Dose, Benefit, and Risk in Medical Imaging

T0179300

SERIES IN IMAGING IN MEDICAL DIAGNOSIS AND THERAPY

Series Editors: Bruce R. Thomadsen and David W. Jordan

Handbook of Small Animal Imaging: Preclinical Imaging, Therapy, and Applications
George Kagadis, Nancy L. Ford, Dimitrios N. Karnabatidis, and George K. Loudos (Eds)

Comprehensive Brachytherapy: Physical and Clinical Aspects
Jack Venselaar, Dimos Baltas, Peter J. Hoskin, and Ali Soleimani-Meigooni (Eds)

Handbook of Radioembolization: Physics, Biology, Nuclear Medicine, and Imaging
Alexander S. Pasciak, PhD., Yong Bradley, MD., and J. Mark McKinney, MD. (Eds)

Monte Carlo Techniques in Radiation Therapy
Joao Seco and Frank Verhaegen (Eds)

Stereotactic Radiosurgery and Stereotactic Body Radiation Therapy
Stanley H. Benedict, David J. Schlesinger, Steven J. Goetsch, and Brian D. Kavanagh (Eds)

Physics of PET and SPECT Imaging
Magnus Dahlbom (Eds)

Tomosynthesis Imaging
Ingrid Reiser and Stephen Glick (Eds)

Ultrasound Imaging and Therapy
Aaron Fenster and James C. Lacefield (Eds)

Beam's Eye View Imaging in Radiation Oncology
Ross I. Berbeco, PhD. (Eds)

Principles and Practice of Image-Guided Radiation Therapy of Lung Cancer
Jing Cai, Joe Y. Chang, and Fang-Fang Yin (Eds)

Radiochromic Film: Role and Applications in Radiation Dosimetry
Indra J. Das (Eds)

Clinical 3D Dosimetry in Modern Radiation Therapy
Ben Mijnheer (Eds)

Observer Performance Methods for Diagnostic Imaging: Foundations, Modeling, and Applications with R-Based Examples
Dev P. Chakraborty (Eds)

For more information about this series, please visit: https://www.crcpress.com/Imaging-in-Medical-Diagnosis-and-Therapy/book-series/CRCIMAINMED

Dose, Benefit, and Risk in Medical Imaging

Edited by
Lawrence T. Dauer
Bae P. Chu
Pat B. Zanzonico

CRC Press
Taylor & Francis Group
Boca Raton London New York

CRC Press is an imprint of the
Taylor & Francis Group, an **informa** business

CRC Press
Taylor & Francis Group
6000 Broken Sound Parkway NW, Suite 300
Boca Raton, FL 33487-2742

First issued in paperback 2020

© 2019 by Taylor & Francis Group, LLC
CRC Press is an imprint of Taylor & Francis Group, an Informa business

No claim to original U.S. Government works

ISBN-13: 978-1-4822-3754-2 (hbk)
ISBN-13: 978-0-367-65696-6 (pbk)

Visit the Taylor & Francis Web site at
http://www.taylorandfrancis.com

and the CRC Press Web site at
http://www.crcpress.com

Contents

SECTION I   Introduction

SECTION II   Optimization and Dose Reduction in Medical Imaging

SECTION III   Radiation Risks in Medical Imaging

SECTION IV   Medical Imaging: Safety Approach

SECTION V   Patient Communications and Shared Decision-Making

SECTION VI   Conclusion

Series Preface

Since their inception over a century ago, advances in the science and technology of medical imaging and radiation therapy are more profound and rapid than ever before. Further, the disciplines are increasingly cross-linked as imaging methods become more widely used to plan, guide, monitor, and assess treatments in radiation therapy. Today, the technologies of medical imaging and radiation therapy are so complex and computer-driven that it is difficult for the people (physicians and technologists) responsible for their clinical use to know exactly what is happening at the point of care, when a patient is being examined or treated. The people best equipped to understand the technologies and their applications are medical physicists, and these individuals are assuming greater responsibilities in the clinical arena to ensure that what is intended for the patient is actually delivered in a safe and effective manner.

The growing responsibilities of medical physicists in the clinical arenas of medical imaging and radiation therapy are not without their challenges, however. Most medical physicists are knowledgeable in either radiation therapy or medical imaging, and expert in one or a small number of areas within their disciplines. They sustain their expertise in these areas by reading scientific articles and attending scientific talks at meetings. In contrast, their responsibilities increasingly extend beyond their specific areas of expertise. To meet these responsibilities, medical physicists periodically must refresh their knowledge of advances in medical imaging or radiation therapy, and they must be pre- pared to function at the intersection of these two fields. How to accomplish these objectives is a challenge.

At the 2007 annual meeting of the American Association of Physicists in Medicine in Minneapolis, this challenge was the topic of conversation during a lunch hosted by Taylor & Francis Publishers and involving a group of senior medical physicists (Arthur L. Boyer, Joseph O. Deasy, C.-M. Charlie Ma, Todd A. Pawlicki, Ervin B. Podgorsak, Elke Reitzel, Anthony B. Wolbarst, and Ellen D. Yorke). The conclusion of this discussion was that a book series should be launched under the Taylor & Francis banner, with each volume in the series addressing a rapidly advancing area of medical imaging or radiation therapy of importance to medical physicists. The aim would be for each volume to provide medical physicists with the information needed to understand technologies driving a rapid advance and their applications to safe and effective delivery of patient care.

Each volume in the series is edited by one or more individuals with recognized expertise in the technological area encompassed by the book. The editors are responsible for selecting the authors of individual chapters and ensuring that the chapters are comprehensive and intelligible to someone without such expertise. The enthusiasm of volume editors and chapter authors has been gratifying and reinforces the conclusion of the Minneapolis luncheon that this series of books addresses a major need of medical physicists.

The series Imaging in Medical Diagnosis and Therapy would not have been possible without the encouragement and support of the series manager, Lou Chosen, Executive Editor at Taylor & Francis. The editors and authors, and most of all I, are indebted to his steady guidance of the entire project.

William R. Hendee
Founding Series Editor

Preface

The application of ionizing radiation and radioactive substances in diagnostic, interventional, and therapeutic procedures in medicine is clearly beneficial for hundreds of millions of people each year worldwide. However, often just mentioning the word "radiation" often evokes fear in patients, families, and health-care professionals alike. Radiation is perceived as a unique hazard and at times the resultant "radiophobia" needs to be recognized and properly managed, especially with regard to medical radiation decision-making.

As public awareness of medical radiation exposure has increased, there has been heightened awareness among patients, physicians, and regulatory agencies of the importance and need for holistic benefit-and-risk evaluations and discussions. Shared decision-making implies a need for patient-provider discussions about the benefits and risks of medical procedures that involve the use of radiation. Such discussions may be challenging. This book provides an overview, by imaging modality, of the current aspects of radiation doses, benefits, and risks to provide information essential for such dialogue.

The radiation protection principles of justification and optimization drive the overall safe use of radiation and radioactive substances in medicine. Selecting the appropriate test and then optimizing the protocols and procedures for appropriate image quality while maintaining low radiation doses are important aims. Such optimization needs to be performed while considering the underlying reasons for the procedures as well as the age, gender, and size of the patient.

This work first lays out essential inquiries related to dose, benefit, and risks. It then spans the topics of radiation dose index monitoring systems (and associated "big data") at national and international levels, diagnostic reference levels (and achievable dose concepts), dental, mammography, fluoroscopically guided interventions, computed tomography applications, nuclear medicine imaging, hybrid imaging platforms, pediatric, and pregnant patient considerations. In addition, this volume addresses the current fundamental radiobiological and radiation epidemiology information, public concerns, and communication strategies.

The target audience includes medical physicists, clinicians, radiation technologists, other related fields, and medical radiation workers. This book benefits from the contributions of nationally and internationally recognized authors who have researched and published in these areas of inquiry and have pursued their integration into the clinical environment. Their extensive knowledge is evident from their comprehensive treatment of the topics, practical examples, suggestions, and references for deeper development. The editors are sincerely thankful for their excellent contributions.

About the Editors

Lawrence T. Dauer, PhD, DABHP is Corporate Radiation Safety Officer, Associate Attending Physicist, and Associate Clinical Member in the Departments of Medical Physics and Radiology at Memorial Sloan Kettering Cancer Center (MSKCC) in New York City. He earned an MS in Health Physics and a PhD in Adult Education. He is certified in comprehensive health physics by the American Board of Health Physics and is past chair of the Radiation Safety Committee of the American Association of Physicists in Medicine (AAPM), past President of the Greater New York Chapter of the Health Physics Society (HPS), Executive Council Member of the Medical Physics Section of HPS, a past member of the Joint Safety Committee of the Society for Interventional Radiology and the American College of Radiology, past council member of the Radiological and Medical Physics chapter of AAPM, and a member of editorial and review boards of several scientific journals. He received the Elda E. Anderson Award from HPS in 2005 and was named an HPS Fellow in 2017. He is a council member and serves on the Board of Directors of NCRP. He also served as a member of the International Commission on Radiological Protection Committee 3 on protection in medicine, a member of the science council for the International Organization for Medical Physics, and was on the program committee for the International Atomic Energy Agency's International Conference on Radiation Protection in Medicine-Setting the Scene for the Next Decade. He serves on the Radiation Advisory Committee of the U.S. Environmental Protection Agency's Science Advisory Board. He has several publications in the topical areas of radiation protection and risks in the fields of detection, radiology, interventional radiology, x-ray imaging, nuclear medicine, and radiation oncology, as well as surgery and medicine.

Bae P. Chu is the Lead Health Physicist at Memorial Sloan Kettering Cancer Center (MSK). She earned her bachelor's degree in Health Science at James Madison University and her master's degree in Environmental and Occupational Health Science from Hunter College. She is a board-certified Radiation Technologist in the state of Maryland. Bae is currently working towards completing her doctorate at the Graduate School of Public Health and Health Policy in New York. She has served on the executive board of the Greater New York Health Physics Society as President, Vice President to Toastmasters at MSK, and member of the American Industrial Hygiene Association. Her current academic research focuses on the perceptions of radiation isolation and interventions related to practice, training, and patient care that would impact engagement, safety, and quality of care.

Pat B. Zanzonico received a BS in Physics from Cooper Union in 1977 and a PhD in Biophysics from the Cornell University Graduate School of Medical Sciences in 1982. He served on the faculty of the Department of Radiology (Nuclear Medicine) of the New York Hospital-Cornell Medical Center and is currently a member and attending physicist at Memorial Sloan Kettering Cancer Center, Co-Head of the center's Small-Animal Imaging Laboratories, and Chairman of its Committee on Radiation. He also serves on the Special Contributing Faculty of the Gerstner Sloan-Kettering Graduate School and is an adjunct professor of Applied Physics and Mathematics at Columbia University. Dr. Zanzonico is Associate Editor of the *British Journal of Radiology* and the *European Journal of Nuclear Medicine* and a member of the Editorial Boards of the *Journal of Nuclear Medicine* and *Medical Physics*. He is also a member of the Medical Internal Radionuclide Dosimetry (MIRD) Committee of the Society of Nuclear Medicine and Molecular Imaging and past member and vice chairman of the NRC's Advisory Committee on Medical Uses of Isotopes (ACMUI), and a past consultant to the International Atomic Energy Agency (IAEA). Dr. Zanzonico has over 120 peer-reviewed publications and over 75 invited presentations. He is actively involved in biomedical research on radionuclide-based methods for detecting and localizing tumor hypoxia, immune effector-cell trafficking, patient-specific dosimetry for radionuclide therapies, and small-animal and molecular imaging.

Contributors

Adam M. Alessio
Department of Radiology and Bioengineering
    and Mechanical Engineering
University of Washington
Seattle, Washington

Wesley E. Bolch
Biomedical Engineering and Medical Physics
Advanced Laboratory for Radiation Dosimetry
    Studies
J. Crayton Pruitt Family Department of
    Biomedical Engineering
Department of Biomedical Engineering
University of Florida
Gainesville, Florida

James A. Brink
Department of Radiology
Massachusetts General Hospital
Boston, Massachusetts

Bae P. Chu
Department of Medical Physics
Memorial Sloan Kettering Cancer Center
New York City, New York

Lawrence T. Dauer
Department of Medical Physics
Department of Radiology
Memorial Sloan Kettering Cancer Center
New York City, New York

Yusuf E. Erdi
Department of Medical Physics
Diagnostic X-Ray Quality Assurance
Memorial Sloan Kettering Cancer Center
New York City, New York

Frederic H. Fahey
Department of Nuclear Medicine/PET Physics
Boston Children's Hospital
Boston, Massachusetts

Amy M. Geyer
Department of Radiation Oncology
St. Luke's Mountain States Tumor Institute
Boise, Idaho

William J. Godwin
Department of Radiation Oncology
Medical University of South Carolina
Charleston, South Carolina

Geoffrey S. Gold
Department of Clinical Psychology
Hofstra University
Hempstead, New York

Jennifer L. Hay
Department of Psychiatry and Behavioral
    Sciences
Memorial Sloan Kettering Cancer Center
New York City, New York

Adam Kesner
Department of Medical Physics
Molecular Imaging and Therapy Physics (Nuclear
    Medicine)
Memorial Sloan Kettering Cancer Center
New York City, New York

Linda Kroger
Department of Radiology
University of California Davis School of
    Medicine
Sacramento, California

Daniel J. Long
Department of Medical Physics
Diagnostic X-Ray Quality Assurance
Memorial Sloan Kettering Cancer Center
New York City, New York

Alan G. Lurie
Department of Oral and Maxillofacial Diagnostic
    Radiology
Department of Diagnostic Imaging and
    Therapeutics
University of Connecticut School of Dental
    Medicine
Farmington, Connecticut

Dustin W. Lynch
Department of Medical Physics
Diagnostic X-Ray Quality Assurance
Memorial Sloan Kettering Cancer Center
New York City, New York

Usman Mahmood
Department of Medical Physics
Diagnostic X-Ray Quality Assurance
Memorial Sloan Kettering Cancer Center
New York City, New York

Colin J. Martin
Department of Clinical Physics and
    Bio-Engineering
University of Glasgow
Glasgow, United Kingdom

Matthew Maynard
Department of Medical Physics
Willis-Knighton Cancer Center
Shreveport, Louisiana

Donald L. Miller
Office of In Vitro Diagnostics and Radiological
    Health
Center for Devices and Radiological Health
U.S. Food and Drug Administration
Silver Spring, Maryland

Tatjana Paunesku
Department of Radiation Oncology
Northwestern University Feinberg School of
    Medicine
Chicago, Illinois

Anand M. Prabhakar
Department of Radiology
Harvard Medical School
and
Department of Emergency Radiology
Department of Cardiovascular Imaging
Massachusetts General Hospital
Boston, Massachusetts

Lawrence N. Rothenberg
Department of Medical Physics
Memorial Sloan Kettering Cancer Center
and
Department of Medical Physics
Columbia University
New York City, New York

Briana Sexton-Stallone
Department of Radiology
Boston Children's Hospital
Boston, Massachusetts

Ferid Shannoun
Secretariat of the United Nations Scientific
    Committee on the Effects of Atomic Radiation
United Nations
Vienna, Austria

David C. Spelic
Office of In Vitro Diagnostics and Radiological
    Health
Center for Devices and Radiological Health
U.S. Food and Drug Administration
Silver Spring, Maryland

Brent K. Stewart
Department of Radiology
University of Washington School of Medicine
Seattle, Washington

Ted S. Treves
Department of Radiology
Harvard Medical School
Boston, Massachusetts

Wolfgang Weiss
United Nations Scientific Committee on the
    Effects of Atomic Radiation
United Nations
New York City, New York

Gayle E. Woloschak
Department of Radiation Oncology and
    Radiology
Northwestern University Feinberg School of
    Medicine
Chicago, Illinois

Pat B. Zanzonico
Department of Medical Physics
Memorial Sloan Kettering Cancer Center
New York City, New York

I

Introduction

1

Essential Inquiries: Dose, Benefit, and Risk in Medical Imaging

Pat B. Zanzonico,
Bae P. Chu, and
Lawrence T. Dauer

1.1 Introduction

The introduction of ionizing radiation in medicine revolutionized the diagnosis and treatment of disease. It has also dramatically improved the quality of health care. The United Nations Scientific Committee on the Effects of Atomic Radiation (UNSCEAR) and the National Council on Radiation Protection and Measurements (NCRP) have identified that medical exposures represent the largest human-made source of radiation exposure (NCRP 2009, 2001; UNSCEAR 2008). The use of nuclear medicine, particularly positron emission tomography (PET), as well as computed tomography (CT) has grown dramatically over the last several decades. For example, as tabulated in a American College of Radiology (ACR) "white paper" published in 2007 (Amis et al. 2007), the annual number of nuclear medicine procedures in the United States increased threefold (from 7 to 20 million), and the annual number of CT procedures increased twentyfold (from 3 to 60 million) between 1985 and 2005. In 2005, more than 1700 PET and PET-CT scanners were in use in the United States, and another 500 essentially all PET-CT units were sold, with more than 1.3 million patients undergoing PET and PET-CT studies (IMV 2004; Czernin et al. 2007; Kincade 2007). Along with broader usage and larger numbers of procedures comes an increasing responsibility for ensuring the many beneficial uses of radiation while preventing or minimizing detrimental radiation effects. The aim is to achieve the clinical objective of medical imaging and therapy with constant vigilance on overall safety.

Accompanying this dramatic growth, concern over the radiogenic risks associated with medical imaging has intensified dramatically, largely due to reports of such risks, some alarmist in tone, in both the scientific and lay media. Reasonable concern is beneficial in terms of promoting critical evaluation of imaging procedures, with technical optimization, elimination of truly unnecessary procedures, and minimization of imaging doses, without compromising the diagnostic information being sought. However, the excessive emphasis on radiogenic cancer risk can create the misconceptions that not only is radiation the *only* risk to be considered in medical imaging, but also that the benefit of imaging procedures may actually be outweighed by the risk (Hendee 2013).

Patients as well as practitioners should, of course, become well-informed about the risks and benefits of all aspects of medical care, radiation-related and otherwise. However, when advising patients about these risks, only a perfunctory statement is generally offered to the effect that any theoretical risks of radiation exposures are far smaller than the direct benefits of the study. Importantly, the word "theoretical" should be used because virtually all risk information is predicated on the "linear no-threshold" (LNT) theory of radiation carcinogenesis. This theory posits that stochastic, or statistical, risks (primarily of cancer) observed in populations exposed to high doses of radiation can be linearly extrapolated to the much lower doses that are encountered in diagnostic medical procedures and certain other circumstances (NRC/NAS 2006). Although the LNT model may be a defensible (i.e., appropriately prudent) basis of radiation protection (e.g., for formulating population-based radiation dose limits), its application to *individual* patients and their medical management is unjustified (ICRP 2007). For example, the LNT-derived risk of a radiogenic cancer associated with a specific diagnostic procedure for an individual patient may conceivably lead the patient and/or the patient's physician to forego the procedure, favoring a sub-optimal procedure with its own, non-radiogenic risks, which may exceed the theoretical radiogenic risk of the original procedure. Dose-response models for radiogenic cancer induction are addressed in further detail below and in subsequent chapters.

This volume is intended to provide a comprehensive review and analysis of the dosimetry and risks of essentially all current radiation-based diagnostic imaging procedures, including plain radiography, dental radiography, mammography, fluoroscopy, computed tomography, planar and tomographic scintigraphy (including PET) and single-photon emission computed tomography (SPECT), and hybrid, or multi-modality, imaging. A point of emphasis will be the optimization of these procedures in terms of minimizing patient dose without compromising the specific diagnostic information being sought. An additional point of emphasis will be the special considerations for diagnostic imaging of the pregnant patient. Importantly, quantification of risk, with critical review of the dose-response models used for such quantification, and of the oft-neglected clinical benefit, will also be considered.

1.2 Radiation Quantities and Units

The lexicon of radiation dosimetry and radiobiology can be confusing, and, in fact, can complicate an understanding of radiobiological effects. Definitions of various quantities used to specify radiation "dose," and of selected related quantities, are therefore presented. A compilation of System Internationale (SI) and conventional quantities and their symbols, units, and conversion factors is presented in International Commission on Radiation Units & Measurements (ICRU) Report No 33 (ICRU 1980) and NCRP Report No 82 (NCRP 1977).

Exposure: Exposure, X, is perhaps the oldest quantity used to express the amount of radiation and is basically the electrical charge produced by radiation per unit mass of air. The SI unit of exposure is the C/kg and the conventional unit, the roentgen (R); 1 R equals 2.58×10^{-4} C/kg and 1 C/kg equals 3,876 R. Strictly speaking, exposure is defined only for XX- and gamma rays in air, but, it may often be convenient to refer to a value of exposure at a point inside a material other than air.

Absorbed dose: Absorbed dose, Dm refers to the amount of energy imparted by ionizing radiation to a material divided by the mass of that material. It is perhaps the most widely used and biologically meaningful quantity for expressing radiation dose. The international unit for absorbed dose is the Gray (1 Gy), which is equal to one joule of energy per kilogram or material (J/kg). In the United States, the rad is also commonly used, with 100 rads equaling one gray (1 rad = 0.01 Gy). The biological effects from radiation exposure depend heavily on the amount of dose delivered, and can be separated into stochastic and deterministic effects.

> *Stochastic and deterministic effects*: The biological effects resulting from radiation exposure are often divided into two general categories: stochastic and deterministic. Stochastic effects refer to those that are probabilistic in nature such that the likelihood of the effect increases with the total radiation dose absorbed by the various organs and tissues of the severity of the effect, however, is independent of the absorbed dose. Deterministic effects are those for which a radiation dose threshold (subject to biological variation) usually exists, with the severity as well as the probability of the effect increasing with the radiation dose (Balter et al. 2010).

Exposure of the body to radiation may cause cell injury or cell death. In the context of diagnostic and other low-level exposures, cancer induction is the stochastic injury of greatest interest for radiation exposure. Deterministic tissue injuries, however, arise when the radiation dose is sufficient to damage a critical number of cells, effectively overwhelming repair and repopulation mechanisms.

> *Linear energy transfer*: The quality as well as the quantity of radiation are important determinants of the frequency and/or severity of radiogenic biological effects. Sparsely ionizing radiations, such as X- and gamma rays intermediate to high-energy electrons (i.e., electrons with energies of the order of 1 keV or greater), and beta particles are characterized as "low-quality" radiations whereas, densely ionizing radiations, such as low-energy electrons (e.g., Auger electrons), protons, neutrons, and alpha particles, are typically characterized as "high-quality" radiations. For the *same* absorbed dose, the frequency and/or severity of biological effects are generally less for sparsely, than for densely ionizing radiations. The quality of radiation is characterized by the "linear energy transfer, L or LET," which is the energy deposited by radiation per unit path length. The SI unit of LET is J/m, and the conventional unit is keV/μm; 1 J/m equals 6.25×10^9 keV/μm, and 1 keV/μm equals 1.60×10^{-10} J/m.

> *Relative biological effectiveness*: The influence of LET on the frequency and/or severity of biological effects is quantified by the "relative biological effectiveness (RBE)," the ratio of the absorbed dose of a reference radiation (typically a widely available sparsely ionizing radiation such as cobalt-60 gamma rays) to that of the radiation whose RBE is being evaluated to produce the same biological effect. The RBE is a ratio of absorbed doses and thus is a dimensionless quantity.

> *Radiation weighting factors and equivalent dose*: A simplified version of the RBE, the "radiation weighting factor, W_R," was devised for purposes of radiation protection. The so-called "equivalent dose, H_T, in tissue or organ T," is related to the radiation weighting factors, W_R, and the mean absorbed doses, $D_{T,R}$, to tissue or organ T due to radiations R:

$$H_T \equiv W_R D_{T,R} \qquad (1.1)$$

The equivalent dose is thus essentially a "scaled" absorbed dose where the scaling factor W_R reflects the biological effective of radiation R.

1.3 Medical Imaging Background[1]

Historically, imaging modalities have often been divided into two general categories, structural (or anatomical) and functional (or physiological). Anatomical modalities, depicting primarily morphology with excellent spatial resolution, include X-rays (plain radiography), magnetic resonance imaging (MRI), CT, and ultrasound (US). Functional modalities, depicting primarily information related to underlying metabolism and biochemistry, include magnetic resonance spectroscopic imaging (MRSI), functional magnetic resonance imaging (fMRI), and (planar) scintigraphy, SPECT, and PET. This traditional distinction between anatomical and functional imaging modalities is increasingly arbitrary

---

[1] Imaging modalities which do not employ ionizing radiation—MRI, US, and optical—are beyond the scope of this book.

and inaccurate because dynamic and/or static MRI, CT, and U.S. imaging may be performed following administration of a blood-flow or molecularly targeted contrast agent and functional images obtained. The functional modalities form the basis of the rapidly advancing field of "molecular imaging," defined as the direct, or indirect, non-invasive monitoring and recording of the spatial and temporal distribution of in vivo molecular, genetic, and/or cellular processes for biochemical, biological, diagnostic, or therapeutic applications (Mankoff 2007). In addition to the foregoing "established" imaging modalities, new modalities, primarily based on optical (i.e., fluorescence) imaging (in intra-operative and laparoscopic settings), are emerging.

Imaging modalities may be further characterized as planar and tomographic. In planar imaging (also known as projection or two-dimensional imaging), structures or imaging signal foci through the full thickness of the patients are effectively summed (i.e., superimposed upon one another) in the image. In tomographic (or three-dimensional [3D]) imaging, relatively thin sections of tissue (typically about 1 to several millimeters in thickness) are mathematically reconstructed from the acquired raw data so that the superposition of structures, or signal foci, is eliminated and their 3D distribution is visualized. The principal advantage of tomography lies in its improved image contrast and greater quantitative accuracy: by eliminating the signal contribution in tissues above and below the section of interest, the target (e.g., tumor)-to-background ratio and the accuracy of image-derived signal levels improve.

Radiography, including plain radiography (X-rays), mammography, fluoroscopy, and CT, is based on the differential transmission of X-rays (typically with energies of the order of 100 keV) among different tissues. In addition, non-tomographic (i.e., planar) imaging, differences in thickness and therefore in the path length of the X-rays through the body also impact overall X-ray transmission and radiographic image contrast. The X-ray stopping power of a medium (expressed, e.g., as the linear attenuation coefficient μ), such as tissue, is determined by its electron density (i.e., the number of electrons per cubic centimeter). This in turn is determined by the effective atomic number and the mass density of the medium. As a result, bone (with a relatively high effective atomic number and electron density due to its calcium content), lung (with a low mass density and electron density due to its air content), and soft tissue can be readily distinguished from one another on planar X-rays as well as CT scans. However, soft tissues are difficult to impossible to resolve due to small differences in electron density among soft tissues. The differential attenuation of X-rays among tissues (particularly among soft tissues) is generally enhanced as the X-ray energy is decreased—primarily as a result of increasing probability of photoelectric versus Compton-scatter interactions with decreasing X-ray energy. For example, utilizing a combination of specialized X-ray tube targetry, low tube voltage, and minimal filtration, mammography is generally performed with relatively low-energy X-rays (with average energies well under 30 keV) to take advantage of the enhanced image contrast at such low energies. Of course, the penetrability of X-rays through tissue decreases as well with decreasing energy, meaning that higher energy X-rays must be used, and some contrast sacrificed, for imaging of the abdomen, pelvis, and other body parts that are thicker than the breast. In fluoroscopy, CT, and certain other radiographic studies, contrast agents (e.g., intravenously administered iodinated agents or orally administered barium-containing solutions) are used to temporarily increase the electron density of different tissues, with the resulting radiographic enhancement of different tissues related to their differential contrast agent content.

In modern CT scanners, slip-ring gantry technology is used, and the X-ray tube rotates at very high speeds emitting a fan beam of X-rays while the patient is translated through the X-ray tube/detector gantry. The X-ray beam thus traces a spiral, or helical, pattern through the patient; such devices are therefore called spiral, or helical, CT scanners. In so-called multi-slice scanners, there may be up to 256 contiguous banks of detectors in the longitudinal direction of the patient, allowing simultaneous acquisition of data for reconstruction of up to 256 transverse sections (or "slices"). Therefore, near whole-body CT studies can be acquired in a matter of seconds to minutes. CT scans can thus be acquired so rapidly that sedation of children and other patients, who might be unable to remain still for the duration of the scan becomes unnecessary; this is partly responsible for the rapid growth in CT over the past several decades, particularly in pediatric populations.

Fluoroscopy is a planar radiographic imaging modality in which a series of images (or frames) are acquired in rapid succession (i.e., dynamically). It is essentially a radiographic motion picture, which allows real-time imaging. Therefore, fluoroscopy can be used to visually position catheters, and other internal devices, in patients. Because fluoroscopy requires a very large number of X-ray exposures, limitations on the X-ray tube performance and the patient radiation doses become important. Historically, fluoroscopy systems included a component known as an image intensifier, which converts the X-ray transmission image to an electronic, and then an optical, image while amplifying the "intensity" of the image. The image amplification allows a relatively low X-ray tube current to be used, thereby avoiding excess heating of the X-ray tube targetry and minimizing the patient dose per frame and for the overall study. Nonetheless, despite this and other technical innovations, patient doses in cardiovascular fluoroscopy, and interventional imaging, are considerably higher than those for other imaging modalities. For example, skin-absorbed doses can be up to several Gy (several hundred rad) can be delivered (Balter et al. 2010; Balter and Miller 2014; Balter et al. 2012; Miller et al. 2003; NCRP 2010a). Therefore, the acute threshold absorbed dose for skin effects approximately 2 Gy (200 rad) may be exceeded (Balter et al. 2010; Balter and Miller 2014). In rare instances, skin doses may be considerably higher than 2 Gy (200 rad) resulting in necrotic damage severe enough to necessitate surgical reconstruction, particularly because fluoroscopically guided procedures have become more complex and beam-on times have increased. It is largely for this reason that in 1994 the United States Food and Drug Administration issued a health advisory titled, "Avoidance of Serious X-ray Skin Injuries to Patients during Fluoroscopically Guided Procedures" (FDA 1994) and established regulatory limits for the entrance dose *rate*, specifically for the so-called air kerma rate (AKR). With certain exceptions, the AKR shall not exceed 88 mGy/minute (10 R/min) at the measurement point. (The foregoing regulatory limit applies specifically to fluoroscopic equipment manufactured on or after May 19, 1995. If the X-ray source is below the patient table, the AKR shall be measured at 1 cm above the tabletop. If the source is above the table, the AKR shall be measured at 30 cm above the tabletop with the end of the beam-limiting device, or spacer, positioned as close as possible to the point of measurement. In a C-arm type of fluoroscope, the AKR shall be measured at 30 cm from the input surface of the fluoroscopic imaging.) The FDA subsequently amended its regulation titled, "Performance Standards for Ionizing Radiation Emitting Products. Fluoroscopic Equipment" (FDA 2015). The additions required that fluoroscopic equipment manufactured after June 10, 2006 display both AKR and cumulative air kerma in order to provide the fluoroscopist with real-time patient radiation dose data, with the intent that such information would result in reduced radiation doses.

Dental X-rays are similar to plain radiographs, and the contrast between teeth (essentially bone) and soft-tissue structures in and around the oral cavity is pronounced. The X-ray doses received by a dental patient are typically very low (~approximately 15 mrem for a full mouth series of up to 18 views, according to the American Dental Association website). Incidental exposure to the patient is reduced by the use of a lead apron and/or lead thyroid collar. Personnel exposure is reduced to negligibly low levels by the dental assistant stepping out of the room, or behind shielding, during the actual X-ray exposure. Panoramic images are extraoral views where the imaging detector is exposed while outside the patient's mouth, yielding, in a single view, the patient's oral anatomy. These images were developed as a rapid alternative to the 18-view full-mouth series. However, the diagnostic accuracy of panoramic images compared to full-mouth series, has been questioned. There is increasing use of CT in dentistry, particularly to plan dental implants, with significantly higher radiation doses to the patient than for conventional dental radiography. Specially designed cone beam CT scanners have been used, which also require careful dosimetric assessments.

Radionuclide imaging, including SPECT and PET, utilizes unsealed sources of radioactivity, usually administered systemically and intravenously, in the form of radiotracers (in a clinical setting, also known as radiopharmaceuticals). Radionuclide imaging, in general, and SPECT and PET, in particular, offer a number of important advantages. First, the specific activity (i.e., activity per unit mass) of radiopharmaceuticals and the detection sensitivity of radionuclide imaging instruments are sufficiently high that the administered activities needed for imaging correspond to non-pharmacologic, non-perturbing mass doses (typically in the sub-nanomole range). This is in contrast to CT and MRI, for example, where the mass doses of various contrast agents are far higher typically in the micromole to millimole range and thus may perturb the subject under study. Second, radionuclide images are quantitative, or at least semi-quantitative, meaning that image "intensity" (i.e., counts) reflects the radiotracer-derived activity concentration. For PET, images are routinely absolutely quantitative and may be parameterized, for example, in terms of activity concentration. For other imaging modalities, the relationship between the contrast agent, or other analyte concentration, and image intensity is typically not as direct. Third, a large number and variety of molecularly targeted and/or pathway-targeted radiotracers (such as metabolites and metabolite analogs, neurotransmitters, drugs, receptor-binding ligands, antibodies and other immune constructs, etc.). have been, and continue to be, developed for increasingly specific characterization of in situ biology. However, radionuclide imaging is not without its drawbacks. These include relatively coarse spatial resolution (5 to 10 mm expressed as the full-width half-maximum of the system line spread function), which is approximately an order of magnitude poorer than the spatial resolution (~approximately1 mm) of CT and MRI.

The basic paradigm of both SPECT and PET imaging includes acquisition of projection images from multiple angles at one to three angular increments around the subject, correction of the acquired data for non-uniform response of the imaging system, and mathematical reconstruction of transverse-section images. The resulting transverse images are essentially contiguous, with no inter-section gaps. Therefore, the reconstructed 3D array of volume elements, or voxels, may be rearranged at any angle relative to the longitudinal axis of the patient and thus yield coronal, sagittal, or oblique as well as transverse images. Another important advantage of emission tomography is the ability to visualize the 3D distribution of activity in situ, that is, ascertain the depths of foci of activity.

Developed in the late 1950s by Hal Anger, the gamma camera, also known as the scintillation or Anger camera, has long been the predominant imaging device for SPECT and single photon (gamma- and X-ray) imaging in general. Almost all gamma-camera scintillation crystals are composed of thallium-doped sodium iodide (NaI[Tl]), with thicknesses of the order of 1 cm; such a crystal stops ~approximately 95% of the 140-kev gamma rays emitted by technetium-99m, the most commonly used non-PET radionuclide. The gamma camera collimator, comprised of a lead plate with holes (apertures) through which radiation must pass to reach the crystal, "directionalizes" the incoming radiation where any radiation traveling at an oblique angle to the axes of the apertures will strike the inter-aperture lead walls (septa) instead of the crystal, thereby allowing only radiation traveling parallel, or nearly parallel, to the aperture axes to reach the crystal and contribute counts to the resulting image. So-called parallel-hole collimators, in which the apertures and septa are parallel to one another, are used almost exclusively. Once the incident radiation passes through the collimator aperture, it strikes and may produce a scintillation (or light "flash") within the crystal. The resulting light signal is distributed among a two-dimensional array of photomultiplier tubes (PMTs) backing the crystal; a PMT is essentially a vacuum tube that converts the light signal to an electronic signal, which is then amplified by virtue of an ~1000 approximately V high voltage. The light intensity reaching each PMT varies inversely with the distance between the scintillation and the respective PMT: the farther the PMT is from the scintillation, the less light it receives and the smaller its output pulse. This inverse relationship is the basis of the Anger position logic circuitry for determining the precise position of a scintillation within the crystal. In older gamma cameras, the x- and y-coordinates were calculated by analog circuitry, that is, using matrices of resistors. In the current system, this is done by digitizing the output signal from each PMT and using digital electronics. For SPECT, the gamma camera assembly actually rotates around the subject to acquire projection images, each typically taking 20 to 30 seconds. Because of the total length of time (20–30 min) thus required for such a study, dynamic SPECT imaging remains largely impractical. However, gamma camera collimators are interchangeable and one may choose to use parallel-hole collimation for either dynamic or static planar imaging.

PET is based on the annihilation coincidence detection of the two co-linear (approximately 180° apart) 511-keV gamma-rays resulting from the mutual annihilation of a positron and an electron [12–14].

Positrons are emitted by radionuclides with an unstably high proton-to-neutron ratio, typically traveling a very short distance (approximately 1 mm or less) in tissue or other media before undergoing annihilation. When both photons from an annihilation event interact simultaneously (actually, within ~10 nanoseconds) with two detectors connected to a coincidence circuit, the circuit is triggered and a coincidence event is generated. In PET, the direction from which an event originated is thus defined electronically, and an important advantage of annihilation coincidence detection is that absorptive collimation (as is used in gamma cameras) is not required. As a result, the sensitivity of PET is two to three orders of magnitude higher than that of SPECT. Current PET scanners typically span a distance of 15 to 20 cm in the patient's longitudinal direction. Thus, a whole-body PET scan will require data acquisition at six to seven discrete bed positions and subsequent merging, or "knitting," of the discrete images into a whole-body image. In contrast to SPECT, the detector assembly does not rotate (as described below), and projection data completely around the patients are simultaneously acquired, thereby making dynamic PET imaging more practical.

Modern PET scanners generally employ a series of stationary rings of discrete, small-area detectors (i.e., scored block detectors, or individual, or pixilated, crystals) encircling the subject. The scintillation detector materials most widely used in PET are: bismuth germanate (BGO, $Bi_4Ge_3O_{12}$), cerium-doped gadolinium oxyorthosilicate (GSO[Ce] or GSO, Gd_2SiO_5:Ce), cerium-doped lutetium oxyorthosilicate (LSO[Ce] or LSO, Lu_2SiO_5:Ce), and cerium-doped lutetium-yttrium oxyorthosilicate (LYSO[Ce] or LYSO, Lu_2YSiO_5:Ce). LSO and LYSO have emerged as the detectors of choice for PET. BGO, GSO, LSO, and LYSO have higher effective atomic numbers and mass densities than NaI(Tl) and thus higher stopping power for 511-keV annihilation gamma rays. Although the stopping power and resulting detection sensitivity of NaI(Tl) are sufficient for the lower energy X- and gamma rays (about approximately 70 to 360 keV) emitted by commonly used single-photon emitters, such as technetium-99m, indium-111, thallium-201, and iodine-131, NaI(Tl) has inadequate stopping power, and therefore sensitivity, for 511-keV annihilation gamma rays.

Image registration and fusion have become increasingly important components of clinical imaging and have led to the development of a variety of pertinent software and hardware tools, including multimodality (e.g., SPECT-CT and PET-CT) devices that "automatically" provide registered and fused 3D image sets. Because information derived from multiple images is often complementary (e.g., localizing the site of an apparently abnormal metabolic process to a pathologic structure such as a tumor), integration of image data in this way may be helpful and even critical. In addition to anatomic localization of signal foci, image registration and fusion provide intra- as well as inter-modality corroboration of diverse images and more accurate and more certain diagnostic and treatment-monitoring information. However, the problem is that differences in image size and dynamic range, voxel dimensions and depth, image orientation, subject position and posture, and information quality and quantity make it difficult to unambiguously co-locate areas of interest in multiple image sets. Therefore, the objectives of image *registration* and *fusion* are (a) to appropriately modify the format, size, position, and even shape of one, or both, image set(s) to provide a point-to-point correspondence between images and (b) to provide a practical integrated display of the images thus aligned.

There are two practical approaches to image registration and fusion, "software" and "hardware" approaches. In the software approach, images are acquired on separate devices, imported into a common image-processing computer platform, and registered and fused using the appropriate software. In the hardware approach, images are acquired on a single, multi-modality device and transparently registered and fused with the manufacturer's integrated software. Both approaches are dependent on software sufficiently robust to recognize and import diverse image formats. Therefore, the availability of industry-wide standard formats, such as the ACR-NEMA DICOM standard (i.e., the ACR and National Electrical Manufacturers Association for Digital Imaging and Communications in Medicine standard), is critical.

All manufacturers of SPECT, PET, and CT scanners market multi-modality scanners, combining PET or SPECT with CT scanners in a single device. These instruments provide near-perfect registration of images of in vivo function (SPECT or PET) and anatomy (CT) using a measured, and presumably fixed, rigid transformation (i.e., translation and rotation in three dimensions) between the image sets. These devices have already had a major impact on clinical practice, particularly in oncology. PET-CT devices, for example, have eliminated "PET-only" scanners from the clinic. Although generally encased in a single seamless housing, the PET and CT gantries in such multi-modality devices are separate; the respective fields of view are separated by a distance of the order of 1 m and the PET and CT scans are performed sequentially.

In addition to PET-CT scanners, SPECT-CT scanners are now commercially available. The design of SPECT-CT scanners is similar to that of PET-CT scanners in that the SPECT and CT gantries are separate and the SPECT and CT scans are acquired sequentially, not simultaneously. In such devices, the separation of the SPECT and CT scanners is more apparent because the rotational and other motions of the SPECT detectors effectively preclude encasing them in a single housing with the CT scanner. PET-MRI scanners have recently been introduced; these allow simultaneous, rather than sequential, acquisition of the PET and MRI studies.

1.4 Assessing the Risks of Medical Imaging

The radiobiologic effects of radiation are often distinguished as either stochastic (i.e., statistical) or non-stochastic (i.e., deterministic). The distinction between stochastic and deterministic effects is perhaps best understood in terms of their respective probability-dose and severity-dose relationships, illustrated in Figure 1.1. A stochastic effect is characterized by the absence of a threshold, meaning that any radiation dose above background is associated with a corresponding finite (or non-zero) increase in the probability above background of the effect occurring. As the dose increases above background, the excess probability also increases. However, the *severity* of the effect does not increase with dose; that is, the severity of a stochastic effect is independent of dose. Stochastic effects include radiation-induced

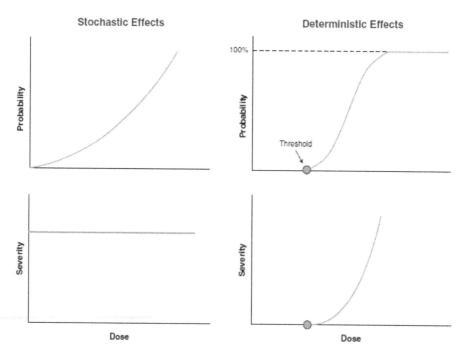

FIGURE 1.1   Radiation effects. Stylized probability-dose and severity-dose relationships for stochastic and deterministic effects. (From Zanzonico, P. et al., *JACC Cardiovasc Imag.*, 9, 1446–1461, 2016.)

carcinogenesis and germ cell mutagenesis and are generally associated with low-level (e.g., diagnostic) exposures. A deterministic effect is characterized by a well-defined threshold dose, meaning that the probability of the effect occurring does not increase above the background probability until the threshold is exceeded. However, once the threshold dose is exceeded, the severity as well as the excess probability of the effect increases with dose, with essentially all irradiated individuals exhibiting the effect (i.e., the probability reaches 100%) at sufficiently high doses; the dose-dependent probability increases in a sigmoidal fashion typical of pharmacological dose-response curves. The range of effects of radiation on skin typifies deterministic effects, as discussed below. Deterministic effects are generally associated with high-level (e.g., therapeutic) radiation exposures.

At the cellular level, stochastic effects presumably result from non-lethal genetic mutations and, in principle, the clonogenic proliferation of a single mutated cell may progress to a tumor. Although an over-simplification (ignoring immunosurveillance and other homeostatic functions), this is mechanistically consistent with the presumed absence of a threshold dose for a stochastic effect such as cancer induction. Induction of a deterministic effect, on the other hand, requires elimination by apoptosis, or other cell killing mechanisms, of a critical mass of cells within one or more functional cell compartments in order to induce a demonstrable clinical effect. This is consistent with a non-zero threshold for such an effect and with the dose dependency of the severity as well as the probability of deterministic effects.

The principal radiation risk of practical concern is the possibility of cancer induction. The scale of this risk appears to differ sharply between pre- and post-natal irradiation. The actual risk, if any, of radiation carcinogenesis at diagnostic and other comparably low doses remains highly controversial, and it has been argued that there is, in fact, a hormetic (or protective) effect against the development of cancer at doses at or below ~approximately 0.1 Sv (10 rem). The literature on radiation carcinogenesis and radiation hormesis is vast and continues to grow; this is addressed in subsequent chapters. However, a brief review of pre- and post-natal radiation carcinogenesis and radiation hormesis is provided in this section.

Pre-natal Irradiation. The increased risk of childhood cancer, a stochastic effect, associated with in utero irradiation is worrisome even at *diagnostic* radiation doses. The NCRP recently published a detailed review and evaluation of preconception and pre-natal radiation exposure health effects (NCRP 2013). Originally, the Oxford Survey of Childhood Cancers in the 1950s suggested an association between an increased risk of childhood cancer, principally leukemia, and exposure in utero to diagnostic X-rays (Stewart and Hewitt 2013). However, the apparent carcinogenicity of in utero irradiation at diagnostic levels remains controversial. For example, in a more recent population-based study (Ray et al. 2010) in Ontario, Canada, from 1991 to 2008, 5,590 mothers underwent diagnostic imaging studies (73% were CT scans and 27% nuclear medicine scans) and 1,829,927 mothers did not. After a median follow-up of 8.9 years, 4 childhood cancers developed in the exposed group (1.13 per 10,000 person-years) and 2539 in the unexposed group (1.56 per 10,000 person-years), yielding a hazard ratio of 0.68 (95% CI: 0.26–1.8) and therefore suggesting that diagnostic imaging in pregnancy is not carcinogenic. Although the *association* between in utero irradiation and childhood cancer appears to be incontrovertible, the question remains whether this is a causative relationship or an example of "reverse causation" (Boice 2015b): because pregnant women were referred for imaging for a medical issue, their children may have been at a naturally increased risk for cancer due to their mother's underlying medical condition rather than as a result of any diagnostic irradiation. However, the preponderance of evidence to many suggests a causative relationship between diagnostic irradiation in utero and an increased risk of childhood cancer. The relevant literature was reviewed in detail by Doll and Wakeford (1997), who concluded that a typical obstetric X-ray examination (corresponding to a fetal radiation doses of ~approximately 10 mGy [1 rad]) results in an increase of approximately 40% in the incidence of childhood cancer, with an excess absolute risk of ~approximately 5%/Gy (0.05%/rad). In light of such a significant risk of childhood cancer associated with (and possibly causatively related to) diagnostic imaging studies, prudence dictates that proceeding with such a study in a pregnant (or possibly pregnant) female should be based on a considered and documented decision as to its medical necessity despite the foregoing risk. This decision should include consideration of both alternative procedures, which do not involve irradiation of the conceptus, and the timeliness (or lack thereof) of the procedure's availability relative to the urgency of obtaining the diagnostic information. The issue of diagnostic imaging studies in the pregnant patient is addressed in further detail in subsequent chapters.

Post-natal Irradiation. Although the possibility of induction of a subsequent cancer in an unborn child from in utero diagnostic irradiation is understandably worrisome, a more frequently encountered concern is the risk of radiation carcinogenesis resulting from imaging studies in pediatric and adult patients. The challenge in assessing this risk is that there are very few reliable data in humans quantifying an increased cancer incidence, if any, following diagnostic radiation doses (i.e., less than approximately 100 mSv [10 rem]). Unfortunately, no prospective epidemiologic studies with appropriate non-irradiated controls have definitely demonstrated either the adverse or hormetic, i.e., beneficial effects of radiation doses less than 100 mSv (10 rem) in man. Furthermore, current estimates of the risks of low-dose radiation suggest that very large-scale epidemiological studies with long-term follow-up would be needed to actually quantify any such risk or benefit; such studies may be logistically and financially prohibitive.

The most creditable dose-response data for radiation carcinogenesis in man mainly involve doses one to two orders of magnitude greater than those encountered in diagnostic imaging studies of the order of 1 Sv (100 rem) and greater including, most notably, the A-bomb survivor follow-up data. However, a handful of high-profile studies have reported cancer risks derived from relatively low-dose exposures. For example, Pierce and Preston (2000) published an analysis of the A-bomb Radiation Effects Research Foundation data on cancer risks among survivors receiving doses less than 0.5 Sv (50 rem), with ~approximately 7000 cancer cases among ~approximately 50,000 low-dose survivors. They concluded that cancer risks are not overestimated by linear risk estimates computed over the dose range 0.05–0.1 Sv%-10 rem), with a statistically significant risk in the range 0–0.1 Sv (0–10 rem) and an upper confidence limit on any possible threshold of 0.06 Sv (6 rem).

Estimation of the excess cancer risk from imaging studies and other low-dose exposures requires mathematical extrapolation of high-dose, dose-response data to the lower diagnostic-dose range. There are at least several distinct dose-response models for radiation carcinogenesis that can be used for this extrapolation: the supra-linear model, the LNT model, the sub-linear (or linear-quadratic) model, and the hormesis model (Figure 1.2).

The supra-linear model implies that the cancer risk *per unit dose* (i.e., the risk per Sv or per rem) is greater at lower than at higher doses. There are no data, or mechanistic considerations, which support such a model, and it is thus not a creditable option for extrapolation of high-dose cancer risk estimates to diagnostic doses.

The LNT model, which implies a uniform cancer risk per unit dose from higher to lower doses, is the model currently recommended for use in radiation protection by authoritative advisory bodies, such as the International Commission on Radiological Protection (ICRP) (2007), the NCRP (1993, 2001), UNSCEAR (2000, 2008), and that has been adopted by regulatory agencies such as the Nuclear Regulatory Commission (NRC) (1991). As its name indicates, also implicit in the LNT model is that there is no threshold dose for radiation carcinogenesis; that is, there is no radiation dose above background below which there is no finite increase in cancer risk. However, it must be understood that at very low doses, the concomitant risks are also very low. The excess absolute risk is the number of excess fatal cancers per number of irradiated individuals (excess above the naturally occurring incidence) predicted by the model in a large (and therefore gender- and age averaged) population that has been exposed to a uniform whole-body dose (or effective dose) of radiation. The excess relative risk (ERR) is the excess absolute risk divided by the naturally occurring incidence of fatal cancer and may be expressed as either a fraction, or a percentage, per unit effective dose. It should be emphasized that risk extrapolations based on an LNT slope estimate cannot be applied reliably to individuals but only

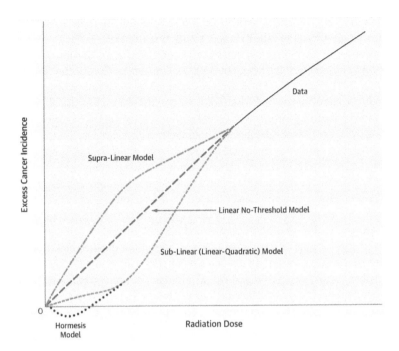

FIGURE 1.2 Dose-response curves for radiation carcinogenesis. Stylized dose-response curves for radiation carcinogenesis for the supra-linear, linear no-threshold (LNT), sub-linear (or linear-quadratic), and hormetic models. Note that for the hormetic model the excess incidence becomes negative at low radiation doses, indicating a cancer incidence less than the naturally occurring incidence and thus a radioprotective effect. (From Zanzonico, P. et al., *JACC Cardiovasc Imag.*, 9, 1446–1461, 2016.)

to large populations (ICRP 2007); that is, populations sufficiently large that differences in radiation sensitivity related to gender, age, diet and other lifestyle effects, and intrinsic biology are effectively averaged out.

The sub-linear, or linear-quadratic, model implies that the excess cancer risk per unit dose is lower than at higher dose and further implies the possibility of at least a practical threshold dose for radiation carcinogenesis, that is, a non-zero dose below which there is no demonstrable increase in cancer incidence.

According to the hormesis model, individuals that have been exposed to low radiation doses actually have a *lower* subsequent risk of cancer than those who have not, presumably as a result of radiogenic upregulation of cellular repair mechanisms or other adaptive response(s) (Vaiserman 2010a, 2010b). Although radiation hormesis had been largely dismissed for many years, there are mounting creditable data in the peer-reviewed scientific literature supporting this phenomenon (Calabrese and Blain 2011). Feinendegen, for example, recently reviewed a number of pre-clinical studies demonstrating radiation hormesis and, in particular, radioprotective adaptive responses to low-dose irradiation and concluded that radiation doses less than about 600 mGy (60 cGy) induced a pronounced (~50%) protective effect against a variety of molecular, cellular, and whole-animal radiation effects (Feinendegen 2016).

It has been argued that: (1) the data and associated analyses supporting the LNT model are further refuted by epidemiologic and experimental studies; (2) this model overstates the risk of radiation carcinogenesis at doses of the order of 100 mSv (10 rem) and less; and (3) this model does not account for creditable evidence for a threshold for cancer induction, that is, a non-zero radiation dose below which there is no increased risk of cancer (Tubiana et al. 2005; Tubiana et al. 2009; Cuttler 2009; Siegel and Stabin 2012). Opposed views contend that the preponderance of data, especially epidemiology data, support, or at least are consistent, with an LNT dose-response model down to the low-dose range (NCRP 2001; Little et al. 2009). The validity, applicability, and utility of the LNT model and alternative models for the purposes of radiation protection remains controversial and is subject to often contentious debate (Cuttler 2009; Calabrese 2009, 2013, 2015a,b). As John Boice, President of the NCRP, has aptly stated (Boice 2015), "LNT is not TNT, but differences in opinions sometimes appear explosive!"

Application with certitude of the LNT risk model with extrapolations of risk at even low dose levels has led to derivation of alarming estimates of excess numbers of cancers as a result of medical imaging. Such exercises largely ignore the considerable uncertainties involved, among other considerations. When the LNT-derived risk overestimates any actual radiogenic risk, it undermines a reasonable benefit-risk analysis and may thus adversely impact the patient's medical management. The LNT model may in fact overestimate the actual risks of diagnostic and other low-level exposures, and there may be creditable evidence for a threshold for radiogenic cancer induction at a radiobiological or cellular level (Tubiana et al. 2005), that is, a non-zero radiation dose below which there is *no* increased risk of cancer. The International Organization for Medical Physics issued a policy statement highlighting the substantial uncertainties in estimating population cancer risk and noting the dangers of extrapolating risk estimates for radiation doses that are less than 100 mSv (10 rem) (Hendee 2013) (99). The use of risk factors to estimate public health consequences from individual or population exposures must be considered in the context of the attendant uncertainties. These include uncertainties related to dosimetry, epidemiology, low statistical power, modeling radiation risk data, and generalization of risk estimates across different populations and dose rates (NCRP 2010). Uncertainties in such risk estimates have been suggested as being up to a factor of 3 lower or higher than the estimate value itself (UNSCEAR 2012). Such large uncertainties render projections of radiation-induced cancers or other detriment, highly susceptible to biases and confounding influences that may be unidentifiable. Several other professional societies and scientific bodies have provided guidance on the assessment of risk at diagnostic and other comparably low doses. The ICRP states, "There is…general agreement that epidemiological methods used for the estimation of cancer risk do not have the power to directly reveal cancer risks in the dose range up to around 100 mSv" (ICRP 2007). The Health Physics Society advises against estimation of health risks below an individual dose of 50 mSv in one year, or a lifetime dose of 100 mSv above that received from natural sources, noting that "…below this level, only dose is credible and statement of associated risks are more speculative than credible" (HPS 2010). Finally, the American Association of Physicists in Medicine (AAPM 2011) has stated that, "…risks of medical imaging at effective doses below 50 mSv for single procedures or 100 mSv for multiple procedures over short time periods are too low to be detectable and may be non-existent."

1.5 Assessing the Benefits of Medical Imaging

Although the point is often made that the benefits of the uses of radiation in medicine are much greater than any theoretical risks, *quantitative* estimates of the benefits are not cited alongside any quantitative estimates of risk. The expression of benefit in purely qualitative terms versus expression of risk in quantitative, and therefore seemingly more certain, terms may well contribute to a skewed sense of the relative benefits and risks of diagnostic imaging among healthcare providers as well as patients. One benefit of a diagnostic imaging procedure may be expressed as the lives saved, that is, the number of lives that have been lost by *not* performing the procedure or by performing an alternative, invasive procedure. (Of course, there may be other metrics of benefit, such as improvements in the quality of life, shortening of hospital stays, and reduction of medical care costs.)

One example of such a quantitative benefit-risk analysis is the use of scintigraphic myocardial perfusion imaging to predict and thereby avoid perioperative cardiac events and associated mortality in non-cardiac surgery (96, 97). The most important cause of perioperative cardiac mortality and morbidity is myocardial infarction due to occult coronary artery disease. In a Veterans Administration (VA) series (98), the incidence and mortality of such events associated with vascular surgery (most commonly, carotid endar-terectomy) was 13% and 40%–70%, respectively. Based on pre-operative dipyridamole thallium-201 (^{201}Tl) imaging, the incidence of perioperative cardiac events was 2% for a severity level and an extent of 0% and 100% for a severity level of 3 and an extent of 5–6, with 22% of patients with reversible perfusion defects. Thus, perfusion imaging was highly accurate for prediction of perioperative cardiac events. The number of vascular surgeries (from the VA database) is approximately 9500 per year. Thus, the number of periop-erative cardiac deaths (i.e., fatal myocardial infarctions) is estimated as $9500 \times 0.13 \times 0.40 = 494$ per year. Of these 494 perioperative cardiac deaths annually, $0.22 \times 494 = 109$ were detectable pre-operatively and therefore avoidable; that is, the gross benefit of preoperative perfusion imaging with ^{201}Tl is 109 lives saved per year in the VA system. The effective dose from the ^{201}Tl study is 24 mSv = 0.024 Sv (2.4 rem) (6, 20). Using the LNT-based lifetime risk factor (i.e., ERR) of 0.05/Sv, a total of $9500 \times 0.05 /Sv \times 0.024 Sv = 11$ excess cancer deaths per year is predicted, yielding a net benefit of pre-operative myocardial perfusion imaging of $109-11 = 98$ lives saved per year. If one considers a rest/stress myocardial perfusion study using ^{99m}Tc-MIBI, the ED is approximately 12 mSv (1.2 rem) (6, 20), leading to only six excess cancer deaths per year and possible greater clinical benefit. Performing cardiac PET with rubidium-82 (^{82}Rb)-rubidium chloride would result in an even lower ED of approximately 7.5 mSv (0.75 rem) (6, 20), thus resulting in a theoretical risk of only 3 cancer deaths and a net savings of 106 lives per year.

The foregoing analysis and similar benefit-risk analyses, discussed in a subsequent chapter, demon-strate that diagnostic imaging saves many thousands of lives per year, whereas, the theoretical, and pos-sibly overestimated, cancer risks predicted by the LNT model are typically much lower.

References

AAPM, 2011. *Position Statement on Radiation Risks from Medical Imaging Procedures*. Policy PP 25-A. Alexandria, VA: American Association of Physicists in Medicine.

Amis ES, Butler PF, Applegate KE et al., 2007. American college of radiology white paper on radiation dose in medicine. *J Am Coll Radiol* 4:272–284.

Balter S, Hopewell JW, Miller DL, Wagner LK, Zelefsky MJ, 2010. Fluoroscopically guided interventional procedures: A review of radiation effects on patients' skin and hair. *Radiology* 254:326–341.

Balter S, Miller DL, 2014. Patient skin reactions from interventional fluoroscopy procedures. *Am J Roentgenol* 202:W335–W342.

Balter S, Miller DL, Schueler BA, 2012. Radiation dose measurements and monitoring for fluoroscopically guided interventional procedures. *J Am Coll Radiol* 9:595–597.

Boice JD, 2015a. Radiation epidemiology and recent paediatric computed tomography studies. *Ann ICRP* 44:236–248.

Boice JD, 2015b. The Boice Report No 40: LNT 101. Bethesda, MD.

Calabrese EJ, 2009. Getting the dose-response wrong: Why hormesis became marginalized and the threshold model accepted. *Archives Toxicol* 83:227–247.

Calabrese EJ, 2013. How the U.S. National Academy of Sciences misled the world community on cancer risk assessment: New findings challenge historical foundations of the linear dose response. *Archives Toxicol* 87:2063–2081.

Calabrese EJ, 2015a. Cancer risk assessment foundation unraveling: New historical evidence reveals that the U.S. national academy of sciences (US NAS), biological effects of atomic radiation (BEAR) committee genetics panel falsified the research record to promote acceptance of the LNT. *Archives Toxicol* 89:649–650.

Calabrese EJ, 2015b. An abuse of risk assessment: How regulatory agencies improperly adopted LNT for cancer risk assessment. *Archives Toxicol* 89:647–648.

Calabrese EJ, Blain RB, 2011. The hormesis database: The occurrence of hormetic dose responses in the toxicological literature. *Regulatory Toxicol Pharmacol* 61:73–81.

Cuttler JM, 2009. Commentary on using LNT for radiation protection and risk assessment. *Dose-Response* 8:378–383.

Czernin J, Allen-Auerbach M, Schelbert HR, 2007. Improvements in cancer staging with PET/CT: Literature-based evidence as of September 2006. *J Nucl Med* 48(Suppl 1):78S–88S.

Doll R, Wakeford R, 1997. Risk of childhood cancer from fetal irradiation. *Br J Radiol* 70:130–139.

FDA, 1994. *Avoidance of Serious X-Ray-Induced Skin Injuries to Patients During Fluoroscopically-Guided Procedures*. Silver Spring, MD: Food and Drug Administration.

FDA, 2015. 21CFR8.1020.32. *Performance Standards for Ionizing Radiation Emitting Products. Fluoroscopic Equipment*. Administration FaD, (Ed.). Silver Spring, MD.

Feinendegen LE, 2016. Quantification of adaptive protection following low-dose irradiation. *Health Phys* 110:276–280.

Hendee WR, 2013. Policy statement of the international organization for medical physics. *Radiology* 267:326–327.

HPS, 2010. Radiation risk in perspective. *Position Statement PS010-2 of the Health Physics Society*, adopted January 1996, revised July 2010. McLean, VA.

ICRP, 2007. Recommendations of the international commission on radiological protection. International commission on radiological protection (ICRP) publication 103. *Ann ICRP* 37:1–332.

ICRU, 1980. Radiation quantities and units. *International Commission on Radiation Units and Measurements (ICRU) Report 33*. Bethesda, MD: International Commission on Radiation Units and Measurements (ICRU).

IMV, 2004. Latest IMV PET census shows fast growth in PET-CT installations [online]. Available at: http://www.imvlimited.com/mid/pdf/News/1104/PET03%20ReleaseV2-Nov%202004.pdf.

Kincade K, 2007. PET-CT moves closer to diagnostic standard of care [online]. Available at: http://www.auntminnie.com/index.aspx?sec=spt&sub=tir&pag=dis&itemID=74571.

Little MP, Wakeford R, Tawn EJ, Bouffler SD, Berrington de Gonzalez A, 2009. Risks associated with low doses and low dose rates of ionizing radiation: Why linearity may be (almost) the best we can do. *Radiology* 251:6–12.

Mankoff DA, 2007. A definition of molecular imaging. *J Nucl Med.* 48:18N, 21N.

Miller DL, Balter S, Cole PE et al., 2003. Radiation doses in interventional radiology procedures: The RAD-IR study: Part II: Skin dose. *J Vasc Interv Radiol* 14:977–990.

NCRP, 1977. Environmental radiation measurements. *National Council on Radiation Protection and Measurements (NCRP) Report 82*. Bethesda, MD: National Council on Radiation Protection and Measurements (NCRP).

NCRP, 1993. Limitation of exposure to ionizing radiation, NCRP Report No 116. Bethesda, MD: National Council on Radiation Protection and Measurement (NCRP).

NCRP, 2001. Evaluation of the linear-non-threshold dose-response model for ionizing radiation, NCRP Report No 136. Bethesda, MD: National Council on Radiation Protection and Measurement (NCRP).

NCRP, 2009. Ionizing Radiation Exposure of the Population of the United States, NCRP Report 160. Bethesda, MD: National Council on Radiation Protection and Measurements (NCRP).

NCRP, 2010a. Radiation dose management for fluoroscopically-guided interventional medical procedures, NCRP Report 168. Bethesda, MD: National Council on Radiation Protection and Measurements (NCRP).

NCRP, 2010b. Uncertainties in internal radiation dose assessment, NCRP Report 164. Bethesda, MD: National Council on Radiation Protection and Measurements (NCRP).

NCRP, 2013. Preconception and prenatal radiation exposure: Health effects and protective guidance, NCRP Report 174. Bethesda, MD: National Council on Radiation Protection and Measurements (NCRP).

NRC, 1991. 21CFR20. *Standards for Protection Against Radiation*. Rockville, MD: Nuclear regulatory Commission (NRC).

NRC/NAS, 2006. *Health Risks from Exposure to Low Levels of Ionizing Radiation: BEIR VII Phase 2.* National Research Council, National Academy of Science, Washington, DC: The National Academies Press.

Pierce DA, Preston DL, 2000. Radiation-related cancer risks at low doses among atomic bomb survivors. *Radiat Res* 154:178–186.

Ray JG, Schull MJ, Urquia ML, You JJ, Guttmann A, Vermeulen MJ, 2010. Major radiodiagnostic imaging in pregnancy and the risk of childhood malignancy: A population-based cohort study in Ontario. *PLoS Med* 7:e1000337.

Siegel JA, Stabin MG, 2012. Radar commentary: Use of linear no-threshold hypothesis in radiation protection regulation in the United States. *Health Phys* 102:90–99.

Stewart A, Hewitt D, 1963. Oxford survey of childhood cancers: Progress report I. Monthly bulletin of the Ministry of Health and the Public Health Laboratory Service, Vol. 22, pp. 182–92, London, UK.

Tubiana M, Aurengo A, Averbeck D, Bonnin D, LeGuen B, Masse R, Monier R, Valleron AJ, de Vathaire F, 2005. Dose-effect relationships and the estimation of the carcinogenic effects of low doses of ionizing radiation. Academy of Medicine (Paris) and Academy of Science (Paris) Joint Report No 2. Paris, France.

Tubiana M, Feinendegen LE, Yang C, Kaminski JM, 2009. The linear no-threshold relationship is inconsistent with radiation biologic and experimental data. *Radiology* 251:13–22.

UNSCEAR, 2000. United Nations Scientific Committee on the Effects of Atomic Radiation (UNSCEAR) 2000 Report: Sources and Effects of Ionizing Radiation. Annex I: Epidemiological Evaluation of Radiation-induced Cancer. New York, United Nations.

UNSCEAR, 2008. Effects of ionizing radiation. UNSCEAR 2006 Report to the General Assembly with Scientific Annexes. Vol. 1. New York: United Nations Scientific Committee on the Effects of Atomic Radiation (UNSCEAR).

UNSCEAR, 2012. United Nations Scientific Committee on the Effects of Atomic Radiation (UNSCEAR). The UNSCEAR 2012 Report to the General Assembly. New York.

Vaiserman AM, 2010a. Radiation hormesis: Historical perspective and implications for low-dose cancer risk assessment. *Dose-Response* 8:172–191.

Vaiserman AM, 2010b. Hormesis, adaptive epigenetic reorganization, and implications for human health and longevity. *Dose-Response* 8:16–21.

Zanzonico P, Dauer L, Strauss HW, 2016. Radiobiology in cardiovascular imaging. *JACC Cardiovasc Imag* 9:1446–1461.

II

Optimization and Dose Reduction in Medical Imaging

2

Radiation Dose Management

Brent K. Stewart

2.1 Introduction

Radiation dose monitoring is an obscure topic, but its ultimate goal is to deliver the appropriate dose at all times under any conditions. This encompasses radiation dose policies and procedures, quality assurance (QA) initiatives in support of the principle of optimization,[1] the concept of ALARA (as low as reasonably achievable), and protocol establishment and review. Such systems and programs have been in place for many years as required by state, federal, and oversight organization regulations. However, more recently, concerns regarding the inherent risk due to the increased administration of radiation in the healing arts have called for improved control of radiation exposure to patients. This has evoked additional regulations regarding the documentation and reporting of radiation dose, as well as the development of information systems to support these requirements.

Just as with the advent of any new system, there was no immediate standard moniker. Such systems have been referred to as clinical/patient exposure/radiation dose recording/monitoring/tracking/management systems. Although it's very tempting to use Radiation Exposure Monitoring System (REMS) due to its meaningful acronym in this context, the American Association of Physicists in Medicine (AAPM) Task Group 257 on Selection of a Patient Dose Monitoring System[2] (tasked with creating a practice guideline on the minimum requirements of patient dose tracking systems based on core patient safety objectives) has termed these systems Radiation Dose Index Monitoring Systems and the abbreviation RDIMS will be used for these systems throughout.

Without the ability to impact ALARA at key touch points of the patient's pathway from order entry to finalized report, an institution cannot truly manage dose. RDIMS is simply a tool relieving medical physicists and imaging physicians from the burdens of data collection (usually from multiple disparate sources), management, and visualization of such an extensive dataset. It is still incumbent on stakeholders to modify/augment local policies, local reporting requirements, periodic review, etc., based on the ability to effectively mine, analyze, and visualize patient dose dataset. RDIMS can contribute to an institution's efforts in providing quality and professional service in the spirit of ALARA, which can bolster confidence in the level of service to referring physicians and patients.

2.2 History of Radiation Dose Index Monitoring Systems and Why Now?

2.2.1 History Leading up to Radiation Dose Index Monitoring Systems

There have been many dose monitoring efforts at institutions around the world previous to the more recent advent of commercially available RDIMS and standards that have helped paved the way for their effective implementation. However, it has not always been so easy to obtain dose data in an automated or semi-automated manner (i.e., from a computer file, Digital Imaging and Communication in Medicine [DICOM] headers, or a database). Beyond mere collection, there is the required analysis, production of dashboard graphics, and presentation to stakeholders to drive dose-reduction initiatives, reduction of inter-examination dose variance, and root cause determination of outliers.

In 2007, the author and colleagues published an article in the *American Journal of Roentgenology* describing this process for surveillance of computed radiography sensitivity values (S-number) at one of our institutions,[3] which was initiated in 2002. Since that time our diagnostic physics group was accessing seven different databases to present dose information at ten different monthly or quarterly quality assurance (QA) meetings at the institutions we serve. Modalities covered included computed tomography (CT), fluoroscopically guided interventional procedures (cardio-vascular/interventional and mobile c-arms), and computed and digital radiography (CR/DR).

Whenever a new source of dose data was identified, we devised a means to access it and use it for ALARA. If not, we instituted, as a last resort due to the possibility of data-entry errors, manual entry of such data, for example, into our Radiology Information System (RIS). In some instances as information systems were discontinued, such as when switching vendors, we had to determine new mechanisms for dose-data extraction. With the implementation of a RDIMS covering our institutions, we no longer resort to mining dose data from many disparate sources, but rather access it from a single authoritative source.

2.2.2 Why the Proliferation of Radiation Dose Index Monitoring Systems at This Time?

The radiation dose administered in medical imaging procedures annually has increased nearly six-fold since the early 1980s, from around 0.5 mSv to almost 3.0 mSv.[4,5] This is due primarily to increased utilization of CT, nuclear medicine, and minimally invasive procedures guided through fluoroscopic imaging. This realization and the very infrequent, but dramatic, gross overadministration of dose-causing patient radiation burns widely related in the press the past six years[6–9] has led to greater patient radiation dose consciousness on behalf of professional societies, the medical community, the public, and accrediting/regulating agencies; for example, the two recent American College of Radiology (ACR) White Papers on Radiation Dose in Medicine[10,11] and both the Image Gently[12] and Image Wisely[13] campaigns focused on patient radiation safety in imaging.

Eventually, comprehensive dose guidelines based on study protocol and patient age and taking into account body habitus will be specified, most likely by the ACR, which has already established basic dose reference levels[14] as well as extensive Appropriateness Criteria for imaging studies.[15] However, at this time, legislation at the state level is driving the rapid adoption of RDIMS.

The potential harm to patients through misadministration of radiation dose from imaging studies has led several states to require X-ray dose recording requirements. For instance, in addition to requiring that all facilities offering CT be accredited and surveyed annually by a qualified medical physicist (QMP), California Senate Bill 1237[16] requires documentation of administered radiation dose in the interpretive report of every CT study (except those used for therapeutic radiation treatment planning or for calculation of attenuation coefficients in nuclear medicine studies), and reporting dose misadministration to the referring physician and patient. As of the end of 2013, only California has enacted reporting requirements legislation, while Texas has enacted regulations to do so not only for CT, but also for fluoroscopic procedures.[17] The Joint Commission, healthcare's patient safety foundation, released in mid-2013, proposed requirements for diagnostic imaging services, effective July 2015, that echo many of the requirements of the California legislation.[18]

This heightened awareness of the possible misadministration of radiation dose and new regulation at the state level, thus far, has in turn led to the rush to implement RDIMS to meet these new requirements. In order to adhere to the ALARA principle to lower medical imaging radiation dose while still providing images appropriate for diagnostic decision-making, one must first record dose information in a database that allows trending and comparison of dose levels over time for various subunits (e.g., different facilities or scanners) and subgroupings (e.g., technologists) of the full dataset, a dose alert mechanism for studies exceeding specified thresholds, and longitudinal tracking of dose indicators for individual patients.

2.3 Where Do Radiation Dose Index Monitoring Systems Reside? What Is Its Domain?

RDIMS typically reside within the IT infrastructure of radiology departments, where there is a wealth of knowledge and experience with dosimetry and dose indices. Radiologists have training in ionizing radiation as applied to diagnostic examinations and are accustomed to seeing dose indices for those exams. Diagnostic medical physicists, usually resident in radiology departments, are experts not only in radiation dosimetry, but also in the analysis of dose indices and their limitations in assigning risk.

RDIMS are ideally deployed within the radiology department IT infrastructure, where there is extensive imaging IT infrastructure, for example, Picture Archiving and Communication System (PACS) and RIS, and knowledge of the standards involved: DICOM,[19] Health Level 7[20] (HL7), and Integrating the Healthcare Enterprise (IHE).[21] Many medical physicists will have experience and are conversant with imaging system IT standards from their involvement in the development, implementation, or utilization of PACS or other radiology IT systems at their institutions. These standards are discussed later herein.

Most, but not all, examinations that are cataloged in RDIMS are from radiology: CT, fluoroscopically guided procedures, mammography, nuclear medicine, and radiographic/fluoroscopic studies. For instance, mobile c-arm fluoroscopic units are common in operating rooms. Fluoroscopically guided interventional procedures are routinely performed in the cardiology catherization laboratory, electrophysiology laboratory, vascular laboratory, urology clinic, and pain management clinic, as well as in the angiography/interventional suites in radiology. In addition, some orthopedic clinics may utilize radiography and mini c-arms in their practice. So, while perhaps centered in radiology, the domain of the RDIMS extends throughout the entire medical enterprise, shadowing the medical physicist's annual survey footprint within the institution.

2.4 How Does a Radiation Dose Index Monitoring System Work?

Because the collected dose data across the enterprise comprises a multi-modality, multi-vendor environment, it is important that, like PACS and RIS, the RDIMS be vendor-agnostic and adhere to a set of standards applicable to all. The IHE initiative has created the radiation exposure monitoring (REM) profile[22] to define how RDIMS should similarly collect, evaluate, and export dose information. The REM profile addresses the modality transfer, storage, reporting, and registry submission of dose data as well as helping to ensure that policies and procedures, protocols, and QA processes are robust and reliably followed.

2.4.1 Elements of a Radiation Dose Index Monitoring System

The functional elements of a RDIMS are simple: a database, a core server with web services and communication servers (Figure 2.1). The ensemble should, of course, scale with the inclusion of multiple sites of practice and with large amounts of accumulated longitudinal data. It should also be obvious that the ensemble, as a medical IT system, should have hardware redundancy planned for and encoded into the software. Scalability and redundancy will be considered later in this section.

The database should, of course, be relational, accessible through Structured Query Language (SQL) requests, and robust enough to handle the large volumes of data expected over its operating lifetime. The web interface should adhere to the most up-to-date standards (e.g., HTML5) and also be viewed not only through popular extant browsers (e.g., Firefox, Chrome, Safari), but also those typically included in medical centers' IT workstation image distribution (e.g., Internet Explorer) applied to computers throughout the medical enterprise.

2.4.2 Communication Servers

A primary means of importing dose data into a RDIMS is through DICOM Radiation Dose Structured Report (RDSR; see Figure 2.1) based on DICOM Part 16 Content Mapping Resource for DICOM Structured Reports[23] and its predecessor Supplement 127: CT Radiation Dose Reporting (Dose SR),[24] which delineates the template identifier (TID) fields for RDSR, such as TID 10011–CT Radiation Dose, TID 1002–Observer Context, TID 10012–CT Accumulated Dose Data, and TID 10013–CT Irradiation Event Data. Each of these TIDs contain many subsidiary content items too numerous to elaborate herein. Principally however, RDSR contains an estimate of the accumulated exposure, acquisition parameters, and effective dose (depending on which dose estimate model is used). Multiple CT radiation

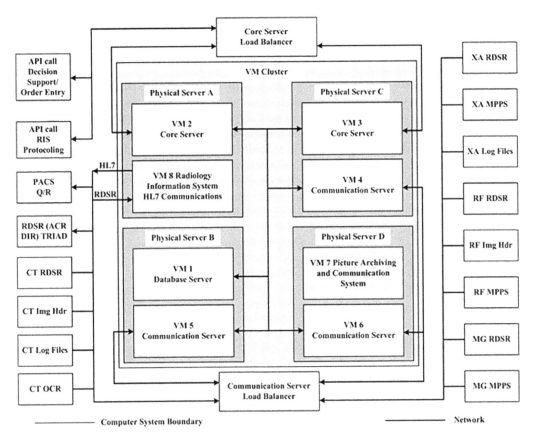

FIGURE 2.1 Schematic diagram of a RDIMS demonstrating the various communication channels between it, imaging modalities and other medical informatics systems (e.g., RIS). RDIMS = Radiation Dose Index Monitoring System, VM = virtual machine, CT = computed tomography, XA = X-ray angiography, MG = mammography, RF = radiographic/fluoroscopic, RIS = radiology information system, ACR = American College of Radiology, DIR = Dose Index Registry, TRIAD = transfer of images and data, HL7 = Health Level 7, RDSR = radiation dose structured report, MPPS = modality performed procedure step, Img Hdr = image header, OCR = optical character recognition, API = application programming interface, PACS = Picture Archiving and Communication System, Q/R = query/retrieve.

dose instances may be generated for a study if, for example, the study workflow is divided into multiple scanning regions, such as chest, abdomen, and pelvis.

DICOM Modality Performed Procedure Step (MPPS) is another means of transmitting dose information to a RDIMS (Figure 2.1). MPPS enables a modality to send reports about an ongoing study, including data about the images acquired, study begin and end times, and minimal information on the dose delivered. However, MPPS messages are transmitted from the modality as the study occurs and are not stored so that they cannot be retransmitted (replayed) as can RDSR. RDSR also contain much more robust dose information for a study. RDSR is thus far preferred over MPPS, but some older imaging devices may only support MPPS.

If RDSR is not available for a modality, it may also be possible to calculate a dose estimate based on information in the DICOM headers for the study image set (Figure 2.1). This necessitates, however, the transmission of the entire study to the RDIMS for DICOM header extraction and dose model calculation. In the case of CT studies, this may involve the transmission of over several gigabytes per study. In addition, both MPPS and image headers lack significant dose details, so utilizing RDSR is by far the preferred method; however, hybrid methods (e.g., both RDSR and DICOM headers) may provide a more comprehensive dataset.

Other, specialized RDIMS dose input methods include optical character recognition (OCR) of a DICOM secondary capture object (screen scrape) dose summary page, analysis of proprietary scanner vendor logs (may only be accessible if the vendor of the scanner and RDIMS are one in the same), and PACS query/retrieve (Figure 2.1). For legacy radiography/fluoroscopic units where none of the above is possible, an ionization chamber may be fitted and the serial output fed into a computer and the resulting dose data then formatted and forwarded to the RDIMS.

To leverage the value of the RDIMS, it is important that it integrates with other medical IT systems, becoming part of daily routine. The ability to update changes to and from other IT systems allows for the creation of a uniform dataset across the institution. The standards involved concern HL7 as well as outbound DICOM RDSR from the RDIMS.

HL7 is a standard for the electronic interchange of clinical, financial, and administrative information among independent healthcare-oriented computer systems. An inbound HL7 interface from a RIS or Hospital Information System (HIS) is important not only for updates of patient demographic information and for examination scheduling to be reflected in RDIMS, but also for correction of the manually entered CT study descriptor, which can be done at completion of the examination by a radiology technologist. In this case, an order message (HL7 message type "ORU") is sent through the HL7 interface to the RDIMS, which updates the study descriptor (Figure 2.1).

DICOM RDSR is the *sine qua non* of RDIMS used not only to import dose information from imaging modalities, but also as an aggregator of such data, to report study dose to other information systems (Figure 2.1). Some of these systems include HIS and RIS, decision support modules within order entry systems, and external dose aggregators, such as the ACR Dose Index Registry (DIR).[25]

The RIS imports RDSR for each performed study to inject specific dose-related elements from the RDIMS into associated RIS database schema fields (Figure 2.1). These dose-related fields may then be automatically inserted into the radiology report, for example, through a pre-defined voice recognition system macro. In the case of decision support modules within order entry systems, the transmitted dose information can be used to alert physicians ordering imaging examination utilizing ionizing radiation of the patient's dose history.

2.4.3 Core Server

The core server not only hosts the web services component of the RDIMS, but also data export, internal dose calculators, an e-mail server for alerts and reports, and lexicon mapping, as well as system and user management facilities.

The basic display screens for the application should include at minimum completed studies, a comprehensive display for each study, including details on patient demographic and high-level examination information, examination acquisition parameters, and dose indices. For each individual patient, there should be cumulative dose by modality and patient dose history in tabular format, as well as a timeline view of all irradiation events. It should also be easy to view dose comparison among database entities (sites, devices, personnel, age range) for standardized study descriptors or protocol names. Tabulations of highest dose studies and cumulative patient dose is helpful for identifying extreme outliers. Our group has found that statistical data is best presented in "box and whisker" plot format so that minimum/maximum, the 25th and 75th percentiles, and median value are easily discerned. This provides a rough representation of the underlying distribution without plotting it in its entirety (Figure 2.2) and is suggested for use by RDIMS vendors.

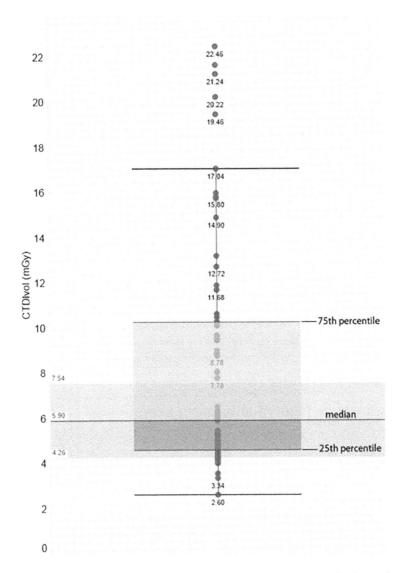

FIGURE 2.2 "Box and whisker plot" for pediatric (0–18-year-old) CT abdomen and Pelvis with Contrast protocol name series CTDIvol (mGy). The "box" (orange) is bounded by the 75th percentile (top) and 25th percentile (bottom) with the median value (boundary between lighter and darker orange). The asymmetry of these levels within the box is an indication of dose distributions typically being non-Gaussian with a long high-dose tail. The interquartile range (IQR) is given by the 75th percentile minus the 25th percentile. The top whisker is bounded by the greatest value data point less than or equal to the 75th percentile plus 1.5 times the IQR. The bottom whisker is bounded by the least value data point greater than or equal to the 25th percentile minus 1.5 times the IQR. Outliers are conventionally those data values outside the whiskers. The darker color bands extending outside the box and centered on the median is the median 95% confidence interval.

A primary function of these systems is the sending of alerts and notifications via e-mail. Alerts are based on thresholds set either manually or through periodic statistical analysis of dose information in the database. In order to provide meaningful alerts, it is necessary to fractionate the data into categories based on a number of parameters. With CT, meaningful parameters for describing the study are via study descriptor, protocol name, or RadLex Playbook ID (RPID). Other parameters meaningful to the sub-categorization of the data include age, weight, anatomical dimension, and the body mass index (BMI).

RadLex[26] is a comprehensive lexicon for the standardized indexing and retrieval of radiology information resources. The RadLex Playbook[27] is a special component of the RadLex-controlled terminology providing a standard, comprehensive lexicon of orderable radiology exams. Each individual "orderable" has a standardized study descriptor and associated RPID. Although initially applied to CT, RadLex now includes orderables for mammography, magnetic resonance, nuclear medicine, ultrasound, and radiography/fluoroscopy exams. RadLex CT orderable terminology has been adopted by the ACR for DIR.

An important function of the core server is to generate reports for specific modalities and specific time periods. These can be generated automatically and sent via e-mail to stakeholders, for example, monthly, or run on demand for a customized time duration. Typically, there are canned reports providing tailored information to various groups of stakeholders, including administrators, physicians, and physicists. Ideally, the ability is provided to create customized reports through SQL queries and dashboard software to generate output in a standard file format such as Adobe PDF.

Once the use of the RDIMS extends beyond a small group of users, for instance, medical physicists and CT technologists, to all physicians in the medical enterprise, it is important for administration of a large number of users that an enterprise-wide authentication for "single sign-on" be employed. This function can be provided by using Lightweight Directory Access Protocol (LDAP).[28] Rather than needing to enter every user directly into the RDIMS, providing them an individual account, the RDIMS passes the user's provided username and password for authentication. In addition, the enterprise user directory also returns the associated user credentials to verify authorization to access the RDIMS as well as the permission level (e.g., standard account or administrator access).

2.4.4 Radiation Dose Index Monitoring System Architecture

As with most IT systems today, the RDIMS should have an architecture that affords high availability. This can be achieved through performance scaling and fault-tolerance. Both are realized through use of virtual machines (VMs), as is common in today's IT world. Virtualization software running on physical server hardware (e.g., x86 or Linux computers) abstracts the physical processors, memory, storage, and resources into multiple VMs. Multiple physical servers can be tied together to create a server cluster with dozens of VMs resident on each producing prodigious processing capacity. A central process outside of the cluster, termed a hypervisor, manages the resource utilization of each VM and for the VM cluster as a whole.

For redundancy, the hypervisor can restart an unresponsive process using a VM on another server or call upon a backup VM allocated for that purpose. Also, server hardware nowadays is generally equipped with a high degree of hardware and software fault-tolerance, such as multiple network cards connected to independent subnets, multiple motherboards, and redundant array of independent disks (RAID) disk drives.

Using the RDIMS as an example, the core (web), database, and communication processes are run on different VMs (Figure 2.3). These VMs could be on the same or different physical server. It is possible to run multiple instances of these processes on several VMs for redundancy, providing a load balancing mechanism is employed. This would also allow for performance scaling should the installation expand beyond current requirements.

Although it is important that users be able to access the RDIMS whenever necessary, the ephemeral nature of MPPS and the onus to collect all dose data dictate that at least some level of redundancy and fail-over be implemented for the communication servers. As such, it is used below as an example of how this may be accomplished. Redundancy and fail-over could similarly be applied to the core and database processes as well.

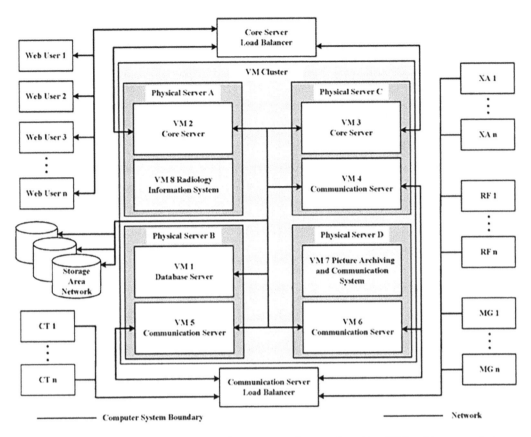

FIGURE 2.3 Schematic diagram of a RDIMS demonstrating the virtual machine (VM) architecture and storage, as well as the load balancers for multiple VMs as they broker the requested services to the various VMs (communication and core services). The communication server load balancer brokers input from the various modalities and queues each association according to the VM communication server best able to field it; likewise for the core server load balancer with each web request. RDIMS = Radiation Dose Index Monitoring System, CT = computed tomography, MG = mammography, RF = radiographic/fluoroscopic, XA = X-ray angiography.

For simple redundancy, each installation should include at minimum two communication servers. The simplest application involves projective load balancing by assigning an equal or nearly equal number of each modality type to each communication server. Vendors usually recommend *a priori* a maximum number of modalities to interface through a solitary VM allocated to communication depending on modality type and mechanism of communication (e.g., MPPS or RDSR). Each individual imaging source is assigned both a primary and backup communication server should the primary fail. However, a better approach is to have the collection of communication servers itself served by a load balancer process that fields communications from all imaging sources and switches each in turn to the communication server deemed best able to handle the incoming request. This simplifies redundancy, as there is no need to setup a backup DICOM destination for each imaging source unit, and also requires no assumptions for *a priori* projection of the load from each modality to specific communication servers. Monitoring the data used by the load balancer (e.g., CPU and RAM utilization) can be used to determine when it is necessary to add an additional VM communication server (e.g., memory saturation or greater than 80% CPU utilization).

2.5 Who Uses Radiation Dose Index Monitoring Systems? How Do They Use It?

The goal of RDIMS implementation is to make dose reduction an integral part of the workflow of existing routines. As such, it is important to examine how the various users of RDIMS might utilize the data and/or reporting from such systems.

2.5.1 Medical Physicists

As mentioned earlier, medical physicists are directly concerned with the safe and efficacious application of radiation to patients in the medical arts. This applied, for example, in making specific dose estimates for pregnant patients or those exhibiting post-fluoroscopically guided procedure erythema. However, medical physicists are also most likely to not only spearhead the implementation and possibly the management of a RDIMS, but also to analyze RDIMS data and present such data in support of institutional QA dose and ALARA processes.

2.5.2 Technologists

When a high-dose alert is issued from the RDIMS, it is the QA or lead technologists that will be tasked to enter commentary directly into the RDIMS as to the underlying cause. There is also important information in the RDIMS regarding the practice of ALARA by individual technologists, for instance, consistently higher dose than their peers or difficulty centering the patient in CT. The supervising technologist can then confer with the technologist regarding best practice to achieve ALARA doses.

2.5.3 Radiologist/Performing Physician

There are several ways radiologists can utilize RDIMS. First, working in conjunction with medical physicists to determine how changes in imaging protocols impact dose as well as image quality and diagnostic information content to maintain doses ALARA. Radiologists will also be the key audience in QA meetings regarding ALARA processes in their institution.

2.5.4 Ordering Physician

It seems unlikely that ordering physicians will have the time or curiosity in most cases to browse the RDIMS for patient dose information. However, one instance of where the just-in-time dose information would have greatest impact for ordering physicians is within the decision support aspect of order entry systems. Patient dose information from the aforementioned application programming interface (API) calls to RDIMS, in conjunction with ACR Appropriateness Criterion and other educational support, can assist the ordering physician in selecting the most appropriate radiological examination for the patient, all things, including any (albeit small) risk due to radiation, considered.

2.5.5 Patient

Although an initial impetus for the deployment of RDIMS, widespread access by individual patients has yet to be realized. Now that the Joint Commission has required documentation of radiation dose in some form or other of the medical record, it remains to be seen whether institutions will provide dose information to patients through traditional paper or electronic medical record document requests, such as the radiology report or through direct or brokered access to the RDIMS.

Radiation and its risks are general poorly understood by the public (as well as many physicians) and potentially controversial. However, as institutions begin leveraging their investment in RDIMS to patients for competitive advantage, relating dose information in a meaningful and most appropriate way to patients, however challenging, is essential.

2.5.6 Professional Societies and Regulatory Agencies

The ACR is a professional society already collecting dose data for CT from institutions across the USA. The ACR DIR is a database of dose indices essential to the creation of radiation dose guidelines and the setting of dose reference levels based on current clinical practice. Such data may also be used to drive standardization of protocols. Regulatory agencies like the Joint Commission will be increasingly interested in the utilization of RDIMS to track outlier cases, especially should any of them prove to be sentinel events. The wide-scale collection of dose data could eventually be useful for research assessing the health impacts of low-dose radiation in the population.

2.6 What Data Are Available? How Can It Be Utilized? How May It Be Presented?

2.6.1 Modalities

Typical modalities currently feeding into RDIMS include CT, fluoroscopically guided intervention, radiographic, mammography devices, and nuclear medicine/positron emission tomography (NM/PET). Dosimetry indices for these modalities are given below, recognizing their limitations as such. Importantly, these dose indices are calculated estimates based on generally non-anthropomorphic phantoms and do not represent actual patient dose in any instance. The emerging consensus is that adding dose estimates received by different parts of the body and for different modalities into a solitary cumulative value to assess a specific patient's risk is flawed.[29] Integration of nuclear medicine dose data into RDIMS is currently hampered due to the paucity of automated transmission of the actually administered radionuclide activity and is not covered in this chapter.

Even though a primary dose indicator associated with a CT study, it is important to realize that the CT dose index ($CTDI_{vol}$) is not the actual dose administered to an individual patient, but rather that absorbed by a standard 16-cm or 32-cm diameter polymethyl methacrylate (PMMA) phantom used by medical physicists to estimate dose. CT scanners use measured $CTDI_{vol}$ from these or similar phantoms at specified technique factors (kVp and mAs) to provide an estimate of the patient $CTDI_{vol}$ in mGy displayed on the modality console prior to scanning.

Another primary CT dose indicator is the dose length product (DLP) in units of mGy-cm, which is a weighted integration of the $CTDI_{vol}$ over each series scan length. Although DLP values are routinely summed over all series and reported as total DLP for CT studies and may be some indication of stochastic risk, not all of that dose may be applied to the same exact region of the body. Similar problems arise when trying to use summed $CTDI_{vol}$ values as a primary dose indicator, as the scan range for the various series may not be equal in magnitude or position. An alternative may be to sum the $CTDI_{vol}$ where the scan ranges overlap to determine the maximum $CTDI_{vol}$ applied to a specific region of the body. This and perhaps the median applied $CTDI_{vol}$ values could be the best simple dose indicators. Of course, the ultimate goal is to estimate organ dose and from this the effective dose for the study exposures. This is addressed later.

The primary dose indicator for fluoroscopic procedures is the air kerma (AK) in units of mGy. The advantage of AK is that it is an estimate that can be related to threshold deterministic effects, which might be observed post-procedure. However, for AK, current estimates displayed at the console are not the entrance skin dose (ESD), but rather calculated for a standard point in space termed the interventional reference point (IRP). It is possible, though difficult to provide an exact figure for the ESD due to variations of geometry, tube angulation, and changes in collimation during a procedure, as well as variation in patient body contour and uncertainty in AK accuracy upwards of ± 35%.[30]

The other dose indicator for fluoroscopic procedures is the dose area product (DAP), also more recently known as the kerma area product (KAP). KAP takes into account variation of collimation during a procedure, so that the magnitude of KAP is the same anywhere along the central line from the X-ray tube focal spot. However, as KAP may be equivalent for high AK with a tight-field collimation and lower AK with wide-field collimation, it is difficult to predict deterministic effects from KAP, but KAP would likely be a better estimator of stochastic effects possibly leading to carcinogenesis. Because deterministic effects are the primary concern for fluoroscopically guided interventional procedures, AK is the preferred dose indicator.

As with fluoroscopically guided studies, the primary dose indicator for projection radiography is the estimated ESD or AK. Usually this is predicted from generator technique factors (e.g., kVp and mAs) at a set distance above the tabletop or from the wall stand cover for an exposure, so would rarely correspond to an individual patient entrance dose. However, what is usually examined in RDIMS is the exposure index (EI) and the deviation index (DI) for the dose delivered to the image detector itself. Formal definitions of EI and DI are given by the AAPM TG 116 report,[31] but essentially EI is proportional to the dose absorbed by the detector for a specified anatomical region. DI relates how much EI varies from the institutionally determined target dose associated for that study's anatomical region and radiographic view (e.g., posterior-anterior or lateral). EI cannot be used to estimate patient dose, but EI and DI are used for QC purposes as immediate post-exposure feedback to the technologist that the exposure was either overexposed (DI > +1), exposed appropriately +0.5 < DI < −0.5, or underexposed and possibly requiring re-exposure (DI < −1). EI and DI may also be used in a QA process to determine if radiographic phototimers are set properly and in the spirit of ALARA if the target EI values might be reduced.

For mammography studies, the estimated average glandular dose (AGD) in units of mGy is the primary dose indicator. AGD is estimated from standardized exposure measurements performed by a medical physicist or vendor field service engineer of the ACR accreditation phantom, taking into account the kVp, beam filtration half-value layer (HVL), and target filter combination, as well as detected breast composition. The mammography device uses this information, scaling by the actual technique factors used for an exposure to calculate the estimated AGD.

2.6.2 Data Analysis and Reliability Concerns

Generation of dose metrics (e.g., median, percentiles) can provide benchmarks against which the effectiveness of dose reduction processes can be compared. Based on the twelve-year experience analyzing dose data at our institution, perhaps the most valid means of analyzing the data is through use of percentiles such as the median or 50th percentile. Calculation of the average allows outliers (extreme or otherwise) to unduly influence the analysis. This is not the case with the median.

Prior to implementing a RDIMS when manual entry data-entry errors (e.g., misplaced decimal points) were possible, dose data were pre-filtered according to how many standard deviations (SD) an outlier was from the median and removed from analysis if greater than 5 SDs away, although utilized for investigation of possible dose misadministration. For fluoroscopically guided procedures, we also looked for unrealistically high dose rate values (e.g., dose rate far greater than 10 R/minute in the absence of high-level control) to discount possible reporting errors (e.g., entering seconds rather than minutes of fluoroscopy time).

Fortunately, with RDIMS, as machine reported dose indices are digitally conveyed, the reliability of the numbers cascading through from source to sink is essentially flawless. The primary concern for the representative nature of the RDIMS-reported dose indices is the accuracy of the ion chambers used to measure the ionization potential of the X-ray beam itself. The ionization chambers (diamentors) used for fluoroscopic systems are required by regulation to only be accurate within ±35% of the actual value (as measured by a calibrated dosimeter).[30] Errors may also be made by service personnel when entering a measured dose value, for example, into the generator of a projection radiography device, which scales this value according to technique factors to estimate the established reference point entrance surface dose in the DICOM header and reported through MPPS or RDSR. It is essential that X-ray-generating devices be surveyed annually by a qualified medical physicist to not only measure output with a calibrated dosimeter but validate that the dose indices reported through the RDIMS are accurate as well.

An important function of analysis is to detect deviations from routine clinical practice and to alert stakeholders to such deviation and to investigate and document the cause as well as follow-up with the patient if there is possibility of deterministic effect. In order to determine and cite deviations, periodic statistical analysis of the dose data is necessary as well as deciding on alert threshold levels. These thresholds may be fixed or statistically determined.

Fixed thresholds can be determined by either analysis of the database at a specific point in time or through the use of virtual target values or diagnostic (or dose) reference levels (DRLs).[32] Virtual target values are set by the institution and could be based on, for instance, the $CTDI_{vol}$ values set in the DoseCheck (NEMA XR 25)[33] software to flag, pre-exposure, significant protocol deviations on CT scanners. In addition, with the ACR DIR, periodic dose indices for CT exams both regionally and nationally based on body part and examination type are available and could be used for virtual target values as a means of benchmarking. DRLs are published recommendations for examination dose or a dose-related indices by professional societies (e.g., ACR), advisory councils, or through regulation.

For statistically determined thresholding, in order to avoid "alert fatigue," it is necessary to set the thresholds so that only those outside two standard deviations or thereabouts are flagged This can be easily done, for example, by setting a warning threshold at the 90th percentile and an alert threshold at the 95th percentile. There are many other methods of threshold setting, for example, based on the inter-quartile range or histogram breakpoint methods, but an interesting new concept is that of acceptable quality dose (AQD).[34]

2.6.3 How May It Be Presented?

In our experience over the decade before the implementation of our RDIMS, we found that it is easy to present too much data at monthly or quarterly QA conferences. It is best to tailor the data presented to the audience (e.g., radiologists) and limit the data presented to actionable occurrences (e.g., statistically significant dose increase for a particular exam). While some analysis variables may prove useful, the primary question is how to best analyze the data and format reports to meet the needs of an institution's dose management policy and dose-reduction plans.

With RDIMS, in addition to the identification of dose outliers, it is possible to interrogate the database in many ways, for instance, breakdown by staff, facility, modality, exam protocol, and device. This provides feedback to enable more effective process controls. Analysis by staff can allow for identification of training needs for specific personnel. Analysis by device can help guide new purchase decisions, identify maintenance needs, and reinforce utilization of low-dose equipment. Viewing trends allows for follow-up on dose reduction program effectiveness.

RDIMS should continuously communicate dose reduction performance data to as many stakeholders (administrators, physicians, physicists, technologists) as possible at their desired level of interaction. For most, this will involve e-mail alerts for examinations exceeding pre-determined thresholds. However, it also may involve the production of periodic reports and dynamic dashboards.

2.7 Advanced Features

2.7.1 Size-Specific Dose Estimate

An advanced featured of a RDIMS is Size-Specific Dose Estimate (SSDE) as delineated by AAPM Task Group 204.[35] SSDE attempts to better estimate the actual patient absorbed dose in CT, rather than in terms of the $CTDI_{vol}$ displayed on the modality console and subsequently reported through RDSR.

The calculation of SSDE is accomplished through the use of conversion factor lookup tables applied to $CTDI_{vol}$ based on the effective diameter of the patient. The effective diameter is estimated by contour image processing of patient anatomy over the scan range based on either single or dual-orthogonal scout images. Generally, should the patient anatomy prove smaller than the PMMA phantom, then the SSDE would be larger than the CTDI and contrariwise (Figure 2.4).

In addition, through information from the images and DICOM header, the patient isocenter shift for CT exams is calculated. Isocenter shift information is important, as patient dose can increase rather dramatically with even seemingly minor offset distances.[36]

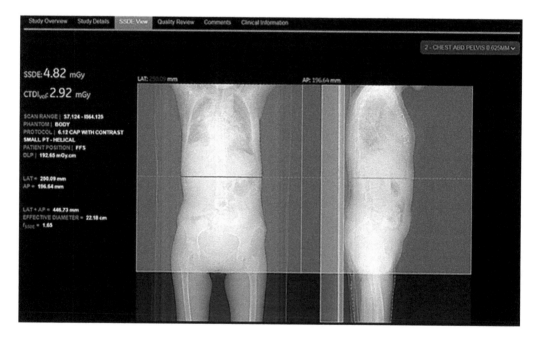

FIGURE 2.4 SSDE view webpage for a thin patient from the RDIMS used at our institution (GE Healthcare DoseWatch). The effective patient diameter was calculated from the machine-estimated patient lateral dimension (PA scout) and AP dimension (lateral scout). In this instance, as the effective patient diameter is far smaller (22 cm) than the PMMA phantom used to measure the CTDIvol (32 cm), the SSDE of 4.82 mGy is far larger (practically a factor of 65% larger) than the reported CTDIvol of 2.92 mGy on the CT scanner console. This demonstrates the value of SSDE towards better, but nowhere exact, estimation of the dose a patient actually receives. SSDE = size-specific dose estimate, RDIMS = Radiation Dose Index Monitoring System, CT = computed tomography, PA = posterior-anterior, AP = anterior-posterior, PMMA = polymethyl methacrylate, CTDIvol = CT Dose Index (volume). Note: screenshot representative of DoseWatch 2.1 current as of January 2017; DoseWatch a trademark of General Electric Company.

2.7.2 Fluoroscopically Guided Intervention Dose Partition by Angulation

Another advanced feature of a RDIMS is display of two-dimensional angulation data for fluoro-scopically guided interventional studies with either single or bi-plane systems. Angulation data for each detector is tabulated during the study and stored in the imaging system and then transmitted to the RDIMS through DICOM RDSR. The RDIMS then bins the run data into a two-dimensional tile grid (Figure 2.5). The two orthogonal axes of this grid are given by the Cranial-Caudal (CRA-CAU) and the Right and Left Anterior Oblique (RAO-LAO) directions. Thirty-degree-by-thirty-degree tiles are typical and the interventional reference point (IRP) dose distributed amongst the tiles based on the geometry of the exposures. Usually the peak accumulated AK is given is as a sum of the tile with the maximum dose and the highest-dose bordering tile. Although this provides much better dose information than simply assuming (in the absence of geometric information) that all the dose was deposited at the same skin location, it still does not provide the patient ESD unless geometric estimations of table height, table and pad attenuation, and collimation (backscatter factor) can be applied.[37]

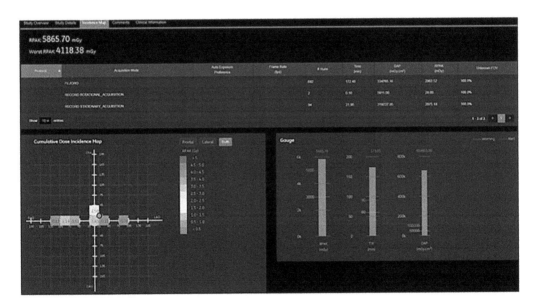

FIGURE 2.5 Fluoroscopically guided intervention dose partition by angulation map from the RDIMS used at our institution (GE Healthcare DoseWatch). Each tile represents the air kerma at the IRP for a 30-degree-by-30-degree solid angle as viewed from the X-ray tube focal spot. In this instance, the total reference point air kerma (RPAK) measured by the system ion chamber (diamentor) is 5865.70 mGy (5.87 Gy). In the absence of this angulation data, the medical physicist might have to assume that this is the peak RPAK, in excess of our institutional substantial radiation dose patient follow up trigger level of 5 Gy (possible permanent erythema), upon which an alert notice is sent to the attending physician. However, examination of how this dose is spread out over solid angle provides a worst or peak RPAK of only 4.12 Gy (sum of dose tiles around the orange dot). In this case, the patient may only experience temporary erythema or temporary epilation. The extra information provided by the 2D dose map is valuable in reduction of false positives and better, but not exact, estimation of patient entrance skin dose. RDIMS = Radiation Dose Index Monitoring System, IRP = interventional reference point. Note: screenshot representative of DoseWatch 2.1 current as of January 2017; DoseWatch a trademark of General Electric Company.

2.7.3 Free-Form Data Analysis

Rather than relying entirely on RDIMS vendors to provide desired dashboards and reports, the medical physicist can dump the data from the RDIMS in some standard format, for example, comma separated variable (CSV) or Microsoft Excel. They can then proceed to analyze the data using a number of tools: SPSS, MATLAB®, etc., which they may have developed prior to RDIMS implementation. However, it would seem to be advantageous for vendors to provide a free-form data analysis component function within the RDIMS, where multivariate analysis may be pursued.

2.7.4 Protocol Management, Correlation/ Visualization of Changes in Protocols

Another advanced feature of RDIMS is protocol management. Protocols are usually created as mainly static documents and made available through institutional intrancts. There is little thought as to enacting change control processes, how to manage version control, who is authorized to modify protocols, or provide an audit trail of modifications.[38] RDIMS can provide these features, but it may also offer within the RDIMS visualization of protocol deviation at the examination level. It might also in the near-term estimate changes in dose for various alterations of protocol technical factors and in the future demonstrate how image quality would be affected by proposed protocol modification.

2.8 Challenges

2.8.1 Study Descriptor Creep

The current state of some CT scanners is such that a technologist manually enters the study descriptor into the console. Unless a strict standard nomenclature is enforced (rare in the author's experience), the number of lexical variants resulting can be quite large. For instance, at our institution last year, the number of unique lexical study descriptors was increasing at a rate of around six per day. The proliferation of lexically unique exam descriptors essentially "diffuses" the data for a specific exam over multiple study descriptor text strings, making it more difficult to assign correct threshold values for alerts.

2.8.2 Advanced Analysis to Compute Patient-Specific Dose

As mentioned previously, the dose estimates currently residing in the RDIMS database do not current allow for patient-specific dose computation. Medical physicists calculate effective dose as an intermediate result in assessing the perceived risk of an exposure. CT effective dose can be estimated through multiplication of the DLP by an anatomical "k" factor.[39] To attempt to generate a patient-specific dose, there must be a means of automatically analyzing image sets to correctly assign anatomical boundaries. It is also necessary to utilize patient-specific, spatially oriented anatomical volumes with Monte Carlo simulation[40] techniques to derive patient-specific dose estimates.[41]

2.8.3 Surface Dose Mapping

The dose displayed for fluoroscopically guided procedures in current RDIMS is not the skin dose. While it is possible to generally approximate the geometries involved into an AK conversion factor to estimate the skin dose,[37] it is better through analysis of vendor-supplied variations of geometry, tube angulation, and changes in collimation during a procedure to create a map on the surface of a 3D regular volume (e.g., elliptic cylinder for torso or ovoid for head; Figure 2.6).[42] Even better would

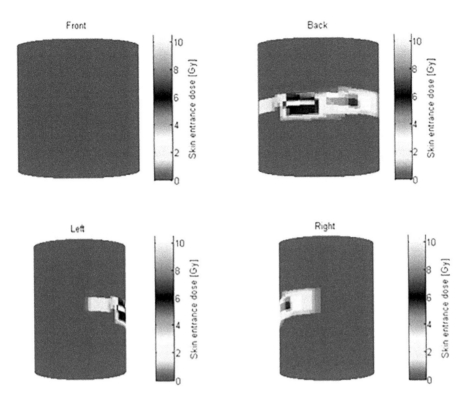

FIGURE 2.6 Surface dose map for a fluoroscopically guided intervention patient at our institution receiving an indicated RPAK in excess of 15 Gy. This caused the author to initiate a sentinel event root cause analysis. All information stored on the angio system for this case was analyzed by the angiography system vendor and projected onto the surface of an elliptical cylinder to determine the peak entrance skin dose. This was determined to be only 10.4 Gy, far lower than the 15 Gy Joint Commission sentinel event threshold. However, it did trigger a greater than 5 Gy threshold alert notice to the attending physician. Surface dose mapping goes beyond 2D solid angle mapping in that the skin dose is shown distributed on the body part avatar, which may be of assistance to the attending in relating the skin area(s) of concern to the patient's healthcare team. Also in this case, the calculation tiles are much smaller than those for the current 2D solid angle mapping, so a better estimate of peak skin dose results. RPAK = reference point air kerma.

be use of scaled 3D regular volumes based on patient dimensions. Whether exact 3D modeling of the patient surface (e.g., from a 3D scanner like the Microsoft Kinect or from CT slice information) would significantly improve the dose estimate remains to be demonstrated.

2.8.4 Real-Time Dose Feedback in the Fluoroscopically Guided Interventional Procedure Suite

For lengthy fluoroscopically guided procedures wherein the estimated surface entrance dose might exceed the threshold dose for deterministic skin damage, it would be valuable to view either the 2D dose grid or 3D surface map in real- or near real-time so that modifications like tube angulations changes could be used to help spare possible skin reaction, possibly for multiple procedures performed over a short time period. Real-time transmission of RDSR messages at the run or frame level would be required, but there are questions on the part of some vendors regarding the requisite bandwidth impacting the clinical operation of their systems.

2.9 Summary

An effective radiation dose management program is much more than simply implementing RDIMS. RDIMS is simply a tool relieving medical physicists and imaging physicians from the burdens of data collection (usually from multiple disparate sources) and management of such an extensive dataset. It is still incumbent on stakeholders to augment local policies, local reporting requirements, annual review, and adjustments, based on the ability to effectively mine, analyze, and visualize this dataset. RDIMS are still relatively new, so many advanced and innovative features to enable successful institution-wide radiation dose management programs are expected to be introduced in the near future.

References

1. ICRP. 2007. Radiological Protection in Medicine. ICRP Publication 105. *Ann ICRP* 37(6): 1–63.
2. Gress, D. A., Dickinson, R. L., Erwin, W. D. et al. 2017. AAPM medical physics practice guideline 6.a.: Performance characteristics of radiation dose index monitoring systems. *J Appl Clin Med Phys* 18: 12–22.
3. Stewart, B. K., Kanal K. M., Perdue J. R., Mann F. A. 2007. Computed radiography dose data mining and surveillance as an ongoing quality assurance improvement process. *Am J Roentgenol* 189: 7–11.
4. Schauer, D. A., Linton O. W. 2009. National council on radiation protection and measurements report shows substantial medical exposure increase. *Radiology* 253: 293–296.
5. Wall, B. F. 2009. Ionising radiation exposure of the population of the United States: NCRP Report No. 160. *Radiat Prot Dosim* 136: 136–138.
6. Domino, D. Settlement reached in Mad River pediatric CT radiation case. *Aunt Minnie.* Available at: http://www.auntminnie.com/index.aspx?sec=ser&sub=def&pag=dis&ItemID=90713. Accessed July 20, 2017.
7. Santa Cruz, N. Cedars-Sinai finds more patients exposed to excess radiation. *The Los Angeles Times.* Available at: http://articles.latimes.com/2009/nov/10/local/me-cedars10. Accessed July 20, 2017.
8. Bogdanich W. After stroke scans, patients face serious health risks. *The New York Times.* Available at: http://www.nytimes.com/2010/08/01/health/01radiation.html. Accessed July 20, 2017.
9. US Food and Drug Administration. FDA unveils initiative to reduce unnecessary radiation exposure from medical imaging. Available at: https://www.fda.gov/radiation-emittingproducts/radiationsafety/radiationdosereduction/ucm199994.htm. Accessed July 20, 2017.
10. Amis, E. S., Butler P. F., Applegate, K. E. et al. 2007. American College of Radiology white paper on radiation dose in medicine. *J Am Coll Radiol* 4: 272–284.
11. Amis, E. S., Butler, P. F. 2010. ACR white paper on radiation dose in medicine: Three years later. *J Am Coll Radiol* 7: 865–870.
12. http://imagegently.dnnstaging.com/. Accessed March 26, 2014.
13. http://www.imagewisely.org/. Accessed July 20, 2017.
14. ACR–AAPM practice guideline for diagnostic reference levels and achievable doses in medical x-ray imaging. 2013. https://www.acr.org/~/media/ACR/Documents/PGTS/guidelines/Reference_Levels_Diagnostic_Xray.pdf?la=en. Accessed July 20, 2017.
15. http://www.acr.org/Quality-Safety/Appropriateness-Criteria. Accessed July 20, 2017.
16. California Senate Bill 1237. http://www.leginfo.ca.gov/pub/09-10/bill/sen/sb_1201-1250/sb_1237_bill_20100929_chaptered.html. Accessed July 20, 2017.
17. 25 TEXAS ADMINISTRATIVE CODE §289.227 Use of Radiation Machines in the Healing Arts (effective May 1, 2013). http://www.dshs.state.tx.us/WorkArea/DownloadAsset.aspx?id=8589975090. Accessed July 20, 2017.

18. The Joint Commission. Revised Requirements for Diagnostic Imaging Services https://www.joint-commission.org/assets/1/18/AHC_DiagImagingRpt_MK_20150806.pdf. Accessed July 20, 2017.

19. DICOM website: http://dicom.nema.org/. Accessed July 20, 2017.

20. Health Level Seven International website: https://www.hl7.org/. Accessed July 20, 2017.

21. Integrating the Healthcare Enterprise (IHE) website: http://www.ihe.net/. Accessed July 20, 2017.

22. O'Donnell, K. 2011. Radiation exposure monitoring: A new IHE profile. *Pediatr Radiol.* Accessed March 26, 2014.

23. Digital imaging communications in medicine (DICOM). 2009. *Part 16: Content Mapping Resource.* NEMA Standards Publications. Rosslyn, VA.

24. Digital imaging communications in medicine (DICOM). 2007. Supplement 127: CT Radiation Dose Reporting (Dose SR). NEMA Standards Publications. Rosslyn, VA. (2007).

25. American College of Radiology Dose Index Registry website: http://www.acr.org/Quality-Safety/National-Radiology-Data-Registry/Dose-Index-Registry. Accessed July 20, 2017.

26. http://www.rsna.org/RadLex.aspx. Accessed July 20, 2017.

27. http://rsna.org/RadLex_Playbook.aspx. Accessed July 20, 2017.

28. The Internet Engineering Task Force (IETF) Request for Comments: 4510. Lightweight Directory Access Protocol (LDAP): Technical Specification. https://tools.ietf.org/rfc/rfc4510.txt. Accessed July 20, 2017.

29. Durand, D. J., Dixon, R. L., Morin, R. L. 2012. Utilization strategies for cumulative dose estimates: A review and rational assessment. *J Am Coll Radiol* 9: 480–485.

30. https://www.accessdata.fda.gov/scripts/cdrh/cfdocs/cfcfr/CFRSearch.cfm?FR=1020.32. Accessed August 1, 2017.

31. Shepard, S. J., Wang, J., Flynn, M. et al. 2009. An exposure indicator for digital radiography: AAPM Task Group 116 (executive summary). *Med. Phys.* 36: 2898–2914.

32. Vassileva, J., Rehani, M. 2015. Diagnostic reference levels. *Am J Roentgenol* 204: W1–W3.

33 http://www.nema.org/standards/Pages/Computed-Tomography-Dose-Check.aspx. Accessed July 20, 2017.

34. Rehani, M. M. 2015. Limitations of diagnostic reference level (DRL) and introduction of acceptable quality dose (AQD). *Br J Radiol* 88(1045): 20140344.

35. Boone, J. M., Strauss, K. J., Cody D. D. et al. 2011. Size-specific dose estimates (SSDE) in pediatric and adult body CT examinations. Report of AAPM Task Group 204, College Park, MD.

36. Li, J., Unni, K., Udayasankar, T. L. et al. 2007. Automatic patient centering for MDCT: Effect on radiation dose. *Am J Roentgenol* 188: 547–552.

37. Dickinson, R. L., Zamora, D. A., Kanal, K. M., Stewart, B. K. 2014. Estimated skin dose look-up tables and their effect on dose awareness in the fluoroscopy-guided imaging suite. *Am J Roentgenol* 203: 630–636.

38. Cody, D. D., Fisher T. S., Gress, D. A. et al. 2013. AAPM medical physics practice guideline 1. a: CT protocol management and review practice guideline. *J Appl Clin Med Phys* 14(5): 3–12.

39. AAPM Report No. 96. 2008. The measurement, reporting, and management of radiation dose in CT. American Association of Physicists in Medicine, College Park, MD.

40. Ding, A., Gao, Y., Liu, H. et al. 2015. VirtualDose: A software for reporting organ doses from CT for adult and pediatric patients. *Phys Med Biol* 60: 5601–5625.

41. Lee, C., Kim, K. P., Bolch, W. E. et al. 2015. NCICT: A computational solution to estimate organ doses for pediatric and adult patients undergoing CT scans. *J Radiol Prot* 35: 891–909.

42. Boujan, F., Clauss, N., Jahn, C. et al. 2012. A new method of real-time skin dose visualisation: Clinical evaluation of fluoroscopy-guided interventions. *Insights Imag* 3(Suppl 1): S380.

3

Nominal and Collective Radiation Dose in Medical Imaging: Critical Review of Survey Studies

David C. Spelic

3.1 Introduction

It is likely that ever since radiation injuries first began appearing soon after the discovery of x-rays, there was a recognized need to know two things[1]:

How much radiation does it take to cause an injury?
How many people were being exposed sufficiently to suffer injury as a result?

Regulatory- and clinical-based efforts to characterize and improve the practice of x-ray-based medical imaging typically rely on responses to these two seemingly simple questions. However, even well into the twentieth century, there was no standardized means to characterize exposure and dose from x-rays. Even today it can be challenging to infer values of dose or even of simpler clinical practice indicators regarding certain populations of radiological interest, such as the number of pediatric computed tomography (CT) examination conducted annually in the U.S. In 1928, the Radiological Congress held at Stockholm introduced the definition of *exposure*, providing a pathway for quantifying not only individual but also collective dose in radiological imaging.[2] Today, there is a wealth of published work to assist with answering these and similar questions. However, nowadays these efforts focus on a different set of questions, such as:

> *How much radiation dose to the patient is sufficient to accomplish the clinical task and provide minimal risk?*
>
> *How many people in an identified population are undergoing radiation-based examinations?*
>
> *How frequently are these examinations resulting in radiation doses that could be lower (or higher) and still satisfy the clinical task?*

Advances in medical imaging technology and clinical practice prompt the need to periodically revisit these questions. This chapter will provide a review of past and recent efforts to answer such questions by means of population surveys.

3.2 Population Surveys

There are numerous benefits to collecting representative data by means of population surveys. These surveys, when done appropriately, can capture a snapshot in time of the state of clinical practice, and also contribute to the broader data collective by which trends over time are documented. Figure 3.1 shows trends in mean glandular dose (to a patient-representative phantom) and image quality in

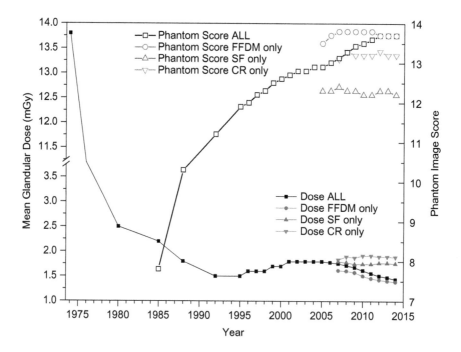

FIGURE 3.1 Trends in mammography dose and image quality. Values are shown for all sites, and separately for screen-film systems (SF), computed radiography (CR) and full-field digital mammography (FFDM). (From U.S. Food and Drug Administration FDA website at http://www.fda.gov/Radiation-EmittingProducts/MammographyQualityStandardsActandProgram/FacilityScorecard/ucm326264.htm., Last updated on 11/29/2017. ScoreYear.)

mammography in the U.S. over a period of several decades. These data were acquired mostly by means of surveys.[3] One notable feature of this graphic is the capture of both dose and indicators of image quality. The principle of "as low as reasonably achievable" (ALARA),[4] while originally directed at the protection of radiation workers, can also guide clinicians' efforts to limit patient doses to levels sufficient to accomplish the clinical task. Assessment of image quality can thereby provide an outcomes-based platform for evaluating the task-appropriate radiation dose. More recently, the concept of optimization of radiation dose in diagnostic x-ray imaging provides a framework for ensuring that patients receive doses that are sufficient for the diagnostic task.[5] For example, diagnostic reference levels provide clinical facilities with exam/procedure-specific target values for patient dose indicators (e.g., values for dose-area product) they can implement as one means of optimizing their program of diagnostic x-ray imaging.

In addition to characterizing dose and image quality, population surveys provide a means of capturing a broad range of clinical practice features, such as quality-control and quality-assurance practices, patient examination volumes, equipment features that can impact dose and image quality such as AEC (automatic exposure control), and staffing descriptors such as training and credentialing. Finally, survey data can provide input into more extensive activities, such as epidemiological studies.[6,7]

A critical component of planning a population survey is the identification of a statistically representative sampling methodology for survey. An important question to ask before developing a sampling strategy is whether the data should capture the state of practice as it exists or document achievable indicators of high-quality practice, such as may be used to develop recommendations for diagnostic reference levels. Depending on the goals of the survey, a number of factors may affect the sampling strategy, such as particular considerations for pediatric examinations, regional differences in the patient population or the clinical practice for the surveyed examination(s), and identification of a sufficient sample size to permit stratification of the survey data across different sub-groups, for example, average dose indicators for hospital sites compared with facilities other than hospitals.

The characterization of dose generally includes assumptions and simplifications, including the capture of suitable radiation measurements to permit one to infer desired dose parameters. For example, a metric of choice to characterize patient doses from radiographic projection examinations is typically entrance surface air kerma. Another important consideration is the method of data collection for conducting dosimetry. Typically, this will involve either the use of patient-representative attenuation phantoms to standardize the measurement procedures or the collection of data from actual patient examinations. Depending on the goals of the survey, the capture of associated peripheral data elements should be considered and may be needed to complete the dosimetry task such as equipment descriptors (e.g., multi-slice format for CT scanners), clinical technique factors, and radiation beam quality (half-value layer). Practice-related parameters that are helpful to characterize radiation dose include the retake frequency, where patients undergo repeated exposures during the examination or receive two different exposures for a complete examination (e.g., a chest radiographic examination comprising a lateral and posteroanterior [PA] projections). In characterizing the doses from CT, for example, collected data should include the fraction of CT examinations that consist of more than one scan.

Finally, the benefits and drawbacks associated with remotely collecting survey data should be considered. Facilities may not be equipped or staffed to take intensive measurements on x-ray equipment but may be able to provide some examination indicators for a small number of patient studies. It is very likely, however, that the success rate of such remote data-gathering methods varies inversely with the volume of requested data: survey sites will generally only provide what is convenient (or required) for them to provide.

3.3 Radiation Dosimetry

Rarely is actual patient dose, such as absorbed dose, captured as part of a survey. More likely, the dosimetric data harvested during these activities are dose *indicators* that can be used to further characterize dose. For example, the dose index values reported to end users or directed to a computer database in newer CT equipment typically include values for volume CTDI ($CTDI_{vol}$). Similarly, for fluoroscopy, a dose indicator that can conveniently be captured is cumulative air kerma to an identified reference point. One program that does collect actual dose (albeit mean glandular dose to a breast phantom of standard size and tissue composition, not to an actual patient) is the U.S. mammography inspection program, mandated by the MQSA (Mammography Quality Standards Act) of 1992.[8] A challenge in survey planning, therefore, is identifying the dosimetric endpoint for the activity, and capturing appropriate and sufficient data to achieve this survey goal. While there are numerous methods to achieve this, generally the capture of dosimetry data will rely either on phantom-based or patient-based methods. We will briefly discuss the advantages and drawbacks of each method.

3.4 Phantom-based Versus Patient-based Dosimetry

One convenient way of inferring patient dose from radiological examinations or procedures is to use phantoms. These devices are typically composed of convenient materials such as polymethylmethacrilate (PMMA) and are designed to provide x-ray attenuation properties similar to that of typical patients.[9] One of the earliest known applications of a device to represent a real patient for a specific x-ray examination was the use of a coconut as a model for the human head![10] For the survey of radiographic equipment, patient-equivalent phantoms are used to drive the AEC system to a level of x-ray output that would typically result for a real patient. Phantom-based methods have the advantage of permitting the capture of patient-representative dose measurements with a single collection event at each site. The use of phantoms also guarantees that the same patient-representative characteristics are employed across all surveyed sites. These devices can also be fitted with image quality test objects for evaluation under typical clinical conditions. This level of convenience also introduces a limitation because phantom-based methods characterize the response of the x-ray equipment only for the particular patient size it was designed to model. Patient-based survey efforts can provide a statistical body of data not possible with standard phantoms, such as characterizing the range of doses for adult body CT examinations performed on newer scanners that are equipped with means for modulating the x-ray tube current based on patient size.

Patient-based surveys typically require more site-level effort depending on the number of clinical examinations requested from each survey site, the frequency with which the examination is conducted for the patient population of interest, and the scope of data sought for in each examination. Further, an examination may be conducted infrequently (e.g., x-ray examinations on pediatric patients) that the collection of an acceptable volume of patient-level data is difficult or impossible to achieve. Dosimetry can be collected from available post-exposure display values on the equipment, for example, $CTDI_{vol}$ and dose-length product (DLP) from newer CT equipment. For equipment not equipped with such dose display and recording features, measurements would need to be acquired for patient examinations. A significant benefit to gathering patient-based data is the ability to conduct statistical studies on captured dose parameters, such as distributions of dose indicators with patient age and size. Further, the datasets of participating facilities can be combined to improve the statistical robustness of data analyses. In 1992 the National Radiological Protection Board (now a part of Public Health England) published a comprehensive protocol for the collection of patient-representative x-ray measurements for diagnostic radiology. The document specifies procedures for identifying representative patients, types of radiographic examinations to report, and dose indicators to be collected.[11]

3.5 Dose Quantities

The appropriate dose indicators for survey collection depends on the imaging modality. A thorough treatment of the various radiation quantities appropriate in medical x-ray imaging is provided in the ICRU Report 74.[12] Dosimetric quantities for diagnostic radiographic examinations include entrance skin air kerma (with or without backscatter) and dose-area product (DAP) (or kerma-area product (KAP)]). For CT, the computed tomography dose index (CTDI) is typically collected, and there are several versions of this particular dose index, such as $CTDI_{100}$, $CTDI_{fda}$, $CTDI_{free\text{-}air}$, $CTDI_w$, and $CTDI_{vol}$.[13]

One dose quantity that has been widely utilized in expressing overall risk is effective dose (ED). A thorough discussion of ED is beyond the scope of this work; Huda and Gkanatsios, and McCollough and Shueler, provide a thorough discussion of this dosimetric parameter.[14,15] Briefly, ED is the absorbed dose to the entire body that results in the same overall stochastic radiation detriment as the ED of the particular dose event under study. ED is typically computed from survey dose quantities and computational resources that are usually modality-specific. Its utility lies in its applicability to comparisons across different modalities. ED is intended to be used for specific populations and is not intended for individual members of the population. While one cannot compare entrance air kerma for a routine PA chest radiograph to the $CTDI_w$ for a routine chest CT scan, one can compare their EDs.

3.6 Survey Challenges

As for any means of characterizing radiation dose for a radiological examination, one must recognize the limitations and sources of errors associated with broad multi-site surveys. In particular, for quantifying indicators of dose, it is essential that all data values are derived using the same methodologies from site to site, such as when requesting clinical sites to collect and report dose indicators. For example, dose values from newer imaging instruments that are capable of reporting selected dose indicators, such as through a digital imaging and communications in medicine (DICOM) object (Radiation Dose Structured Report, or RDSR[16]), should be validated to ensure they are based on the same computational methodology and reflect similar practice characteristics. The American College of Radiology (ACR) has made a concerted effort to standardize the collection of dosimetry and related data for CT, including efforts to standardize the reporting of the CT examination names, which often vary broadly across institutions and scanners.[17] In cases where there are likely to be differences, these should be identified prior to the survey and addressed during the data analyses.

Measurements with instrumentation-related differences should also be addressed early to avoid the challenges of validating data post-survey. For example, commercially available instrumentation for the measurement of air kerma is now often configured with a solid-state-based detector, as opposed to the reference-standard ionization chamber-based technologies. These two technologies in combination with their associated electronics may respond differently during the x-ray event, not just with regard to the spectral character of the beam but also with regard to the beam modulation and means of triggering the measurement event. Image quality assessment can pose a similar challenge, such as when images of test objects are assessed on site under differing viewing conditions or by different readers across multiple sites. Proper survey planning can either minimize or eliminate a number of these limitations.

3.7 Image Quality

Whenever a patient is exposed to ionizing radiation, efforts should be made to ensure that a clinical benefit is realized at the lowest possible dose. Therefore, the results for patient doses captured during survey activities can, and should, be interpreted in the context of the clinical benefit. A convenient means of characterizing this aspect of the risk-benefit scenario is by capturing indicators of image quality. Unfortunately, even for a given radiographic modality, there are likely no broadly accepted methodologies for evaluating image quality. However, there is a general consensus that a small number of indicators for image quality merit evaluation, among them are spatial detail, low-contrast detectability, and noise. Commercially available methods and tools exist to evaluate these image-quality metrics. An alternative to image-quality assessment by means of imaging non-clinical test objects is to directly assess the quality of clinical images, such as the methodology used by Muhogora et al.[18] Such methods unfortunately seldom permit comparison across different survey activities. For this reason, we will limit the balance of our discussion to dose surveys. Whatever the means and methods used to assess image quality, it is important to consider the distinction between the assessment of image quality and characterizing actual clinical benefit, a much more challenging endpoint to characterize.

3.8 General Radiography

Advances in technology continue to influence radiological imaging, including the practice of projection radiography. Soon after the x-ray was introduced as a routine clinical diagnostic tool, efforts were made to determine exposure conditions that would result in a radiograph of good quality. During the first half of the twentieth century, a popular means for configuring a proper x-ray exposure setting was by use of a slide rule calculator, and several x-ray equipment companies produced these including Eastman Kodak and Philips.[19] In 1942 a method for automatically controlling x-ray exposure was published.[20] Today, most radiographic equipment is equipped with means for automatically determining the optimal exposure technique factors. Ironically, this technology now typically referred to as AEC that removed much decision-making from the selection of appropriate radiographic technique factors also made the determination of typical patient doses more challenging to measure. Whereas previously one would configure all the x-ray equipment technique factors for the specific examination, the AEC feature would now determine one or more exposure parameters based on the level of x-ray attenuation by the patient. To measure patient-representative doses from x-ray equipment equipped with AEC technology, radiographic phantoms are employed to drive the AEC system to the output that would be delivered to a patient represented by the phantom. Published results of patient doses based on those phantoms should clarify the specific patient characteristics the phantom(s) represent (e.g., five-year-old child vs. an average size adult).

3.9 United States Radiological Survey Efforts

While there were likely earlier attempts to characterize population exposures to radiographic examinations, a number of significant population-based studies took place in the U.S. during the 1940s, and continued into the 1970s. A study of the overall practice of radiology conducted by the ACR in the late 1940s found that approximately 25 million x-ray examinations are performed annually by radiologists.[21] Focusing on chest radiography, typically the most frequently performed radiographic examination, an estimated 7.1 million chest examinations were conducted annually at the time of survey.* Significant health events can dramatically change practice over short periods of time. For example, a survey of Public Health Service medical facilities by Moeller et al. in the early 1950s found that approximately 15 million chest examinations were conducted in 1950 to screen for tuberculosis. Those exams mostly used photofluorography and were likely the highest contributor to U.S. population dose from medical x-rays at the time, according to the study authors.[22]

---

* Our estimate assumes that chest examinations were conducted either as a radiographic film examination or as a photofluorographic examination. The percentages for these two types of examinations among all x-ray-based examinations from the ACR survey were 51.9% and 33.6%, respectively. The study also finds from a limited subset of surveyed facilities that chest radiography comprises approximately 33% of all radiographic studies. Then the number of chest radiographs is estimated as $0.33 \times (0.336 + 0.519) \times 25$ million x-ray examinations annually = 7.1 million chest examinations annually.

There is a lack of concise, examination-specific dose estimates for the studies discussed above. Data from the study of Moeller et al.[10] indicate a reasonable estimate for patient exposure from chest radiography (including photofluorography) of approximately 1R. More broadly, the authors estimate an overall figure for patient exposure from all x-ray imaging examinations to be approximately 11R per examination, including 2.7R for radiographic examinations, 1.0R for photofluorographic examinations, and 65.0R for fluoroscopic examinations. Dental x-rays were taken with a typical patient exposure of approximately 5R, likely owing in part to the direct-film method of image capture.

Following the studies discussed above, the interest in characterizing the state of practice of radiology in the U.S. via large-scale surveys accelerated, with a number of such studies taking place in the 1960s and 1970s. The National Health Survey, conducted by the (then) U.S. Department of Health, Education, and Welfare, U.S. Public Health Service in cooperation with the U.S. Census Bureau, periodically conducted surveys of a sample of U.S. households for data regarding indicators for the state of health in the general population. During the period 1960–1961, the U.S. National Health Survey collected, by means of visitations to approximately 38,000 households, data regarding medical and dental x-ray visits, including health, social, and demographic data for members of visited households.[23] Among their findings, there were an estimated 82 million visits to medical facilities in the U.S. for a diagnostic x-ray examination and 49 million visits to dental offices for x-ray examinations. Notably, the routine chest x-ray examination was identified as the most frequently performed procedure: an estimated 51 million such examinations were conducted during the survey time frame. Overall, there were 48 visits to clinical facilities for medical x-rays for every 100 persons in the population. Lacking in this significant survey effort was any data permitting patient dose estimates. The subsequent Health Interview survey conducted in 1970 also captured data on x-ray usage in the U.S. While comprehensive in scope, the missing survey element, indicators for patient dose, would not be collected until the later surveys by the U.S. Public Health Service in the 1960s and 1970s, beginning with the 1964 X-ray Exposure Study.

The motivation for the 1964 and 1970 X-ray Exposure Studies came from a recognition that a significant contribution to the exposure of the general population to radiation came from medical and dental radiological imaging.[24,25] This was the first large-scale U.S. effort to characterize both population-level x-ray examination statistics and also gather sufficient data to permit the inference of patient doses from specific imaging examinations. The Bureau of the Census trained approximately 80 interviewers to visit 10,029 U.S. households identified for survey (of which 376 did not participate), and gather data covering 31,289 individuals regarding x-ray visits and related items. Interestingly, the 1964 survey captured data for x-ray systems equipped with photo-timers. This was done by introducing a 14-inch by 17-inch sheet of lead of thickness 0.015 inches to approximate a typical adult chest. The lead sheet was incorporated into a film packet that was sent to participating clinical sites to be exposed on appropriate radiographic equipment for the indicated examination.

Following the success of the 1964 survey, a second survey was conducted in 1970. This survey essentially repeated the 1964 survey, capturing data for 67,000 persons from 21,500 households.

Table 3.1 summarizes findings from the surveys discussed above. We provide statistics for overall rates for all x-ray procedures (excluding therapeutic procedures) and also for the chest examination. We chose to highlight chest radiography because (1) it is likely the most frequently performed radiographic examination, and (2) the procedure is highly standardized: nearly every patient (69% + 27% = 95%) has one (36,484/53164 or 69%) or two (14,340/53,164 or 27%) films.*

---

* Percentages for chest examination consisting of either one or two films taken from 1964 X-ray Exposure Study report, pg 163.

Dose, Benefit, and Risk in Medical Imaging

TABLE 3.1   Trends in X-ray Exam Rates in the U.S

Year of Survey	References	U.S. Population (thousands)	No. of All X-ray Exams (thousands) (rate/100)	No. of Chest X-ray Exams (thousands)/ (rate/100)
Late 1940s–1950	(1) Dade Moeller, James Terrill, Samuel Ingraham. Radiation Exposure in the United States. *Public Health Reports* 68(1), 1953: 57–65 (2) SW Donaldson. The Practice of Radiology in the United States: Facts and Figures. *Am J Roent* 66:929, 1951	150,700	25,400[a] (notes)/17	15,000/10
1960–1961	Public Health Service Publication No. 584-B38, October 1962	178,000	82,000/46	51,000/29
1964	U.S. Department of Health, Education and Welfare, Public Health Service. Population Exposure to X-rays U.S. 1964. Public Health Service publication No. 1519, 1966	186,700	90,700[b]/49	53,200/29
1970	U.S. Department of Health, Education and Welfare, Public Health Service. Population Exposure to X-rays U.S. 1970. DHEW publication (FDA) 73–8047, November 1973	199,800	141,700/71	65,200/33
2001	D Spelic et al. Nationwide Surveys of Chest, Abdomen, Lumbosacral Spine Radiography, and Upper Gastrointestinal Fluoroscopy: A Summary of Findings. *Health Phys* 98(3) 2010: 498–514.	284,800	NA	155,500/55

[a] Calculated as 83,000 exams/day in U.S. (Donaldson et al. *Am J Roent* 1951) × 306 working days (Donaldson et al.) = 25.4 million exams.

[b] This number is the stated number of x-ray visits (93 million) minus the number of therapy visits (2.28 M).

The two U.S. Public Health Service (PHS) surveys of 1964 and 1970 merit some comment regarding their execution. The sampling strategy for these two surveys was based on the same methodology used for the Health Statistics survey activities. In this method of sampling, the U.S. is divided into discrete geographical regions, and small clusters of households within each region are randomly identified for survey. These surveys were also conducted in two phases: a household visitation followed by a response package mailed to clinical facilities that provided information on the x-ray examinations to members of the surveyed households. Interestingly, the data needed to conduct the dosimetry for these two surveys were gathered by mail; no on-site measurements were taken from the x-ray equipment. Clinical staff exposed a sensitometric film packet with the identified x-ray technique factors, and upon receipt survey staff processed the film and used the sensitometric response curve to infer corresponding exposure. These data were then used to infer values for gonadal doses for the surveyed radiographic and fluoroscopic examinations based on laboratory measurements from an anthropomorphic phantom.[26]

We reviewed several additional survey efforts conducted by the U.S. government. A limited survey of U.S. hospitals was conducted in 1980, capturing data regarding diagnostic imaging procedures (including nuclear medicine and ultrasound), and again included data regarding patient demographics.[27] This is likely the first major U.S. survey to include CT as an identified imaging modality for survey. A first-ever large-scale survey of mammography was conducted in the later 1970s. The *Breast Exposure Nationwide*

Trends (BENT) surveys collected data directly from clinical service providers regarding patient doses and included a follow-up phase to those facilities found to be administering high or low doses (among other findings). This follow-up effort assisted these facilities with optimizing their mammography practice, particularly focusing on quality control and quality assurance procedures.[28,29] Survey staff found patient exposures to be as high as 14R, noting that at the time of survey, there was still a significant number of sites using direct film and xeromammography. Facilities identified for follow-up were later revisited to verify whether recommended changes were instituted. Almost coincidentally with the BENT program came DENT, *Dental Exposure Normalization Technique* program. Noting that studies have shown patient exposures from dental radiographs to be as high as 5R, the survey planners developed recommendations for optimal exposure ranges for dental radiographs and provide recommendations for improved quality control and quality assurance.[30] For this particular survey program, state radiological health personnel were trained to conduct site visits while the U.S. Bureau of Radiological Health provided equipment and logistical support. Although these two survey activities directed additional efforts at not only capturing the state of practice but also at improving the practice, it is unfortunate that sufficient data were not gathered to infer population-level statistics, such as the number of examinations in the U.S. at the time of survey.

Likely the longest running national survey effort is the *Nationwide Evaluation of X-ray Trends* (NEXT) program, a joint effort by the U.S. Food and Drug Administration (FDA) and State radiological health programs through the Conference of Radiation Control Program Directors (CRCPD).* The U.S. then Bureau of Radiological Health provided logistical and equipment support, and post-survey data analyses, while state radiological health personnel conducted the site visits to a randomly identified sample of clinical sites likely to perform the examination under survey. The NEXT program has collected radiological survey data on selected x-ray examinations since 1972 and presently continues to periodically conduct surveys of x-ray studies of current public health interest. Initially the program surveyed 12 commonly performed examinations, including chest, abdomen, lumbar spine, thoracic spine, dental bitewing, and selected extremities. The advent of AEC technology complicated the collection of data for inferring patient exposures; therefore, beginning in 1984, NEXT focused on one particular examination for survey but employed patient-equivalent phantoms to capture x-ray measurements on AEC-equipped x-ray systems.[9] These later NEXT surveys also collected indicators of image quality using a set of test objects that are radiographed and evaluated under typical clinical conditions at the survey site. Figure 3.2 provides a summary of findings from past NEXT surveys for selected x-ray examinations. Figure 3.2a shows patient entrance air kerma for routine PA chest, anteroposterior (AP) abdomen, and lumbosacral (LS) spine projections. These three examinations have been repeatedly surveyed, and so provide a good example of the ability to observe trends in the practice with time. Obviously as technology has improved over the decades, patient doses have decreased. Figure 3.2b shows the rate of examinations per 100 persons of the U.S. population covering the period from the 1964 U.S. PHS study to the most recent NEXT surveys for the indicated examinations between 2001 and 2003. The figure shows that the rates for abdomen radiography and routine upper gastrointestinal (UGI) fluoroscopy are steady compared to the substantial increases for chest and lumbosacral spine radiography. The increased use of CT may partly explain these observations. The NEXT survey of CT conducted in 2000 found that routine CT examinations of the adult abdomen and combined abdomen plus pelvis comprised nearly 30% of all CT examinations (at least for the CT scanner most frequently used at the site for the surveyed examinations).[31]

---

* Conference of Radiation Control Program Directors, Inc. Frankfort KY. www.CRCPD.org (last accessed on April 17, 2014)

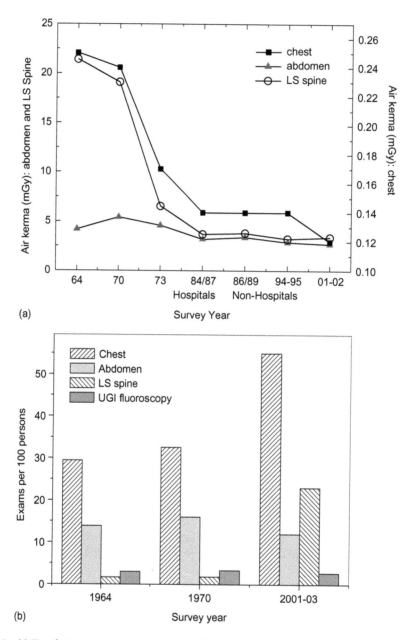

(a)

(b)

FIGURE 3.2   (a) Trends in patient entrance air kerma for U.S. survey activities. (b) Trends in per capita radiographic exam rates in the U.S. (From Spelic, D et al., *Health Phys.*, 98, 498–514, 2010.)

Additional studies of the exposure of the U.S. population to ionizing radiation from medical imaging have been conducted, including a broad study by Fazel et al. between 2005 and 2007. This retrospective study of the imaging procedures conducted on approximately 655,613 health-care enrollees documented population dose quantities such as annual effective dose per person and distributions of effective doses across patient age ranges.[32] Finally, Mettler et al. published in 2008 a collective catalog of effective doses for medical imaging procedures, pulling together data from a large number of sources including survey activities.[33] Tables 3.2 and 3.3 provide summaries of patient exposure (air kerma) and effective dose from selected surveys for radiographic examinations of the chest, abdomen, and lumbosacral spine. These examinations were selected for tabulation because of their relatively high examination rates compared to other radiographic examinations.

TABLE 3.2　Patient Entrance Air Kerma for Chest, Abdomen, and Lumbosacral Spine Radiography by Survey. Unless Otherwise Indicated Values Are in mGy

Author(s)	Country/ Year Covered	Publication	Reference Citation	Adult PA Chest	Adult AP Abdomen	Adult AP Lumbosacral Spine
A Aroua et al.	Switzerland, 1998	Dosimetric aspects of a national survey of diagnostic and interventional radiology in Switzerland	*Med Phys* 29(10), October 2002	0.18		
WE Muhogora et al.	2005–2006, selected Asia, Africa, E. Europe	Patient doses in Radiographic Examinations in 12 countries in Asia, Africa, and Eastern Europe: Initial results from IAEA projects	*Am J Roent* 190, June 2009; 1453–1461. [ref 18]	0.33[a]	4.07[a]	3.64[a]
PC Shrimpton et al.	1984/1995, UK[c]	Diagnostic medical exposures in the UK.	*Appl Rad Isotopes* 50(1999); 261–269	0.23/0.16[b]	8.2/5.6[b]	9.6/6.1[b]
CJ Tung et al	1998 Taiwan	Determination of guidance levels of dose for diagnostic radiography in Taiwan	*Med Phys* 28(5), May 2001	0.52	5.91	4.77
A Servomaa	1991–1996, various Health Care Level 1 countries	International Atomic Energy Agency. Proceedings of an International Conference held in Malaga Spain March 26–30, 2001; IAEA 2001.	IAEA publication, 2001.	0.11/0.81[d]	2.35/20.4[d]	3.34/22.8[d]
DC Spelic et al.	U.S. 1964, 1970	Nationwide Surveys of Chest, Abdomen, Lumbosacral Spine Radiography, and Upper Gastrointestinal Fluoroscopy: A Summary of Findings. tabulated values are 1964/1970 U.S. public health studies	*Health Phys* 98(3):498–514; 2010	0.25/0.24	4.2/5.4	21.4 19.1
DC Spelic et al.	U.S. mid-1990s and 2001–2002	Nationwide Surveys of Chest, Abdomen, Lumbosacral Spine Radiography, and Upper Gastrointestinal Fluoroscopy: A Summary of Findings. tabulated values are 1994(CH), 95(ABD,LS)/2001 (CH), 2002 (ABD,LS)	*Health Phys* 98(3):498–514; 2010	0.14/0.12[e]	2.9/2.7[e]	3.2/3.4[e]

[a] ESD after implementation of Quality Control program.
[b] ESD.
[c] Results labeled 1995 include data collected since 1988 (see page 264 of paper).
[d] Values (X/Y) are lowest/highest of tabulated values.
[e] Values (X/Y) are for 1994/2001 (chest) and 1995/2002 (abdomen and lumbosacral spine).

TABLE 3.3 Effective Dose by Exam and Survey. Unless Otherwise Indicated, Values Are in mSv

Author(s)	Country/Year Covered	Publication	Reference Citation	Adult Chest	Adult Abdomen	Adult Lumbosacral Spine
A. Aroua et al.	Switzerland, 1998	Dosimetric aspects of a national survey of diagnostic and interventional radiology in Switzerland	*Med Phys* 29(10), October 2002	0.052 (male); 0.063 (female)	1.34	3.4
PC Shrimpton et al.	1984/1995	Diagnostic medical exposures in the UK. Values are (1984/1995).	*Appl Rad Isotopes* 50(1999); 261–269	0.024/0.017[a]	1.1/0.7[a]	1.09/0.69[a]
A Servomaa	1991–1996	International Atomic Energy Agency. Proceedings of an International Conference held in Malaga Spain March 26–30, 2001; IAEA 2001. cell values are lowest/highest among tabulated countries (Health Care Level I)	IAEA publication 2001	0.02/0.3[b]	0.24/2.5[b]	1.1/4.33[b]
Spelic et al.	NEXT 1980s–2001–2002	Nationwide Surveys of Chest, Abdomen, Lumbosacral Spine Radiography, and Upper Gastrointestinal Fluoroscopy: A Summary of Findings	*Health Phys* 98(3):498–514; 2010	0.027/0.030/0.026[a]	0.60/0.51/0.48[a]	0.58/0.50/0.54[a]
F Mettler et al.	Various	Effective Doses in Radiology and Diagnostic Nuclear Medicine: a Catalog	*Radiology* 248(1): 254–263; July 2008	0.02 mSv (PA only)	0.7	1.5
E Samara et al.	Switzerland 2008	Exposure of the Swiss population by medical x-rays: 2008 review.	*Health Phys* 102(3): 263–270; 2012	0.051		

[a] Values (X/Y/Z) are PA chest: (1984–1986/1994/2001); AP abdomen: (1987–1989/1995/2002); AP lumbosacral spine: (1987–1989/1995/2002).
[b] Cell values (X/Y) are lowest/highest of tabulated countries.

3.10 Survey Efforts Outside the United States

A number of surveys have been conducted outside the U.S. and we discuss several studies here. The United Kingdom National Radiological Protection Board (NRPB), now with the Health Protection Agency (HPA), Public Health England, conducted several large-scale surveys including studies going back to 1957 and 1977 of the resident population. Selected results for two such U.K. studies are provided in Tables 3.2 and 3.3. A notable activity is the establishment of a central database for the collection of patient dose measurements and associated data.

UNSCEAR (United Nations Scientific Committee on the Effects of Atomic Radiation) conducts periodic studies of the usage of medical imaging procedures.[34] Two comprehensive reports pull together a vast collection of survey data to detail global usage and patient doses for a broad range of medical imaging procedures.[35,36] Also, Muhogora et al. provide findings from an International Atomic Energy Agency (IAEA) initiative to survey both image quality and patient dose for common radiographic examinations and to introduce quality assurance into the clinical practice of participating agencies and sites.[18] Selected findings are also presented in Table 3.2.

The investigators of these various survey activities relate a common observation: a broad range of patient doses in radiographic imaging, largely regardless of the particular examination or procedure surveyed. Muhogora et al. note for examination that one concern is "the significant variation in dose levels to patients of similar size undergoing the same type of radiographic examination." Further, their data demonstrated the positive impact that quality assurance can have on both patient dose and image quality, noting the "potential for dose reduction without affecting the quality of radiographic images in this study..." This is certainly a worthy goal in the best interest of the population of patients undergoing such radiographic examinations even if the individual dose is low compared to other imaging modalities, notably CT.

3.11 Mammography

In the United States, mammography is regulated under the MQSA of 1992.[37] This law specifies standards of quality for the practice including quality standards for image quality, patient dose, quality control and quality assurance practices, and mandates requirements for clinical staff education, training, and experience. Prior to the MQSA, several x-ray survey activities, including the BENT program described earlier, highlighted the need for further improvement in patient dose and image quality for mammography. Three timely NEXT surveys conducted in 1985, 1988, and 1992 characterized observations similar to those described above regarding general radiography: broad ranges of values for technical factors in mammography, including patient dose (as inferred from dose to a standard phantom), image quality, film processing quality, and darkroom film fog levels.[38] At the time of these surveys, xeromammography was still in clinical use and was found to have significantly higher patient dose than film-based imaging. Figure 3.1 displays trends in mammography dose and image quality over nearly 40 years. The majority of dose and image quality data were captured using standard phantoms, providing a means for observing trends with time. It is notable that patient dose has decreased with time while image quality has improved. These improvements over time are likely due to efforts to optimize the imaging chain as well as the institution of effective quality control and quality assurance practices, including significant improvement in film processing quality during the first decade of MQSA requirements. Today approximately 99% of certified mammography facilities in the U.S. are using digital-based mammography equipment, and approximately 52% of U.S. certified facilities have digital breast tomosynthesis (DBT) systems.[39]

Outside the U.S. comparable efforts have been made to document the state of clinical practice in mammography. The Irish National Cancer Screening Service invites women of age 50–64 to participate in their breast screening program. With the large majority of facilities using digital mammography equipment, surveys are performed to assess patient dose. In 2009 a study was conducted of patient dose focusing on digital mammography equipment.[40] The investigators analyzed the data from approximately 100 patient examinations and inferred doses for the craniocaudal and mediolateral oblique views. Their findings are summarized in Table 3.4. Similarly, the National Health Service

TABLE 3.4 Mammography Dose Indices by Survey Study

Author(s)	Country, Year of Survey	Publication	Reference	Dose (mGy)	Avg Breast Thickness (mm)	Comments
Baldelli et al.	Ireland, 2009	Comprehensive Dose Survey of Breast Screening in Ireland	*Rad Prot Dosim* 145(1), 2011: 52–60 [ref 40]			Dosimetry based on study of approximately 2910 patient exams, and includes contributions from extra images and retakes;
		Craniocaudal view		1.27	63	
		Mediolateral oblique view		1.34	60.5	
		Examination		2.72	61.4	
Young et al.	UK, 1997–1998, 2001–2002	Radiation doses received in the UK Breast Screening Programme in 2001 and 2002	UK: KC Young, A Burch, JM Oduko *BJR* 78, March 2005: 207–218 [ref 41]			Authors report dose for analyses of both patient exam data and for a standard phantom; We inferred the dose value for 1997–98 based on authors' observation of 6% increase between 1997–98 and 2001–02 values.
			Craniocaudal view	1.86/1.96	51.5/54.1	
			Oblique view	2.36/2.23	54.3/56.8	
			Standard breast (40 mm PMMA phantom)	1.34/1.42		
Jamal et al.	Malaysia 1999–2001	A study of mean glandular dose during diagnostic mammography in Malaysia and some of the factors affecting it	*BJR* 76, 2003: 238–245.	1.23	NA	Craniocaudal view, dose measured using ACR protocol, Radiation Measurements Inc. (RMI) 156 phantom
Conway et al.	USA, 1985, 1988, and 1992	National Survey of Mammographic Facilities in 1985, 1988, and 1992	*Radiology* 191, 1994: 323–330. [ref 38]			Doses are for craniocaudal view; 1985 survey used only the RMI 152 phantom (4.7 cm breast); Later surveys used RMI 156 phantom (4.2 cm breast), 50% glandular, 50% adipose tissue equivalence.
			1985	2.2	NA	
			1988	1.8	NA	
			1992	1.5	NA	
Spelic et al.	U.S. 1995, 2006	United States radiological health activities: inspection results of mammography facilities	*Biomed Imaging Interv J* 2007; 3(2):e35 [ref 3]			Doses are based on standard phantom equivalent to 4.2 compressed breast, 50% glandular, 50% adipose tissue equivalence.
			1995	1.51	NA	
			2006	1.78	NA	

Breast Screening Program in the UK provides screening mammography to women of age 50–64 years. KC Young et al. analyzed dose data regarding a large, representative sample of patient examinations conducted in 2001 and 2002.[41] In addition to patient-specific data, they inferred dose to a standard breast (45 mm thickness, comprised of 50% glandular, 50% adipose tissue) by means of a phantom consisting of 40 mm of PMMA. At the time of the survey, the objective of this breast screening program was for dose to the standard breast to be 2 mGy or less. They found standard breast dose to be 1.42 mGy and observed this to be slightly (6%) higher than that for 1997–98. Interestingly, Figure 3.1 showing trends in dose and image quality in the U.S., shows a similar, slight increase in dose between 1995 and 2001. Noting that at this period of time mammography in the U.S. was predominantly a film-based modality,[3] one possible explanation for these increases in dose is the realization by the clinical community that higher film optical densities can provide higher image quality, as Figure 3.1 seems to indicate.

3.12 Fluoroscopy

Characterizing patient doses from fluoroscopy is challenging for several reasons. For most routine fluoroscopic examinations, there are two components of dose delivery to the patient: fluoroscopy and radiography. Even for a well-defined fluoroscopic examination the contributions to patient dose from these two features can vary broadly. In addition, fluoroscopic examinations are dynamic in nature, not only from the standpoint of examination descriptors, such as total fluoroscopy time or the number of image acquisitions acquired, but also from the extent of patient anatomy exposed and the beam angulations that can be employed during an examination. In the U.K., D. Hart et al. reported data representing 4,122 patients undergoing the barium meal and swallow examination and found fluoroscopy times to range (minimum to maximum) between 38 seconds and 2118 seconds.[42] They also found the number of digital spot images per examination to vary broadly: for the percutaneous transluminal coronary angioplasty (PTCA) (1 stent) procedure, the number of spot acquisitions varied (minimum to maximum) between 1 and 90 images. S. Balter describes methodologies for characterizing patient skin dose from interventional fluoroscopic procedures,[43] and Miller et al. describe results of a prospective study of interventional fluoroscopy, including a comparison between different technical indicators such as cumulative fluoroscopy time, dose-area product, and cumulative dose at the interventional reference point for a number of clinical procedures.[44]

We will limit the following discussion to general diagnostic fluoroscopy, and as was done for radiography; we will focus on an examination of relatively high frequency: the UGI fluoroscopy examination. We will briefly mention some past survey efforts to characterize patient dose and population examination rates for this diagnostic procedure. In the U.S., the previously discussed surveys by the U.S. Public Health Service in 1964 and 1970 documented rates of diagnostic x-ray examinations including fluoroscopy. UGI fluoroscopy comprised 56% of all fluoroscopy examinations in 1964 and 44% of all fluoroscopy examinations in 1970. A similar survey was conducted in 1980 of hospitals in the U.S. conducting a number of diagnostic x-ray examinations. The most recent NEXT survey of the UGI examination was conducted in 2003.[45] A summary of findings from these survey activities is presented in Table 3.5 along with comparable results for surveys in the U.K. While the per capita examination rate (per 1000) in the U.S. remained steady between 1964 and 2003, the U.K. surveys observed a substantial drop in rate between 1983 and 1997–1998 (Tanner et al.).[46] Many reasons may explain this decrease, including the increased use of CT and endoscopy as noted by Tanner et al. The report by D. Hart et al.[42] illustrates the advantage of capturing data regarding actual patient examinations versus the use of patient representative phantoms as routinely implemented in the NEXT survey program. Such a methodology can permit the capture of a variety of dosimetric indicators for specific patient groups, including total fluoroscopy time, dose-area product, number of digital spot radiographs per examination, and associated technique factors as applicable. Finally, such data collection methodologies can support efforts to develop recommendations for reference levels, and a detailed discussion of these efforts is provided in Chapter 25.[47]

TABLE 3.5 Exam Rates and Per Capita Rates for the Upper Gastrointestinal Fluoroscopy Exam in the U.S. and U.K.

Author(s)	Country, Year of Survey	Publication	Citation	Total Annual No. Exams (×1000)	Exam Rate (per thousand popul)	Comments
(1) U.S. Public Health Service; (2) Spelic et al.	U.S. 1964/1970	(1) Population Exposure to X-rays U.S. 1970; (2) Nationwide Surveys of Chest, Abdomen, Lumbosacral Spine Radiography, and Upper Gastrointestinal Fluoroscopy: A Summary of Findings	(1) U.S. Dept of Health Education and Welfare, Public Health Service; DHEW Publication (FDA) 73–8047; (2) *Health Phys* 98(3):498–514; 2010	5880/5625	30/33	Total exam figures are from PHS publication; Exam rates are from Spelic et al.
	U.S. 1970			5625	33	
Mettler	U.S. 1980	Diagnostic Radiology: Usage and Trends in the United States, 1964–1980	*Radiology* 162, 1987: 263–266	7600	33[a]	
Spelic et al.	U.S. 2003	Nationwide Surveys of Chest, Abdomen, Lumbosacral Spine Radiography, and Upper Gastrointestinal Fluoroscopy: A Summary of Findings	*Health Phys* 98(3):498–514; 2010	7800	27	
	UK 1983 (barium meal)			601[b]	11[c]	RJ Tanner et al. Frequency of medical and dental examinations in the UK 1997–98. (2000).
Tanner et al.	UK (NHS, England) 1997–1998 (barium meal)	Frequency of Medical and Dental X-ray Examinations in the UK—1997/1998	National Radiological Protection Board (NRPB) publication NRPB-R320, 2000	80.5[c]	1.4[d]	

[a] Estimated using figure for total annual no. of exams from Mettler (1987) and figure for U.S. population in 1980: approximately 227,157,000 (also from Mettler 1987, Table 9).

[b] Estimated from Tanner et al. [ref #46]: Pg 28 provides graphical depiction of number of diagnostic (medical) x-ray exams per 1000 population. Population of UK in 1983 is estimated at 56,316,000 [ref #]; Then total number of medical x-ray exams in UK in 1983 is 485 exams × 56,316 = 27.6 million exams. The number of UGI exams is then 0.022 (percentage of UGI of all tabulated medical exams, page 25 of Tanner) × 27.6 million exams = approximately 600,900.

[c] Exam rate per thousand is based on estimated number of exams divided by the resident (UK) population for 1983: 601,000/56,316,000 = 0.011 or 11 per 1000 population.

[d] Value is directly from Tanner et al. (2000) [ref #46]

[e] Value for exam rate per thousand population is estimated as total number of UGI exams divided by resident (UK) population: 83,514 exams/58,395,000 = 0.00138 or 1.4 exams per thousand.

3.13 Computed Tomography

Computed tomography (CT) has seen explosive growth in usage during the past two decades, with annual CT examination volumes in the U.S. nearly tripling from 22.6 million examinations in 1996 to 62 million examinations in 2006.[48] In 2006 CT was found to have the largest modality-specific contribution to the U.S. collective effective dose (440,000 person-Sv/899,000 person-Sv or approximately 50%) from medical sources of radiation, and the second largest contribution (24%) for all sources after radon and thoron background sources (37%).[46] Cone-beam computed tomography (CBCT) is increasingly being used in new areas of clinical practice including dental and maxillofacial applications. A NEXT survey of dental x-ray imaging conducted in 2014–2015 found there to be approximately 5,500 CBCT systems installed in U.S. dental facilities, providing an estimated 4.2 million examinations of adult and pediatric patients at the time of survey.[49]

There are a number of challenges associated with large-scale surveys of CT, and this in part arises from the large number of parameters that need to be captured in order to determine dosimetric quantities such as $CTDI_w$, $CTDI_{vol}$, DLP, and ED. A more recent challenge is the technological feature of tube current modulation on newer CT scanners, in which, similar to the AEC feature on radiographic equipment, the CT scanner modulates the x-ray tube current in response to patient size. Therefore, simply capturing a static value for tube current or tube current-time product may not be representative of actual clinical practice.

One source for data regarding the state of clinical practice for CT (as well as other diagnostic imaging modalities) is the series of benchmark reports by the market research company, International Marketing Ventures (IMV). These reports gather data from a substantial percentage of clinical sites identified for survey and periodically repeat surveys to capture trends.[50] Among the survey outcomes are facility examination/procedure workloads, staffing characteristics, and estimates for total U.S. CT annual procedure rates. These surveys do not report any indicators or estimates for patient dose. The NEXT program also conducted periodic surveys of computed tomography, the most recent ones conducted in 2000 and in 2005–2006.[28,51] Similar to the NEXT surveys discussed previously, these surveys comprised an onsite visit including measurements on CT equipment to gather sufficient data to allow dosimetric quantities to be computed. Examination-specific and total facility workloads were also gathered for the determination of total U.S. CT examination rates at the time of survey. The ACR administers the National Radiology Data Registry, a warehouse of individual ACR registries including the Dose Index Registry, allowing clinical CT sites to compare dose index values from their clinical practice with those derived from a database representative of current regional and U.S. state of practice.

Table 3.6 provides a summary of selected findings from several survey efforts. Similar surveys conducted outside the U.S. regarding the practice of CT are summarized here. Shrimpton et al. captured via questionnaires data regarding twelve common CT examinations, covering approximately one-quarter of the population of CT scanners in the U.K.[52] Dose indicators such as $CTDI_w$, $CTDI_{vol}$, DLP, and ED were inferred from collected data that were combined with available scanner-specific $CTDI_w$ coefficients (mGy-mA^{-1}s^{-1}). The authors also use their findings to derive recommendations for national reference doses. Brix et al. characterized the doses to the German population in 2002 from multi-slice CT and compared to the results of a previous survey conducted on single-slice CT scanners installed in the later 1990s.[53] Their data were collected by means of questionnaires to both hospital and private practice sites and covered 14 CT examinations. Muhogora et al. conducted a study of dose from CT body examinations in 18 countries in Africa, Asia, and Eastern Europe as part of an IAEA initiative in developing countries.[54] Data were collected between 2005 and 2008 regarding 5 body examinations performed at 73 clinical sites. Given that the adult abdomen and pelvis CT examination is typically found during surveys to be one of the most frequently performed examinations, it is notable that the investigators here collected data separately for CT scanning of the abdomen and the pelvis. This observation highlights the challenges that can occur when comparing findings across different surveys when data collection methodologies are sufficiently different.

TABLE 3.6 Dose Index Values for CT Exams from Selected Surveys

Author(s)	Country, Year of Survey	Publication	Citation	Dose Quantity/Comments	Adult Head	Adult Abdomen and Pelvis
Conference of Radiation Control Program Directors, Inc. (CRCPD)	U.S. 2000	Tabulation and Graphical Summary of 2000 Survey of Computed Tomography	CRCPD Publication E-07-2, February 2007 [23]	$CTDI_{vol}$ (mGy) / Eff dose per exam (mSv)	63 (brain and Post F)* / 2.1 (brain and Post F)*	17** / 13.7**
Conference of Radiation Control Program Directors, Inc. (CRCPD)	U.S. 2005–2006	Nationwide Evaluation of X-ray Trends, Computed Tomography 2005–2006; Preliminary Summary	CRCPD publication E-15-3, 2015 [51]	$CTDI_{vol}$ (mGy) / Eff dose per exam (mSv)	66 / 2.5	20.5 / 15.9
Brix et al.	Germany 2002	Radiation exposure in multi-slice versus single-slice spiral CT: results of a survey	*Eur Radiol* 13 (2003); 1979–1991	$CTDI_{vol}$ (mGy) [multi-slice/single-slice] / Eff dose per exam (mSv): [multi-slice/single-slice]	Brain: 60.6/56.1 / Brain: 2.8/2.8	12.6/13.9 / 14.4/17.2
Shrimpton et al.	UK 2003	National survey of doses from CT in the UK: 2003	*BJR* 79(2006): 968–980	$CTDI_{vol}$ (mGy) / Eff dose per exam (mSv)	Cerebrum: 49 / Head: 1.5	11 / 7.1
Muhogora et al.	Various, 2005–2008	Patient doses in CT examinations in 18 countries: initial results from International Atomic Energy Agency projects	*Rad Prot Dosimetry* 136(2) 2009: 118–126	Africa: $CTDI_w$ (mGy) [range] / Asia: $CTDI_w$ (mGy) [range] / E. Europe: $CTDI_w$ (mGy) [range]	NA / NA / NA	11.9–22.7/7.3–26 / 11.7–21.6/16.8–28.4 / 10.2–21.2/8.3–21.8

* Values for 2000 survey for Adult Head are for axial mode of scanning.
** Values for 2000 survey for Adult Abdomen and Pelvis are for helical mode of scanning.

3.14 Summary and Conclusions

Optimizing the use of ionizing radiation for medical imaging depends in part on access to information that reliably and accurately characterizes the state of practice. Population surveys can provide these data as a snapshot of the present state of clinical practice as well as a record of trends with time as technology and clinical methods progress. The data summarized in this chapter are by no means a complete and comprehensive library of survey activities but are intended to provide a sample of efforts taken to document the state of clinical practice in diagnostic radiology. One common theme that can likely summarize the findings from most surveys across the spectrum of diagnostic x-ray examinations is the broad range of values typically observed for patient dose and related indicators. In the U.S. a number of early survey activities discussed above have documented the impact that the observed broad range of quality in film processing has had on patient dose and image quality.[2,43] A survey of chest and abdomen radiography in Malaysia found values for selected clinical kVp for the chest examination to range (minimum to maximum) from 60 to 125 kV, and an accompanying broad range for entrance surface dose.[55] Arguably one of the best approaches to reducing this broad spectrum of patient dose indicators and optimizing the clinical practice is the institution of quality control and quality assurance practices. Muhogora et al.[15] demonstrated well the achievements of these quality activities, and the U.S. MQSA inspection program also has demonstrated improvements largely driven by the institution of quality control and quality assurance practices.[56]

As imaging technology continues to develop, such as the conversion from film-based to digital-based radiographic imaging and the broad establishment of CBCT, there will be a need to characterize and optimize the practice of these and other developing areas in diagnostic radiology. Periodic surveys support efforts to not only characterize the state of practice but also provide a platform for the development and periodic revision of recommendations for quality indicators such as diagnostic reference levels (Chapter 4). In the U.S. the National Council on Radiation Protection and Measurements has published such recommendations based largely on available survey data.[57] The European Commission also published guidance for developing diagnostic reference levels (DRLs) based on either phantom or patient data.[58] Population surveys support these and similar radiological health efforts directed at improving medical x-ray-based imaging.

References

1. A Brodsky and RL Kathren. Historical development of radiation safety practices in radiology. *RadioGraphics* 1989; 9(6):1267–1275.
2. WR Hendee and FM Edwards. *Health Effects of Exposure to Low-Level Ionizing Radiation*. Institute of Physics Publishing, London, UK, 1996.
3. DC Spelic, RV Kaczmarek, M Hilohi, S Belella. United States radiological health activities: Inspection results of mammography facilities. *Biomed Imaging Interv J* 2007; 3(2):e35.
4. The ALARA principle is described in numerous publications from professional organizations such as the IAEA: see https://www-pub.iaea.org/MTCD/Publications/PDF/Pub1145_web.pdf (last accessed August 8, 2018).
5. ICRP. Radiation protection in medicine. ICRP Publication 105. Ann ICRP 2007; 37(6):1-63.
6. AB de González, M Mahesh, K-P Kim, M Bhargavan, R Lewis, F Mettler, C Land. Projected cancer risks from computed tomographic scans performed in the United States in 2007. *Arch Intern Med* 2009; 169(22):2071–2077.
7. AB de González, S Darby. Risk of cancer from diagnostic X-rays: Estimates for the UK and 14 other countries. *The Lancet* 2004; 363(9406):345–351.
8. Further information regarding the MQSA inspection program, including summaries of findings from facility inspections, is available from the U.S. Food and Drug Administration: http://www.fda.gov/Radiation-EmittingProducts/MammographyQualityStandardsActandProgram/default.htm (last accessed February 22, 2014)

9. BJ Conway et al. Beam quality independent attenuation phantom for estimating patient exposure from x-ray automatic exposure-controlled chest examinations. *Med Phys* 1984; 11(6):827–832.

10. DW Moeller. Reflections of an old scout-exposures from various sources of ionizing radiation. *Health Phys News* 2011; 39(6):18–20.

11. Dosimetry Working Party of the Institute of Physical Sciences in Medicine. National Protocol for Patient Dose Measurements in Diagnostic Radiology. National Radiological Protection Board, 1992.

12. ICRU. ICRU Report 74; Patient Dosimetry for X-Rays used in Medical Imaging. Journal of the ICRU, Oxford University Press, Vol 5 No 2, 2005.

13. See for example: Radiation Exposure in Computed Tomography. European Coordination Committee of the Radiological and Electromedical Industries. Edited by Hans Dieter Nagel. 2nd Ed: October 2000.

14. W Huda, N Gkanatsios. Effective dose and energy imparted in diagnostic radiology. Med *Phys* 1997; 24(8):1311–1316.

15. CH McCollough, BA Shueler. Calculation of effective dose. *Med Phys* 2000; 27(5): 828–837.

16. For further information on radiation dose structured report see for examination, David Clunie. Radiation Dose Structured Reports. Presented at 2013 annual meeting of the American Association of Physicists in Medicine, Indianapolis IN, August 4–8, 2013. http://www.dclunie.com/papers/ AAPM_2013_RDSR_Clunie.pdf (last accessed February 16, 2014)

17. The American College of Radiology. Dose Index Registry; information available at: http://www. acr.org/Quality-Safety/National-Radiology-Data-Registry/Dose-Index-Registry (last accessed February 5, 2014).

18. WE Muhogora et al. Patient doses in radiographic examinations in 12 countries in Asia, Africa, and Eastern Europe: Initial results from IAEA projects. *Am J Roentgenol* 2008; 190: 1453–1461.

19. JE Burns. Radiographic exposure slide rules. *Br J Radiol* 1999; 72: 48–54.

20. RH Morgan. A photoelectric timing mechanism for the automatic control of Roentgenographic exposure. *Am J Roentgenol Radium Ther* 1942; 40: 220–228.

21. SW Donaldson. The practice of radiology in the United States: Facts and figures. *Am J Roent* 1951; 66: 929.

22. D Moeller, J Terrill, S Ingraham. Radiation exposure in the United States. *Pub Heal Rep* 1953; 68(1): 57–65.

23. U.S. Department of Health, Education, and Welfare, U.S. Public Health Service. Health Statistics from the U.S. National Health Survey. Public Health Service Publication No. 584–B38, October 1962.

24. U.S. Department of Health, Education and Welfare, Public Health Service. Population Exposure to X-rays U.S. 1964. Public Health Service publication No. 1519, 1966.

25. U.S. Department of Health, Education and Welfare, Public Health Service. Population Exposure to X Rays U.S. 1970. DHEW publication (FDA) 73-8047, November 1973.

26. U.S. Department of Health, Education, and Welfare, Public Health Service. Population Dose from X-Rays U.S. 1964. Public Health Service publication 2001, October 1969.

27. Radiation Experience Data (RED), Documentation and Results of the 1980 Survey of U.S. Hospitals. U.S. Department of health and Human Service, Public Health Service, Rockville MD. HHS Publication FDA 86-8253, September 1985.

28. J Jensen, P Butler. Breast exposure: Nationwide trends; a mammographic quality assurance program results to date. *Rad Technol* 1978; 50(3): 251–257.

29. R Jans, PF Butler, JL McCrohan, WE Thompson. The status of film/screen mammography, results of the BENT study. *Radiology* 1979; 132: 197–200.

30. U.S. Department of Health and Human Services, Public Health Service. Dental Exposure Normalization Technique "DENT" Instruction manual. HEW publication (FDA) 76-8042; May 1976.

31. Conference of Radiation Control Program Directors. Nationwide Evaluation of X-Ray Trends (NEXT): Tabulation and Graphical Summary of 2000 Survey of Computed Tomography. CRCPD publication E-07-2, 2007.

32. R Fazel et al. Exposure to low-dose ionizing radiation from medical imaging procedures. *N Engl J Med* 2009; 361(9): 849–857.

33. F Mettler, W Huda, T Yoshizumi, M Mahesh. Effective doses in radiology and diagnostic nuclear medicine: A catalog. *Radiology* 2008; 248(1): 254–263.

34. For further information on the activities of UNSCEAR, visit http://www.unscear.org/unscear/en/about_us.html (last accessed August 8, 2018).

35. United Nations Scientific Committee on the Effects of Atomic Radiation: UNSCEAR 2008 Report to the General Assembly, with scientific annexes; Volume I: Sources of Ionizing Radiation. Report to the General Assembly, Scientific Annexes A and B; United Nations, New York. Available at www.unscear.org (last accessed August 6, 2018).

36. United Nations Scientific Committee on the Effects of Atomic Radiation: UNSCEAR 2008 Report to the General Assembly, with scientific annexes, Sources and Effects of Ionizing Radiation, Volume I: Sources. Report to the General Assembly, Scientific Annexes A and B; United Nations, New York. Available at www.unscear.org.

37. Mammography Quality Standards Act of 1992: Public Law 102-539, 102d Congress; October 27, 1992.

38. BJ Conway et al. National survey of mammographic facilities in 1985, 1988, and 1992. *Radiology* 1994; 191(2): 323–330.

39. Figure for the percentage of digital mammography sites in the U.S. was computed from the numbers for mammography facilities identified by the Food and Drug Administration as having digital mammography equipment and for all certified mammography facilities as of October 2017. Numbers were acquired on August 10, 2018 at https://www.fda.gov/Radiation-EmittingProducts/MammographyQualityStandardsActandProgram/FacilityScorecard/ucm113858.htm.

40. P Baldelli, J McCullagh, N Phelan, F Flanagan. Comprehensive dose survey of breast screening in Ireland. *Rad Prot Dosimetry* 2011; 145(1): 52–60.

41. KC Young, A Burch, JM Oduko. Radiation doses received in the UK breast screening programme in 2001 and 2002. *Br J Radiol* 2005; 78: 207–218.

42. D Hart, MC Hillier, BF Wall. Doses to Patients from Radiographic and Fluoroscopic X-ray Imaging Procedures in the U.K. 2005 Review. Health Protection Agency Publication HPA-RPD-029, Didcot, OX, 2007.

43. S Balter. Methods for measuring fluoroscopic skin dose. *Pediatr Radiol* 2006; 36(Suppl 2): 136–140.

44. DL Miller et al. Radiation doses in interventional radiology procedures: The RAD-IR study. Part I: Overall measures of dose. *J Vasc Interv Radiol* 2003; 14: 711–727.

45. D Spelic, R Kaczmarek, M Hilohi, A Moyal. Nationwide surveys of chest, abdomen, lumbosacral spine radiography, and upper gastrointestinal fluoroscopy: A summary of findings. *Health Phys* 2010; 98(3): 498–514.

46. RJ Tanner et al. Frequency of medical and dental X-ray examinations in the UK- 1997/98. National Radiological Protection Board publication NRPB-R320, Didcot, OX, 2000.

47. D Miller, M Hilohi, D Spelic. Patient radiation doses in interventional cardiology in the U.S.: Advisory data sets and possible initial values for U.S. reference levels. *Med Phys* 2012; 39(10): 6276–6286.

48. National Council on Radiation Protection and Measurements. Ionizing Radiation Exposure of the Population of the United States. NCRP Report No. 160, March 2009. Data drawn from p. 91 and pp. 11–12.

49. K Farris, D Spelic. Nationwide Evaluation of X-Ray Trends: Highlights of the 2014-15 NEXT Dental Survey. In: *Proceedings of the 47th Annual Conference on Radiation Control. Conference of Radiation Control Directors*. Publication E-15-4, Frankfort, KY, August 2015.

50. IMV Medical Information Division, Des Plaines IL. For further information on IMV Benchmark Reports, visit www.imvlimited.com.

51. Conference of Radiation Control Program Directors. Tabulation and Graphical Summary of 2005–2006 Survey of Computed Tomography. CRCPD Publication E-15-3; 2015.

52. PC Shrimpton, MC Hillier, MA Lewis, M Dunn. National survey of doses from CT in the UK: 2003. *Br J Radiol* 2006; 79: 968–980.

53. G Brix et al. Radiation exposure in multi-slice versus single-slice spiral CT: Results of a nationwide survey. *Eur Radiol* 2003; 13: 1979–1991.

54. WE Muhogora et al. Patient doses in CT examinations in 18 countries: Initial results from international atomic energy agency projects. *Rad Prot Dosimetry* 2009; 136(2): 118–126.

55. A Hambali et al. Entrance surface dose and image quality: Comparison of adult chest and abdominal x-ray examinations in general practitioners' clinics, public and private hospitals in Malaysia. *Rad Prot Dosim* 2009; 133(1): 25–34.

56. OH Suleiman, DC Spelic, JL McCrohan, GR Symonds, F Houn. Mammography in the 1990s: The United States and Canada. *Radiology* 1999; 210: 345–351.

57. National Council on Radiation Protection and Measurements. Reference Levels and Achievable Doses in Medical and Dental Imaging: Recommendations for the United States. NCRP Report 172, September 2012.

58. European Commission. Radiation Protection 109: Guidance on Diagnostic Reference Levels (DRLs) for Medical Exposures, 1999. http://ec.europa.eu/energy/nuclear/radiation_protection/doc/publication/109_en.pdf (last accessed June 2, 2014).

4

Diagnostic Reference Levels

Donald L. Miller

4.1 Introduction

The International Commission on Radiological Protection (ICRP) has defined three basic principles of radiation protection: justification, optimization of protection, and the application of dose limits (ICRP 2007a). In simplest terms, justification in medical imaging means that the benefits of radiation exposure should exceed the risks. Justification is not discussed further here. Justification and optimization apply to all radiation exposures. The principle of the application of dose limits does not apply to medical exposures, because limiting radiation use can interfere with the practice of medicine, may hamper efforts at diagnosis or treatment, and may result in more harm than good (ICRP 2007a). Instead, ICRP emphasizes the use of justification and optimization to protect patients.

Optimization of protection is usually referred to in the United States simply as "optimization" when used in the context of medical imaging. The National Council on Radiation Protection and Measurements (NCRP) prefers the term "as low as reasonably achievable," or "the ALARA principle" to "optimization" (NCRP 1993). For medical exposures, optimization means that radiology equipment (e.g., radiography, fluoroscopy, and computed tomography [CT] systems) is appropriately designed and manufactured, appropriately selected by the medical facility, installed and calibrated correctly, and maintained in good working order. It also means that appropriately trained individuals operate this equipment correctly, using appropriate protocols. The goal is to keep patient radiation doses "as low as reasonably achievable, economic and social factors taken into account" (the ALARA concept) (ICRP 2007a). As applied to medical imaging, optimization is best described as management of the radiation dose to the patient to be commensurate with the clinical purpose (ICRP 2007b). This chapter discusses diagnostic reference levels (DRLs) as a tool for optimization.

In the context of an imaging examination, how can one determine if practices at a facility result in patient radiation doses that are ALARA? Radiation dose metrics at the facility must be compared to some external standard. How is such a standard determined? One approach is to compare dose metrics at a facility to values of the same dose metric at other facilities, with the logic that values at one facility that are high relative to other facilities are a good indicator that patient radiation doses at that facility are not ALARA. This is the method recommended by the ICRP. The indicator used for comparison is called a DRL (ICRP 1996). DRLs are a form of investigation level, where an examination of the cause should occur if the level is exceeded (ICRP 1991). (The term "reference level" should not be used in lieu of DRL, because in ICRP usage, "reference level" is a separate and distinct concept [ICRP 2017]). Use of DRLs as a tool for optimization has been termed the "DRL process" (ICRP 2017).

4.2 Diagnostic Reference Levels

DRL values are determined for specific imaging examinations (e.g., head CT, posteroanterior [PA] chest radiograph). The value of a DRL is based on values of specific dose metrics (also called DRL quantities) observed in clinical practice, most commonly determined from a survey of a number of clinical sites. Each site contributes data from that procedure for a group of procedures and patients; the average value of specified dose metrics for each procedure is contributed to a central registry. Both mean and median values have been used as the "average" value (IPEM 2004; NCRP 2010), but ICRP now recommends use of the median value (ICRP 2017).

The DRL value is determined as a specific percentile of the distribution of the observed doses, most commonly the 75th percentile (Aroua et al. 2007). A rounded value of the 75th percentile (Hart et al. 2002; NRPB 1999) and the 80th percentile of the dose distribution have been used in the past to specify DRL values (Gray et al. 2005). Where radiation doses are very variable, the upper confidence limit of the 75th percentile has also been suggested as an appropriate value (Marshall et al. 2000). Previously, ICRP has not recommended any specific percentile, but both ICRP and NCRP now recommend use of the 75th percentile (ICRP 2017, NCRP 2012). Because dose distributions for most imaging examinations typically are not normally distributed (Hart et al. 2002), and frequently are log-normal (Kwon et al. 2011; Tsai et al. 2007), the 75th percentile separates the relatively high-dose "tail" of the distribution from the remainder of the distribution regardless of the type of examination or the dose metric used (Figure 4.1).

Interestingly, a 2014 survey of CT dose data in Scottish hospitals demonstrated a different dose distribution compared to the log-normal distributions seen previously, with the majority of the dose data clustered around a position just below the national DRL value (Sutton et al. 2014) This may be a result of the approach to optimization used in the UK, where medical physicists are regularly

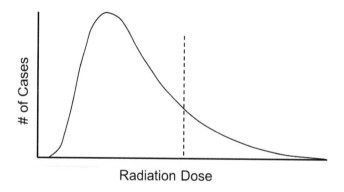

FIGURE 4.1 Shape of a typical log-normal distribution of data. The 75th percentile is shown as a dashed line. Note that the left end of the distribution is not at zero.

involved in optimization efforts, the DRL process has been in use for several decades, and national DRL values are available.

DRLs are a quality assurance and quality improvement tool, used as part of a systematic approach to evaluate local practice. They are a guide to help determine when practices at a facility warrant investigation. To do this, the median radiation dose for a specific imaging procedure at the facility is compared to the DRL value. If it consistently exceeds the DRL value, the facility should conduct an investigation of its equipment and clinical practices in order to reduce radiation doses (NRPB 1990; Wall 2001). The phrase "consistently exceeds" is interpreted by ICRP as meaning in a majority of cases, as opposed to over a period of time (ICRP 2017). The goal of this process is to reduce average doses below the DRL value. When this process is adopted widely, the ultimate result is to shift the right side of the curve in Figure 4.1 to the left. Optimization of practice may not shift the left edge of the curve further to the left, as there is a minimum radiation dose necessary to produce images that are adequate to supply the needed diagnostic information. Adoption of a technological advance that permits adequate quality images to be obtained with lower radiation doses, such as iterative reconstruction as a replacement for filtered back projection in CT, will move the left edge of the curve further to the left. It will not move it to zero, as some radiation dose will always be necessary.

DRLs apply only to groups of patients. *They do not apply to individual patients.* They are intended to be a reasonable indication of dose for average-sized patients, and to provide guidance on what is achievable with current good practice, rather than optimum performance (IAEA 1996). A facility where average patient dose metrics are below the DRL value should not assume that its practices are optimized. A different indicator, the achievable dose (AD), is a better guide to whether dose is optimized (NCRP 2012). The AD is discussed in Section 4.3.

Regardless of whether DRLs or ADs are used, the sequence for investigating practices that result in high doses is the same: first, investigate the equipment, then investigate procedure protocols, and finally investigate operator performance (Balter et al. 2011; Vañó and Gonzalez 2001). This sequence is recommended because it is relatively easy to investigate equipment, but it can be quite difficult to evaluate operator performance.

Use of the DRL process is supported by national and international advisory bodies (Amis et al. 2007; Gray et al. 2005; Hirshfeld et al. 2004; ICRP 2000, 2007b, 2017; Interagency Working Group on Medical Radiation 2014; NCRP 2012). Use of the DRL process may be mandatory in some jurisdictions, but numerical values for DRLS are determined by professional bodies and are advisory (ICRP 2001, 2017). The DRL concept allows flexibility in the selection of examinations to be evaluated and implementation of the DRL process. A number of different organizations have provided guidelines on measuring radiation dose in clinical practice and setting DRL values (CRCPD 2003; CRCPD/CDRH 1992; Hart et al. 2012; ICRP 1991, 1996, 2007a, 2017; IAEA 1996; NCRP 2012; Shrimpton et al. 2014; Wall and Shrimpton 1998).

DRL values are not a guide to or an indicator of image quality. Imaging procedures must supply the diagnostic information required for the clinical task. Radiation doses at a facility that are above or below a particular value do not indicate that images are adequate or inadequate for a particular clinical purpose. *Image quality must be assessed separately from radiation dose* (European Commission 1999; ICRP 2017). Patient radiation doses that are so low that image quality is inadequate are as bad as, or worse, than patient radiation doses that are not ALARA. When image quality is inadequate for the clinical purpose, the administered radiation provides no clinical benefit. Further, the examination must be repeated, and the patient will receive additional radiation from the repeated examination.

4.3 History

The use of patient dose measurements for optimization of protection has been reviewed in detail (Wall and Shrimpton 1998). Beginning in the 1950s, local surveys of dose metrics from diagnostic x-ray examinations were performed, and some relatively general radiation dose information was published

(Billings et al. 1957; Moeller et al. 1953; Norwood et al. 1959; Ritter et al. 1952; Wall and Shrimpton 1998). National surveys of radiation dose metrics began in the United States in the 1960s (Gitlin et al. 1964; U.S. Public Health Service 1966). In the 1970s, the Nationwide Evaluation of X-Ray Trends (NEXT) surveys began in the United States (Center for Devices and Radiological Health 1984), and in the 1980s the National Radiation Protection Board (NRPB, now Public Health England) conducted surveys in the United Kingdom (Shrimpton et al. 1986).

The results of national surveys of dose metrics were the basis for exposure recommendations, first developed in the United States for dental radiography and mammography (CRCPD/CDRH 1992; Jans et al. 1979; Travis and Hickey 1970; Travis et al. 1973; Wall and Shrimpton 1998), then in the United Kingdom (NRPB 1990), and subsequently elsewhere in Europe (European Commission 1996a, 1996b). These exposure recommendations were referred to variously as exposure guides, guideline doses, guidance levels, and reference doses. For example, the international Basic Safety Standards issued by the International Atomic Energy Agency (IAEA) in 1996 (IAEA 1996) included "guidance levels" that had the same meaning as DRLs.

ICRP (1996) began to use the term *diagnostic reference level*, and subsequently published an additional Supporting Guidance 2 (ICRP 2001). A summary of ICRP guidance for DRLs was also included in its Publication 105, "Radiation Protection in Medicine" (ICRP 2007b). In Europe, DRLs were formally introduced in Council Directive 97/43/EURATOM (European Commission 1997, 1999).

An early demonstration of the effectiveness of DRLs for optimization of protection in medical exposures was the Breast Exposure: Nationwide Trends (BENT) mammographic quality assurance program in the U.S. (Jensen and Butler 1978). An initial survey, using phantoms, collected data on entrance exposure from facilities in 19 states. On the basis of these data, trained surveyors visited facilities with high or low exposures and made recommendations to improve technique. At one-year follow-up, there was a substantial decrease in mean exposure and a decrease in the standard deviation of the dose distribution.

In the United Kingdom, where radiation dose data have been collected approximately every five years since the mid-1980s, DRLs determined from the results of the 2005 survey were 16% lower than the corresponding values in the 2000 survey, and approximately half of the corresponding values in a mid-1980s survey (Hart et al. 2009). The value of this tool was recognized in the European Commission's 1997 Medical Exposure Directive (European Commission 1997). As a result of this directive, member States of the European Union were obligated to promote the establishment and use of the DRL process as a strategy for optimization. This requirement was reiterated in European Commission (2013). The 2013 directive also requires that DRL values be updated periodically and that appropriate corrective action be taken without undue delay whenever DRLs are consistently exceeded. As noted above, the phrase "consistently exceeds" is interpreted by ICRP as meaning *in a majority of cases*, as opposed to *over a period of time*.

In the United States, the Conference of Radiation Control Program Directors (CRCPD), an organization of state radiation regulators, issued exposure guides in 1992 and updated them in 2003 (CRCPD 2003; CRCPD/CDRH 1992). Values were based on the results of NEXT surveys (see Chapter 3). The American Association of Physicists in Medicine (AAPM) Task Group on Reference Values for Diagnostic X-Ray Examinations published DRL values for a variety of radiographic, CT, fluoroscopic, and dental examinations (Gray et al. 2005), based on 80th percentile values from earlier NEXT surveys. In 2012, NCRP published Report No. 172, "Reference levels and achievable doses in medical and dental imaging: Recommendations for the United States" (NCRP 2012), which included recommended DRL values from NEXT surveys and other published data. The 2015 revision of Part F of the CRCPD Suggested State Regulations (Diagnostic X-Rays and Imaging Systems in the Healing Arts), Section F.3.a., requires that nationally recognized DRL values be utilized when applicable (CRCPD 2015).

The concept of an AD was introduced in 1999 by the NRPB (1999), which noted that once a facility had adjusted its practices so that patient dose metrics did not exceed DRL values, the next step should be to "reduce doses to those achievable by standard techniques and technologies in widespread use, without compromising adequate image quality." NRPB proposed the concept of AD and suggested that in the long term it should be possible to replace DRLS with ADs. The NRPB suggested that AD values could be defined as those "likely to be attained when complying with accepted recommendations" for good imaging practice, and proposed values for AD that were based, with rounding, on the mean doses observed for a selected sample of departments that met European Commission recommendations on technique (NRPB 1999). NCRP Report No. 172 subsequently proposed an alternative definition for AD, use of the median value obtained from a dose survey (NCRP 2012). The values for ADs given in Report No. 172 were determined primarily by using 50th percentile values of dose metrics obtained from NEXT survey data. ICRP (2017) does not use the term AD, preferring instead "median values of the national survey distribution."

4.4 Determining Diagnostic Reference Level and Achievable Dose Values

Numerical values for DRLs and ADs are determined from the results of surveys of clinical sites. These may be conducted by recording values of dose metrics obtained from patient examinations or by using radiation dose data obtained with phantoms. The use of phantoms has been common in the U.S. and played a major role in most NEXT surveys, while patient dose data have been used in both the U.S. and Europe. Both methods have advantages and disadvantages, but patient surveys are preferred for most medical imaging examinations (ICRP 2017).

Phantoms are used most often for evaluation of general radiography and mammography. Data acquired from phantoms can easily be used to compare equipment within a facility or among facilities. A single measurement is adequate to characterize equipment performance with respect to dose. However, while data acquired using phantoms are adequate to characterize equipment performance, they are not sufficient to evaluate operator performance or clinical practices.

Data acquired from patient examinations include the effect of clinical practices and operator performance but are more difficult to acquire and more difficult to use for comparisons among different imaging rooms or different facilities. For most imaging examinations, the dose required to obtain an adequate image depends on body part thickness and therefore on patient size and weight. Surveys that rely on patient dose data typically specify a relatively narrow range of patient weights (e.g., for adults, 65–75 kg, a mean of 70 ± 3 kg) or use some method to correct for patient body habitus (Chapple et al. 1995; Hart et al. 2002; Lindskoug 1992; Marshall et al. 2000). Each facility must collect data for multiple patients and submit average values for the dose metric of interest. Methods that select patients on the basis of weight reduce the amount of data available, in part because some patients are outside the acceptable weight range, but more frequently because the patient's weight is not known (Hart et al. 2002). Despite these difficulties, patient dose data are generally preferable to phantom-derived data, because they represent actual clinical practice and incorporate the effects of facility practices and routine operator performance. Use of data from patients is also more consistent with ICRP guidance that the numerical value of the DRL should be linked to "defined clinical and technical requirements for the medical imaging task" (ICRP 2001, 2017). This implies that data should come only from similar procedures at all the participating facilities.

When patient dose data are used, it is important to obtain sufficient data from each facility. The choice of sample size is a balance between reducing uncertainty and having a realistic and workable goal. In the United Kingdom, the recommendation is that data be collected for at least 20 patients for each specific imaging procedure being evaluated and preferably at least 30 for fluoroscopic examinations, where variability is greater (Institute of Physical Sciences in Medicine 1992). ICRP (2017) also provides guidance. When data are collected from a limited number of patients, it should be collected only from those examinations where the patients meet the size or weight criteria. For interventional fluoroscopy procedures, where there is no "standard" procedure, a current recommendation is that radiation dose data be collected for all patients (Balter et al. 2011; Miller et al. 2012; NCRP 2010). Data collection can be laborious if manual methods are used, but recent developments in automated data collection and data mining have the potential to simplify the process dramatically and allow collection of data from large numbers of patient examinations (Bhargavan-Chatfield and Morin 2013; Cook et al. 2011; Charnock et al. 2013; Ikuta et al. 2012; ICRP 2017; Kanal 2017; Sodickson et al. 2012). When data from large numbers of patients are available, patient weight has only a minimal effect on the final DRL values (Hart et al. 2002) and data for all patients undergoing the procedure should be used (ICRP 2017).

The dose metric used to develop DRL values depends on the type of examination and the availability of the dose metric. It is more useful to measure exposure than patient dose. The metric should be one that is easily measured, such as volume CT dose index ($CTDI_{vol}$) for CT, administered activity for nuclear medicine, and entrance surface air kerma (ESAK) for radiography (ICRP 2001, 2017). There is value to recording and developing DRLs and ADs for more than one dose metric (ICRP 2017). For example, if both $CTDI_{vol}$ and dose-length product (DLP) are used to develop DRL values, a facility might observe that, for a specific examination, the average $CTDI_{vol}$ is below the DRL value for $CTDI_{vol}$, but the average DLP value is above the DRL value for DLP. This suggests that the scan length used at the facility for this particular CT examination is not optimized. This might not have been as obvious if only $CTDI_{vol}$ or only DLP were evaluated. It has been recommended that all readily available dose metrics should be used to develop DRL values (Miller et al. 2009, 2012; NCRP 2010; Vañó and Gonzalez 2001).

4.5 Specific Imaging Modalities

Specific considerations for DRL and AD development and use are discussed below for radiography, CT, interventional fluoroscopy, nuclear medicine, and pediatric applications. When the DRL process is used to optimize protection as part of the quality improvement process, it is imperative to evaluate image quality at the same time. Assessment of image quality can be subjective, but objective criteria are available for radiography in adults and children and for CT to assist in this evaluation (European Commission 1996a, 1996b, 2000).

4.5.1 Radiography

In the U.S., DRL values have been based on data obtained by using standardized phantoms (NCRP 2012). In Europe, where DRL development has been ongoing for a longer period of time, patient dose data are available and used for determination of DRL values. In the U.S., DRLs for radiography, obtained using phantoms, have been defined using incident air kerma, also called entrance skin exposure in air (Gray et al. 2005). This dose metric does not include backscatter (Rosenstein 2008). The most recent compendium was published in 2012, but the current DRL and AD values are based on the results of earlier surveys (NCRP 2012).

In the U.K., DRL values derived from patient data have been measured with thermoluminescent dosimeters (TLDs) and reported for individual images as ESAK, also called entrance surface dose (ESD), which includes backscatter (Hart et al. 2002; Rosenstein 2008). Kerma-area product (KAP) is used as the dose metric for complete examinations (Wall and Shrimpton 1998). Dose metrics are recorded from examinations performed on a representative sample of 10 or so patients with a mean weight close to 70 kg. This method has been used for several decades (Institute of Physical Sciences in Medicine 1992; Wall and Shrimpton 1998); DRL values for radiography from the 2010 survey are available (PHE 2016). The European Commission has adopted the approach used in the U.K., with the stipulation that "at least 10 patients should be included for each examination at the facility in order to determine an ESD value for comparison with the DRL" (European Commission 1996b).

ICRP recommends using patient data rather than phantoms. Both ESAK and KAP are recommended for radiography. For mammography, ICRP (2017) recommends the use of one or more of incident air kerma, ESAK, and mean glandular dose.

4.5.2 Computed Tomography

The most commonly used dose metrics in CT, CTDI (as either $CTDI_{vol}$ or $CTDI_w$) and DLP, are indicators of radiation output from the CT scanner rather than patient dose (McCollough et al. 2011a). Modern CT scanners display and output both $CTDI_{vol}$ and DLP values. If these values are obtained by scanning phantoms, it is essential to be certain that the same phantom is being used, as there are two different standard phantoms (16 and 32 cm diameter). ICRP (2017) recommends that DRL values be determined for both $CTDI_{vol}$ and DLP. Size-specific dose estimates (SSDE) can be a further source of information for optimization.

It is preferable to use patient data. In Ireland, DRLs for nine common CT examinations have been determined by collecting data through a survey of multiple hospitals (Foley et al. 2012). $CTDI_{vol}$ and DLP data for a minimum of 10 patients were collected from each examination type at each facility, and DRL values were based on the rounded 75th percentile of the dose distribution for each examination. A similar approach was used in Taiwan (Tsai et al. 2007). U.K. DRL values for $CTDI_{vol}$ and DLP from the 2011 survey have been published (PHE 2016).

In the U.S., 75th percentile values for CT dose metrics are available from NEXT surveys (CRCPD 2007) and the American College of Radiology (ACR) CT Accreditation Program (McCollough et al. 2011b). More recently, the ACR Dose Index Registry (DIR) has collected dose metric data ($CTDI_{vol}$, DLP, and SSDEs) on 5.5 million CT examinations, including 1.5 million adult and 80,000 pediatric CT examinations performed at 465 facilities in the first six months of 2013 (Bhargavan-Chatfield and Morin 2013). This wealth of data permits determination of the 75th percentile values of dose metric distributions for a wide variety of CT examinations. The DIR supplies these data to each participating site semi-annually, along with a summary of each site's own dose metric distributions. This markedly simplifies the process of dose optimization at individual facilities, as dose data are collected automatically and data analysis is done at a central facility. These data have also been used to provide U.S. national DRL and AD values for 10 adult CT examinations (Kanal 2017).

Notification Alerts, a component of the CT Dose Check standard (NEMA 2010), are intended to protect patients from inadvertent exposure to excessively high radiation doses. CT DRL values have sometimes been used to set values for these notifications. This is not appropriate (AAPM 2011). DRL values apply to *groups* of patients. Notification Values apply to *individual* patients—if the Notification Value is exceeded, a pop-up notification appears on the operator's console that prompts the technologist to review the scan settings for that patient before proceeding with the examination. Also, DRL values are based on examination of an average-sized patient. CT examinations of large patients require more radiation than examination of average-sized patients to achieve adequate image quality, so use of DRL values as Notification Values will result in a large number of inappropriate notifications when large patients are scanned. This may lead users to ignore relevant notifications due to these false-positive notifications. Ideally, notifications should occur in fewer than 5% of CT examinations (Howard et al. 2014).

4.5.3 Interventional Fluoroscopy

DRL values for diagnostic imaging procedures are derived from data collected for standardized examinations performed on a standard-size patient or phantom (European Commission 1999). Interventional procedures differ from diagnostic imaging procedures in that interventional procedures are not standardized. Interventional procedures demonstrate substantial variability in radiation dose among individual cases, owing to patient, operator, and equipment factors (ICRP 2001). Phantoms are less useful in establishing DRLs for interventional procedures, as their use provides information on equipment performance but does not reflect the effects of variations in patient factors (anatomy, lesion, and disease variations) or operator performance (Balter et al. 2011, NCRP 2010).

In principle, the DRL process could be used for dose management for interventional procedures but is difficult to implement because of the very wide distribution of patient doses, even for the same procedure performed at the same facility (ICRP 2007b; Padovani and Quai 2005). However, the DRL methodology used for diagnostic imaging procedures can be modified to help manage the substantial variability in radiation dose for individual cases seen in interventional procedures.

Radiation doses from interventional procedures are strongly affected by procedure complexity. Complexity reflects variations in patient anatomy, lesion characteristics, and disease severity (Bernardi et al. 2000; IAEA 2009; Peterzol et al. 2005; Vehmas 1997). Because of the marked variability in patient doses, it has been suggested that data on at least 50 patients per facility should be collected for each interventional procedure (Vano et al. 2008). Additionally, Marshall and colleagues (Marshall et al. 2000) have recommended using the upper confidence limit of the 75th percentile of the dose distribution for procedures with wide variation in patient dose resulting from complexity, in order to avoid excessively stringent DRL values caused by sampling limitations in the data.

One way of dealing with the variability in patient dose metrics due to various patient-specific factors is to compensate for the complexity of the procedure (ICRP 2007b). This has been done, for example, for percutaneous coronary artery interventions, where individual cases have been determined to be simple, medium, or complex based on factors such as the number of vessels treated, the number of vessels with severe tortuosity, and the number of bifurcation stents placed (Balter et al. 2008; Bernardi et al. 2000; IAEA 2009). Similar, but preliminary, complexity classification schemes have been proposed for cardiac electrophysiology procedures (Padovani et al. 2008) and some interventional radiology proedures (Ruiz Cruces et al. 2016). However, since assessing procedure complexity requires substantial clinical data that are difficult to collect and often are not available, many published studies have presented DRL values for interventional fluoroscopy procedures without consideration of procedure complexity (Balter et al. 2008; Miller et al. 2009, 2012; Neofotistou et al. 2003; Peterzol et al. 2005; Vano et al. 2009). Similarly, the effects of case-to-case variability in complexity outweigh the effects of patient weight. Patient weight can generally be ignored when DRL values are determined without the use of complexity information (Miller et al. 2009).

Another method can be used to characterize and analyze patient radiation dose for interventional procedures, without the need for the clinical data needed to determine complexity (Balter et al. 2011; ICRP 2017; NCRP 2010). This method requires collection and analysis of data from a greater number of cases than those used to determine DRL values for diagnostic imaging, and ideally uses data from all cases of a specific procedure. This method requires information on the full dose distribution and the uncertainty of the dose distribution (Marshall et al. 2000). It provides a benchmark in the form of an advisory data set (ADS), which consists of the relevant dose metrics for all of the cases of that procedure performed in a large number of facilities (Balter et al. 2011; IAEA 2009). As noted earlier, automated data mining has the potential to allow collection of data from large numbers of patient examinations.

European DRL values generally use KAP (Bleeser et al. 2008; Brambilla et al. 2004; D'Helft et al. 2009; Vano et al. 2009; Zotova et al. 2012), although some published recommendations also include one or more of cumulative air kerma, fluoroscopy time, and the number of images (Brnic et al. 2010; Padovani et al. 2008; Vano et al. 2008). The most recent U.K. DRL values are presented as KAP and fluoroscopy time (PHE 2016). In the U.S., published recommendations typically include all four of these dose metrics (Miller et al. 2009, 2012; NCRP 2012). As noted above, evaluation of facility performance using multiple dose metrics can yield useful information. For example, a facility where median cumulative air kerma values are acceptable but kerma-area product values are high probably has an opportunity to improve its performance by promoting better attention to collimation. ICRP now recommends that DRL values be established KAP, cumulative air kerma, fluoroscopy time, and the number of radiographic images (ICRP 2017). NCRP report No. 172 provides U.S. DRL values for all four dose metrics for selected interventional fluoroscopy procedures (NCRP 2012).

The DRL process is designed to optimize radiation use in order to minimize the risk of a stochastic effect. It is not intended as a means to control the risk of a tissue reaction, such as skin injury. Other methods have been developed to address the risk of tissue reactions (ICRP 2013a; NCRP 2014).

4.5.4 Nuclear Medicine

Nuclear medicine procedures differ from procedures performed using x-rays in that the radiation dose results from administered radiopharmaceuticals. The approach used for radiography, fluoroscopy, and CT, measuring radiation output from the imaging device, is not applicable to nuclear medicine procedures. ICRP (1996) Publication 73, the EC Medical Exposure Directives and guidance (European Commission 1997, 1999, 2013) and NCRP (2012) Report No. 172 recommend that DRL values in nuclear medicine should be expressed in terms of the administered activity of the radiopharmaceutical used for each specific clinical examination. ICRP now observes that the ideal DRL value for a specific clinical task is the administered activity per body weight of the specific radionuclide (ICRP 2017). The recommended administered activity is not based on the 75th percentile of a dose survey, but on the administered activity necessary for an adequate image during a standard procedure (the "optimum" value) (European Commission 1999). As a result, in nuclear medicine the goal for standard procedures is to approach the DRL value as closely as possible without exceeding it. This is different from imaging procedures using x-rays, where the goal is to stay well below the DRL value.

DRLs for nuclear medicine procedures have largely been determined through surveys of practitioners and facilities conducted by professional societies such as the Society of Nuclear Medicine and Molecular Imaging. Collated recommended values for the U.S. are available (NCRP 2012). Data for myocardial perfusion imaging derived from ACR surveys were published in 2016 (Becker et al. 2016). The ACR DIR has provided data on the CT radiation dose component of positron emission tomography (PET)-CT examinations (Alessio et al. 2015). Professional societies have also published recommendations for administered activity (European Association of Nuclear Medicine 2017; Society of Nuclear Medicine and Molecular Imaging 2015).

4.5.5 Pediatrics

Since DRL values for adult examinations are based on dose metrics for an average-sized adult, they are not suitable for children. Initial European surveys of pediatric dose metrics included 3 age groups: infant (10 months), 5 years, and 10 years (European Commission 1996a). More recent U.K. and other publications typically provide DRL values for pediatric patients at 5 standard ages that correspond to standard-sized mathematical phantoms: 0 (newborn), 1, 5, 10, and 15 years (Hart et al. 2009, 2002; Pages et al. 2003). In the U.K. this has been done by normalizing dose metrics from examinations of individual children to those for the nearest standard-sized age (Hart et al. 2000).

Age does not correlate well with size or weight in children. Radiation dose correlates better with a child's weight or size than with the child's age (Järvinen et al. 2011). Current recommendations for designing CT protocols for children are based on patient size or weight, rather than patient age (Frush et al. 2002; ICRP 2013b; Nievelstein et al. 2010; Verdun et al. 2008; Watson and Coakley 2010). DRLs for children should be constructed using weight ranges (ICRP 2017). A few weight-based recommendations for DRL values, based on surveys, have been published for CT (Goske et al. 2013) and nuclear medicine (Gelfand et al. 2011). The data collected by the ACR DIR has provided insight into radiation doses used for pediatric CT (Marin et al. 2015; Strauss et al. 2017). The IMPACT registry (https://www.ncdr.com/webncdr/impact/), a part of the American College of Cardiology's National Cardiovascular Data Registry, has begun collecting radiation dose data along with the clinical data collected for interventional fluoroscopy procedures performed on pediatric and adult patients with congenital heart disease, and should be able to support creation of DRL values for these procedures.

4.6 Conclusion

In medical imaging with ionizing radiation, the principle of optimization of protection (the ALARA principle) means that the imaging study should be performed with the least amount of radiation required to provide adequate image quality. DRLs and ADs are tools that assist in optimization of protection by helping a facility to identify practices that may result in radiation doses to patients that are not optimized. They can be considered investigation levels for medical imaging. They apply to groups of patients and are not to be used to evaluate radiation doses to individual patients.

The DRL process for optimization of protection was developed as a means to provide guidance on what is achievable with current good practice rather than optimum performance. ADs are an additional tool to help reduce doses to those achievable with standard techniques and technologies. The DRL process has been shown to an effective tool. Its use is supported by numerous national and international organizations and is mandatory in Europe.

DRL values are typically set at the 75th percentile of the dose distribution observed from a survey of multiple facilities. In the U.S., NCRP recommends that AD values be set at the 50th percentile of the dose distribution. The dose metric used to develop DRL and AD values should be one that is easily measured. The specific dose metric used depends on the type of examination and the availability of the dose metric. For many imaging modalities, there is value in developing DRL and AD values for more than one dose metric.

DRLs and ADs are effective tools for one component of the optimization process for radiation protection and management of radiation dose. However, they are not a guide or indicator of the other component of optimization, image quality adequate to provide the necessary diagnostic information. Radiation doses at a facility that are above or below a particular value do not indicate whether the resultant images are adequate or inadequate. Image quality must be assessed separately from radiation dose. Optimization must manage the radiation dose to the patient without sacrificing the image quality necessary to achieve the clinical purpose.

References

AAPM. 2011. AAPM recommendations regarding notification and alert values for CT scanners: Guidelines for use of the NEMA XR 25 CT dose-check standard. American Association of Physicists in Medicine, Last Modified April 27, 2011. Available at http://www.aapm.org/pubs/CTProtocols/documents/NotificationLevelsStatement.pdf. Accessed May 28, 2014.

Alessio, A., D. Sengupta, M. Bhargavan-Chatfield, P. Butler, K. Kanal, and F. Fahey. 2015. Survey of CT radiation dose levels during PET/CT from ACR CT Dose Index Registry. *The Journal of Nuclear Medicine*, 56 (suppl 3): 1696.

Amis, E. S., P. F. Butler, K. E. Applegate, S. B. Birnbaum, L. F. Brateman, J. M. Hevezi, F. A. Mettler et al. 2007. American college of radiology white paper on radiation dose in medicine. *Journal of the American College of Radiology* 4 (5):272–284.

Aroua, A., H. Rickli, J. C. Stauffer, P. Schnyder, P. R. Trueb, J. F. Valley, P. Vock, and F. R. Verdun. 2007. How to set up and apply reference levels in fluoroscopy at a national level. *European Radiology* 17 (6):1621–1633.

Balter, S., D. L. Miller, E. Vano, P. Ortiz Lopez, G. Bernardi, E. Cotelo, K. Faulkner, R. Nowotny, R. Padovani, and A. Ramirez. 2008. A pilot study exploring the possibility of establishing guidance levels in x-ray directed interventional procedures. *Medical Physics* 35 (2):673–680.

Balter, S., M. Rosenstein, D. L. Miller, B. Schueler, and D. Spelic. 2011. Patient radiation dose audits for fluoroscopically guided interventional procedures. *Medical Physics* 38 (3):1611–1618.

Becker, M. D., P. F. Butler, M. Bhargavan-Chatfield, B. A. Harkness, D. Metter, C. R. MacFarlane, M. Ghesani, P. Wilcox, and M. E. Oates. 2016. Adult gamma camera myocardial perfusion imaging: Diagnostic reference levels and achievable administered activities derived from ACR accreditation data. *Journal of the American College of Radiology: JACR* 13 (6):688–695.

Bernardi, G., R. Padovani, G. Morocutti, E. Vano, M. R. Malisan, M. Rinuncini, L. Spedicato, and P. M. Fioretti. 2000. Clinical and technical determinants of the complexity of percutaneous trans-luminal coronary angioplasty procedures: Analysis in relation to radiation exposure parameters. *Catheterization and Cardiovascular Interventions* 51 (1):1–9; discussion 10.

Bhargavan-Chatfield, M., and R. L. Morin. 2013. The ACR computed tomography dose index registry: The 5 million examination update. *Journal of the American College of Radiology* 10 (12):980–983.

Billings, M. S., A. Norman, and M. A. Greenfield. 1957. Gonad dose during routine roentgenography. *Radiology* 69 (1):37–41.

Bleeser, F., M. T. Hoornaert, K. Smans, L. Struelens, N. Buls, D. Berus, P. Clerinx, L. Hambach, F. Malchair, and H. Bosmans. 2008. Diagnostic reference levels in angiography and interventional radiology: A Belgian multi-centre study. *Radiation Protection Dosimetry* 129 (1–3):50–55.

Brambilla, M., G. Marano, M. Dominietto, A. R. Cotroneo, and A. Carriero. 2004. Patient radiation doses and references levels in interventional radiology. *Radiologica Medica* 107 (4):408–418.

Brnic, Z., T. Krpan, D. Faj, D. Kubelka, J. P. Ramac, D. Posedel, R. Steiner et al. 2010. Patient radiation doses in the most common interventional cardiology procedures in Croatia: First results. *Radiation Protection Dosimetry* 138 (2):180–186.

Center for Devices and Radiological Health. 1984. *Nationwide Evaluation of X-Ray Trends (NEXT) Eight Years of Data (1974–1981)*. Washington, DC: U.S. Department of Health and Human Services.

Chapple, C. L., D. A. Broadhead, and K. Faulkner. 1995. A phantom based method for deriving typical patient doses from measurements of dose-area product on populations of patients. *British Journal of Radiology* 68 (814):1083–1086.

Charnock, P., B. M. Moores, and R. Wilde. 2013. Establishing local and regional DRLs by means of elec-tronic radiographical x-ray examination records. *Radiation Protection Dosimetry* 157 (1):62–72.

Cook, T. S., S. L. Zimmerman, S. R. Steingall, A. D. Maidment, W. Kim, and W. W. Boonn. 2011. Informatics in radiology: RADIANCE: An automated, enterprise-wide solution for archiving and reporting CT radiation dose estimates. *Radiographics* 31 (7):1833–1846.

CRCPD. 2003. *Patient Exposure and Dose Guide—2003. CRCPD Publication E-03-2*. Frankfort, KY: Conference of Radiation Control Program Directors. Available at http://www.crcpd.org/Pubs/ESERePublishedApr03.pdf. Accessed May 28, 2014.

CRCPD. 2007. *Nationwide Evaluation of X-ray Trends (NEXT). Tabulation and Graphical Summary of 2000 Survey of Computed Tomography*. Frankfort, KY: Conference of Radiation Control Program Directors. Available at http://www.crcpd.org/Pubs/NEXT_docs/NEXT2000-CT.pdf. Accessed May 28, 2014

CRCPD. 2015. *Part F. Medical Diagnostic and Interventional X-Ray and Imaging Systems*. Suggested State Regulations for Control of Radiation. Available at http://crcpd.org/SSRCRs/default.aspx. Accessed May 16, 2016.

CRCPD/CDRH. 1992. *Average Patient Exposure Guides—1992. CRCPD Pub. 92-4*. Frankfort, KY: Conference of Radiation Control Program Directors.

D'Helft, C. J., P. C. Brennan, A. M. McGee, S. L. McFadden, C. M. Hughes, J. R. Winder, and L. A. Rainford. 2009. Potential Irish dose reference levels for cardiac interventional examinations. *British Journal of Radiology* 82 (976):296–302.

European Association of Nuclear Medicine. 2017. *Guidelines*. Vienna, Austria: European Association of Nuclear Medicine. Available at http://www.eanm.org/publications/guidelines/index.php?navId=37 &PHPSESSID=okn9uumdi20mmfjuop6edr7j87. Accessed April 13, 2017.

European Commission. 1996a. *European Guidelines on Quality Criteria for Diagnostic Radiographic Images in Paediatrics. EUR 16261*. Luxembourg, UK: Office for Official Publications of the European Communities. Available at http://bookshop.europa.eu/en/european-guidelines-on-quality-criteria-for-diagnostic-radiographic-images-in-paediatrics-pbCGNA16261/. Accessed May 28, 2014.

European Commission. 1996b. *European Guidelines on Quality Criteria for Diagnostic Radiographic Images. EUR 16260.* Luxembourg, UK: Office for Official Publications of the European Communities. Available at http://bookshop.europa.eu/en/european-guidelines-on-quality-criteria-for-diagnostic-radiographic-images-pbCGNA16260/. Accessed May 28, 2014.

European Commission. 1997. Council Directive 97/43/Euratom of June 30, 1997 on health protection of individuals against the dangers of ionising radiation in relation to medical exposure. *Official Journal of the European Community* L180:22–27. Available at http://ec.europa.eu/energy/nuclear/radioprotection/doc/legislation/9743_en.pdf. Accessed May 28, 2014.

European Commission. 1999. *Radiation Protection 109. Guidance on Diagnostic Reference Levels (DRLs) for Medical Exposures.* Luxembourg, UK: European Commission. Directorate-General Environment, Nuclear Safety and Civil Protection. Available at http://ec.europa.eu/energy/nuclear/radiation_protection/doc/publication/109_en.pdf. Accessed June 5, 2014.

European Commission. 2000. *European Guidelines on Quality Criteria for Computed Tomography. EUR 16262.* Luxembourg, UK: Office for Official Publications of the European Communities. Available at http://bookshop.europa.eu/en/european-guidelines-on-quality-criteria-for-computed-tomography-pbCGNA16262/. Accessed May 28, 2014.

European Commission. 2013. Council Directive 2013/59/Euratom of 5 December 2013 laying down basic safety standards for protection against the dangers arising from exposure to ionising radiation, and repealing Directives 89/618/Euratom, 90/641/Euratom, 96/29/Euratom, 97/43/Euratom and 2003/122/Euratom. *Official Journal of the European Community* L13:1–73. Available at http://eur-lex.europa.eu/LexUriServ/LexUriServ.do?uri=OJ:L:2014:013:FULL:EN:PDF. Accessed May 28, 2014.

Foley, S. J., M. F. McEntee, and L. A. Rainford. 2012. Establishment of CT diagnostic reference levels in Ireland. *British Journal of Radiology* 85 (1018):1390–1397.

Frush, D. P., B. Soden, K. S. Frush, and C. Lowry. 2002. Improved pediatric multidetector body CT using a size-based color-coded format. *AJR American Journal of Roentgenology* 178 (3):721–726.

Gelfand, M. J., M. T. Parisi, and S. T. Treves. 2011. Pediatric radiopharmaceutical administered doses: 2010 North American consensus guidelines. *Journal of Nuclear Medicine* 52 (2):318–322.

Gitlin, J. N., C. R. Hayman, E. G. Proctor, and P. L. Roney. 1964. Field trial of a system to obtain data on population exposure to X-rays. *Public Health Reports* 79:185–192.

Goske, M. J., K. J. Strauss, L. P. Coombs, K. E. Mandel, A. J. Towbin, D. B. Larson, M. J. Callahan et al. 2013. Diagnostic reference ranges for pediatric abdominal CT. *Radiology* 268 (1):208–218.

Gray, J. E., B. R. Archer, P. F. Butler, B. B. Hobbs, F. A. Mettler, R. J. Pizzutiello, B. A. Schueler, K. J. Strauss, O. H. Suleiman, and M. J. Yaffe. 2005. Reference values for diagnostic radiology: Application and impact. *Radiology* 235 (2):354–358.

Hart, D., B. F. Wall, P. C. Shrimpton, D. R. Bungay, and D. R. Dance. 2000. *Reference Doses and Patient Size in Paediatric Radiology. NRPB-R318.* Chilton, UK: National Radiation Protection Board.

Hart, D., M. C. Hillier, and B. F. Wall. 2002. *Doses to Patients from Medical X-ray Examinations in the UK—2000 Review. NRPB-W14.* Chilton, UK: National Radiation Protection Board. Available at http://webarchive.nationalarchives.gov.uk/20140714084352/http://www.hpa.org.uk/webc/HPAwebFile/HPAweb_C/1194947421571. Accessed April 12, 2017.

Hart, D., M. C. Hillier, and B. F. Wall. 2009. National reference doses for common radiographic, fluoroscopic and dental X-ray examinations in the UK. *British Journal of Radiology* 82 (973):1–12.

Hart, D., M. C. Hillier, and P. C. Shrimpton. 2012. *Doses to Patients from Radiographic and Fluoroscopic X-ray Imaging Procedures in the UK—2010 Review. HPA-CRCE-034.* Chilton, UK: Health Protection Agency. Available at https://www.gov.uk/government/uploads/system/uploads/attachment_data/file/342780/HPA-CRCE-034_Doses_to_patients_from_radiographic_and_fluoroscopic_x_ray_imaging_procedures_2010.pdf. Accessed April 12, 2017.

Hirshfeld, J. W., S. Balter, J. A. Brinker, M. J. Kern, L. W. Klein, B. D. Lindsay, C. L. Tommaso, C. M. Tracy, and L. K. Wagner. 2004. ACCF/AHA/HRS/SCAI clinical competence statement on physician knowledge to optimize patient safety and image quality in fluoroscopically guided invasive cardiovascular procedures: A report of the American College of Cardiology Foundation/American Heart Association/American College of Physicians Task Force on Clinical Competence and Training. *Journal of the American College of Cardiology* 44 (11):2259–2282.

Howard, M. E., C. H. McCollough, S. Leng, L. Yu, and M. R. Bruesewitz. 2014. Use of CT dose notification and alert values in routine clinical practice. *Journal of the American College of Radiology* 11 (5):450–455.

IAEA. 1996. *International Basic Safety Standards for Protection Against Ionizing Radiation and the Safety of Radiation Sources. Safety Series No. 115.* Vienna, Austria: International Atomic Energy Agency.

IAEA. 2009. *Establishing Guidance Levels in x ray Guided Medical Interventional Procedures: A Pilot Study. Safety Reports Series No. 59.* Vienna, Austria: International Atomic Energy Agency. Available at http://www-pub.iaea.org/MTCD/publications/PDF/pub1346_web.pdf. Accessed May 28, 2014.

ICRP. 1991. 1990 Recommendations of the international commission on radiological protection. Publication 60. *Annals of the ICRP* 21 (1–3):1–201.

ICRP. 1996. Radiological protection and safety in medicine. ICRP Publication 73. *Annals of the ICRP* 26 (2):1–47.

ICRP. 2000. Avoidance of radiation injuries from medical interventional procedures. ICRP Publication 85. *Annals of the ICRP* 30 (2):7–67.

ICRP. 2001. Radiation and your patient: A guide for medical practitioners. Supporting Guidance 2. *Annals of the ICRP* 31 (4):5–31.

ICRP. 2007a. The 2007 recommendations of the international commission on radiological protection. ICRP publication 103. *Annals of the ICRP* 37 (2–4):1–332.

ICRP. 2007b. Radiation protection in medicine. ICRP Publication 105. *Annals of the ICRP* 37 (6):1–63.

ICRP. 2013a. Radiological protection in cardiology. ICRP Publication 120. *Annals of the ICRP* 42 (2):1–125.

ICRP. 2013b. Radiological protection in paediatric diagnostic and interventional radiology. ICRP Publication 121. *Annals of the ICRP* 42 (2):1–63.

ICRP. 2017. Diagnostic reference levels in medical imaging. ICRP Publication 135. *Annals of the ICRP* 46 (1):1–143.

Ikuta, I., A. Sodickson, E. J. Wasser, G. I. Warden, V. H. Gerbaudo, and R. Khorasani. 2012. Exposing exposure: Enhancing patient safety through automated data mining of nuclear medicine reports for quality assurance and organ dose monitoring. *Radiology* 264 (2):406–413.

Institute of Physical Sciences in Medicine. 1992. *National Protocol for Patient Dose Measurements in Diagnostic Radiology.* Chilton, UK: National Radiological Protection Board. Available at http://www.hpa.org.uk/Publications/Radiation/NPRBArchive/MiscellaneousNRPBReports/rad60misc_pub_NationalProtocol/. Accessed May 28, 2014.

Interagency Working Group on Medical Radiation. 2014. Federal guidance report No. 14. radiation protection guidance for diagnostic and interventional X-Ray procedures. Washington, DC: Environmental Protection Agency. Available at https://www.epa.gov/sites/production/files/2015-05/documents/fgr14-2014.pdf. Accessed May 16, 2016.

IPEM. 2004. *Guidance on the Establishment and Use of Diagnostic Reference Levels for Medical X-ray examinations. Report 88.* New York: Institute of Physics and Engineering in Medicine.

Jans, R. G., P. F. Butler, J. L. McCrohan, and W. E. Thompson. 1979. The status of film/screen mammography: Results of the BENT study. *Radiology* 132 (1):197–200.

Järvinen, H., K. Merimaa, R. Seuri, E. Tyrväinen, M. Perhomaa, P. Savikurki-Heikkilä, E. Svedström, J. Ziliukas, and M. Lintrop. 2011. Patient doses in paediatric CT: Feasibility of setting diagnostic reference levels. *Radiation Protection Dosimetry* 147 (1–2):142–146.

Jensen, J. E., and P. F. Butler. 1978. Breast exposure: Nationwide trends; a mammographic quality assurance program—results to date. *Radiologic Technology* 50 (3):251–257.

Kanal, K. M., P. F. Butler, D. Sengupta, M. Bhargavan-Chatfield, L. P. Coombs, and R. L. Morin. 2017. US Diagnostic reference levels and achievable doses for 10 adult CT examinations. *Radiology* 284 (1):120–133.

Kwon, D., M. P. Little, and D. L. Miller. 2011. Reference air kerma and kerma-area product as estimators of peak skin dose for fluoroscopically guided interventions. *Medical Physics* 38 (6):4196–4204.

Lindskoug, B. A. 1992. The reference man in diagnostic radiology dosimetry. *British Journal of Radiology* 65 (773):431–437.

Marin, J. R., D. Sengupta, M. Bhargavan-Chatfield, K. M. Kanal, A. M. Mills, and K. E. Applegate. 2015. Variation in pediatric cervical spine computed tomography radiation dose index. *Academic Emergency Medicine* 22 (12):1499–1505.

Marshall, N. W., C. L. Chapple, and C. J. Kotre. 2000. Diagnostic reference levels in interventional radiology. *Physics in Medicine and Biology* 45:3833–3846.

McCollough, C. H., S. Leng, L. Yu, D. D. Cody, J. M. Boone, and M. F. McNitt-Gray. 2011a. CT dose index and patient dose: They are not the same thing. *Radiology* 259 (2):311–316.

McCollough, C., T. Branham, V. Herlihy, M. Bhargavan, L. Robbins, K. Bush, M. McNitt-Gray et al. 2011b. Diagnostic reference levels from the ACR CT accreditation program. *Journal of the American College of Radiology* 8 (11):795–803.

Miller, D. L., C. M. Hilohi, and D. C. Spelic. 2012. Patient radiation doses in interventional cardiology in the US: Advisory data sets and possible initial values for US reference levels. *Medical Physics* 39 (10):6276–6286.

Miller, D. L., D. Kwon, and G. H. Bonavia. 2009. Reference levels for patient radiation doses in interventional radiology: Proposed initial values for US practice. *Radiology* 253 (3):753–764.

Moeller, D. W., J. G. Terrill, and S. C. Ingraham 2nd. 1953. Radiation exposure in the United States. *Public Health Reports* 68 (1):57–65. Available at http://www.ncbi.nlm.nih.gov/pmc/articles/PMC2024075/. Accessed May 28, 2014.

NCRP. 1993. Limitation of exposure to ionizing radiation. NCRP Report No. 116. Bethesda, MD: National Council on Radiation Protection and Measurements.

NCRP. 2010. Radiation dose management for fluoroscopically guided interventional medical procedures. NCRP Report No. 168. Bethesda, MD: National Council on Radiation Protection and Measurements.

NCRP. 2012. Reference levels and achievable doses in medical and dental imaging: Recommendations for the United States. NCRP Report No. 172. Bethesda, MD: National Council on Radiation Protection and Measurements.

NCRP. 2014. Outline of administrative policies for quality assurance and peer review of tissue reactions associated with fluoroscopically-guided interventions. Statement No. 11. Bethesda, MD: National Council on Radiation Protection and Measurements.

NEMA. 2010. Computed tomography dose check. NEMA Standards Publication XR 25-2010. Rosslyn, VA: National Electrical Manufacturers Association.

Neofotistou, V., E. Vano, R. Padovani, J. Kotre, A. Dowling, M. Toivonen, S. Kottou et al. 2003. Preliminary reference levels in interventional cardiology. *European Radiology* 13 (10):2259–2263.

Nievelstein, R. A., I. M. van Dam, and A. J. van der Molen. 2010. Multidetector CT in children: Current concepts and dose reduction strategies. *Pediatric Radiology* 40 (8):1324–1344.

Norwood, W. D., J. W. Healy, E. E. Donaldson, W. C. Roesch, and C. W. Kirklin. 1959. The gonadal radiation dose received by the people of a small American city due to the diagnostic use of roentgen rays. *American Journal of Roentgenology, Radium Therapy, and Nuclear Medicine* 82:1081–1097.

NRPB. 1990. Patient dose reduction in diagnostic radiology. Documents of the NRPB 1 (3). Chilton, UK: National Radiological Protection Board.

NRPB. 1999. Guidelines on patient dose to promote the optimisation of protection for diagnostic medical exposures. Documents of the NRPB 10 (1). Chilton, UK: National Radiological Protection Board.

Padovani, R., and E. Quai. 2005. Patient dosimetry approaches in interventional cardiology and literature dose data review. *Radiation Protection Dosimetry* 117 (1–3):217–221.

Padovani, R., E. Vano, A. Trianni, C. Bokou, H. Bosmans, D. Bor, J. Jankowski et al. 2008. Reference levels at European level for cardiac interventional procedures. *Radiation Protection Dosimetry* 129 (1–3):104–107.

Pages, J., N. Buls, and M. Osteaux. 2003. CT doses in children: A multicentre study. *British Journal of Radiology* 76 (911):803–811.

Peterzol, A., E. Quai, R. Padovani, G. Bernardi, C. J. Kotre, and A. Dowling. 2005. Reference levels in PTCA as a function of procedure complexity. *Radiation Protection Dosimetry* 117 (1–3):54–58.

PHE. 2016. National Diagnostic Reference Levels (NDRLs). Public Health England. Available at https://www.gov.uk/government/publications/diagnostic-radiology-national-diagnostic-reference-levels-ndrls/national-diagnostic-reference-levels-ndrls. Accessed March 3, 2016.

Ritter, V. W., S. R. Warren, Jr., and E. P. Pendergrass. 1952. Roentgen doses during diagnostic procedures. *Radiology* 59 (2):238–251.

Rosenstein, M. 2008. Diagnostic reference levels for medical exposure of patients: ICRP guidance and related ICRU quantities. *Health Physics* 95 (5):528–534.

Ruiz Cruces, R., E. Vañó, F. Carrera-Magariño et al. 2016. Diagnostic reference levels and complexity indices in interventional radiology: A national programme. *European Radiology* 26:4268–4276.

Shrimpton, P. C., B. F. Wall, D. G. Jones, E. S. Fisher, M. C. Hillier, G. M. Kendall, and R. M. Harrison. 1986. *A National Survey of Doses to Patients Undergoing a Selection of Routine X-ray Examinations in English Hospitals*. NRPB-R200. London, UK: HMSO.

Shrimpton, P. C., M. C. Hillier, S. Meeson, and S. J. Golding. 2014. *Doses from Computed Tomography (CT) Examinations in the UK—2011 Review*. PHE-CRCE-013. Chilton, UK: Public Health England. Available at https://www.gov.uk/government/uploads/system/uploads/attachment_data/file/349188/PHE_CRCE_013.pdf. Accessed April 12, 2017.

Society of Nuclear Medicine and Molecular Imaging. 2015. Procedure standards. Society of Nuclear Medicine and Molecular Imaging. Reston, VA. Available at https://www.snmmi.org/ClinicalPractice/content.aspx?ItemNumber=6414. Accessed April 13, 2017

Sodickson, A., G. I. Warden, C. E. Farkas, I. Ikuta, L. M. Prevedello, K. P. Andriole, and R. Khorasani. 2012. Exposing exposure: Automated anatomy-specific CT radiation exposure extraction for quality assurance and radiation monitoring. *Radiology* 264 (2):397–405.

Strauss, K. J., M. J. Goske, A. J. Towbin et al. 2017. Pediatric chest CT diagnostic reference ranges: development and application. *Radiology* 284 (1):219–227.

Sutton, D. G., S. McVey, D. Gentle, A. J. Hince, N. MacDonald, and S. McCallum. 2014. CT chest abdomen pelvis doses in Scotland: Has the DRL had its day? *British Journal of Radiology* 87:20140157.

Travis, K. L., and J. E. Hickey. 1970. A state program for reducing radiation exposure from dental x-ray machines. *American Journal of Public Health and the Nation's Health* 60 (8):1522–1527.

Travis, K. L., J. R. Stanton, J. E. Hickey, and B. M. Burnett. 1973. Dental technique normalization: Development and pilot test. IRPA 3, Washington, DC September, 1973. Available at http://www.irpa.net/irpa3/cdrom/VOL.3B/W3B_45.PDF. Accessed May 28, 2014.

Tsai, H. Y., C. J. Tung, C. C. Yu, and Y. S. Tyan. 2007. Survey of computed tomography scanners in Taiwan: Dose descriptors, dose guidance levels, and effective doses. *Medical Physics* 34 (4):1234–1243.

U.S. Public Health Service. 1966. Population Exposure to X-rays U.S. 1964. Public Health Service Publication No. 1519. Washington, DC: U.S. Department of Health, Education, and Welfare.

Vañó, E., and L. Gonzalez. 2001. Approaches to establishing reference levels in interventional radiology. *Radiation Protection Dosimetry* 94 (1–2):109–112.

Vano, E., H. Järvinen, A. Kosunen, R. Bly, J. Malone, A. Dowling, A. Larkin et al. 2008. Patient dose in interventional radiology: A European survey. *Radiation Protection Dosimetry* 129 (1–3):39–45.

Vano, E., R. Sanchez, J. M. Fernandez, J. J. Gallego, J. F. Verdu, M. G. de Garay et al. 2009. Patient dose reference levels for interventional radiology: A national approach. *Cardiovascular and Interventional Radiology* 32 (1):19–24.

Vehmas, T. 1997. Radiation exposure during standard and complex interventional procedures. *British Journal of Radiology* 70:296–298.

Verdun, F. R., D. Gutierrez, J. P. Vader, A. Aroua, L. T. Alamo-Maestre, F. Bochud, and F. Gudinchet. 2008. CT radiation dose in children: A survey to establish age-based diagnostic reference levels in Switzerland. *European Radiology* 18 (9):1980–1986.

Wall, B. F. 2001. Diagnostic reference levels—the way forward. *British Journal of Radiology* 74 (885):785–788.

Wall, B. F., and P. C. Shrimpton. 1998. The historical development of reference doses in diagnostic radiology. *Radiation Protection Dosimetry* 80 (1–3):15–19.

Watson, D. J., and K. S. Coakley. 2010. Paediatric CT reference doses based on weight and CT dosimetry phantom size: Local experience using a 64-slice CT scanner. *Pediatric Radiology* 40 (5):693–703.

Zotova, R., J. Vassileva, J. Hristova, M. Pirinen, and H. Jarvinen. 2012. A national patient dose survey and setting of reference levels for interventional radiology in Bulgaria. *European Radiology* 22 (6):1240–1249.

5

Optimization and Dose Reduction in Dentomaxillofacial Imaging

Alan G. Lurie

5.1 Introduction

Performing dentistry without contemporary imaging is unimaginable. Most dental pathology either hides in the bone and soft tissues not seen with the eye or moves into these places from visible intraoral structures where their deeper effects are often hardly evident or not evident at all. Thus, for most dental and maxillofacial diagnoses, imaging has become an absolutely essential part of the diagnosis and treatment planning for every patient.

Doses from dentomaxillofacial (DMF) imaging tend to be very small, with accompanying minimal, almost negligible, risks to individual patients. However, doses from large-volume cone-beam computed tomography (CBCT) imaging can approach, and even surpass, doses from multi-detector CT (MDCT). The fact that CBCT is dentistry's first volumetric imaging modality, delivering multiplanar views as well as aesthetically attractive 3D emulations, may lead to overuse, particularly in younger, more radiation-sensitive patients. Much attention has been given to risks from DMF imaging; however, the small doses result in risk estimation being population based, and extremely small and difficult to characterize for individual patients.

While there are many modalities and views that are used in dentistry, they can be categorized as conventional intraoral imaging, panoramic imaging, cephalometric imaging, and CBCT. The vast majority of DMF imaging in the United States is conventional and consists of bitewing and periapical intraoral projections. These images provide highly detailed depictions of the teeth, supporting structures, and are the first line of imaging for the most pervasive dental pathologies—dental caries and periodontal disease. Figure 5.1 shows bitewing and periapical images of patients with normal dentition and supporting structures.

Panoramic imaging, introduced in the 1960s and also known as pantomography, is also widely used as it generates variable-thickness, curved-surface tomographic images of the mandible, maxilla, dentoalveolar structures, temporomandibular joints, and maxillary sinuses. Panoramic imaging is used for overall baseline evaluation of new patients, planning of third-molar extractions, evaluation of developing mixed dentition, mandibulofacial trauma, and a variety of other dental issues. Figure 5.2 shows a panoramic image of a patient with normal dentition and supporting structures.

Cephalometric imaging is used primarily by orthodontists and oral and maxillofacial surgeons. These images are similar to skull imaging techniques but differ with focus around delicate facial bone structures and soft tissues to avoid burnout by lowering radiation exposure that would otherwise render the cephalometric image non-usable. Cephalometric images are used to evaluate patients for dental versus skeletal basis of malocclusion, monitoring craniofacial growth and development, and planning for orthodontic treatment and some orthognathic surgical procedures. Figure 5.3 shows a lateral cephalometric image of a normal patient being planned for orthodontic treatment.

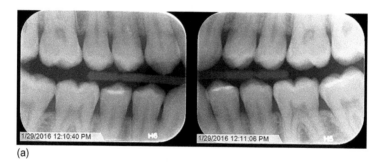

(a)

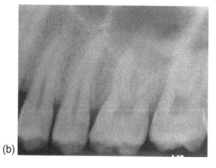

(b)

FIGURE 5.1 (a) Right/left pair of bitewing images of a patient with normal dentoalveolar structures. They show the crowns and coronal root segments of both maxillary and mandibular teeth and are the most widely used images for the detection of dental caries on approximal surfaces of teeth, as well as the detection of early periodontal bone loss between the teeth. Approximal surfaces are shown, as are the indental cortical and cancellous bone. (b) Left maxillary premolar periapical image of a patient with normal dentoalveolar structures. It shows the crowns and roots of the teeth and the supporting bone structures from either maxilla or mandible. It is widely used to detect periapical and furcational pathology, as well as to evaluate more pronounced marginal bone loss.

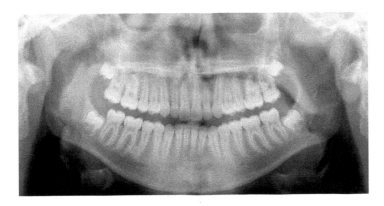

FIGURE 5.2   A panoramic image of a young patient, free of pathology. Four third molars ("wisdom teeth") are still developing. This image is a variable thickness, pantomograph of the dentition and skeletal facial and gnathic structures. As such, it is subject to distortions and non-imaging of objects not in the image layer. Panoramic imaging is widely used as a baseline intake image and is also very useful in third molar evaluation and initial imaging of temporomandibular joint pain, orthodontic patients, osseous lesions, and suspected developmental abnormalities.

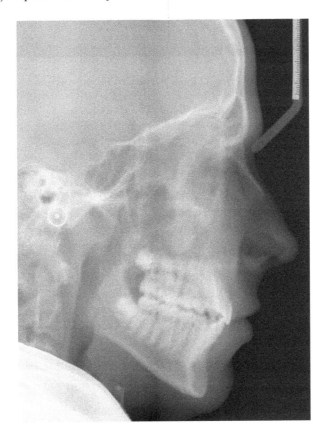

FIGURE 5.3   A lateral cephalometric image of a 15-year-old patient being treatment planned for orthodontic care. Lateral cephalometric images are used primarily in orthodontic treatment planning, but also, in concert with PA cephalometric images, useful in the initial stages of treatment planning for complex orthognathic surgery. Thyroid shielding is used for orthodontic patients, as in this image, but not in orthognathic surgical planning as the soft tissues of the neck must be evaluated. The ruler is attached to the naso-frontal suture ("nasion") and is used in superimposing subsequent intra- and post-treatment lateral cephalometric images on the same patient.

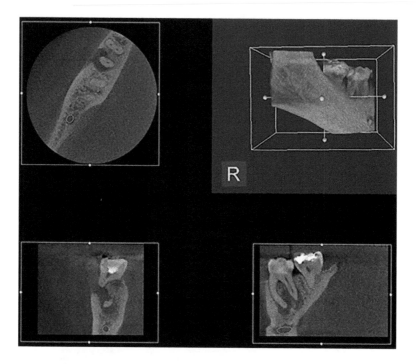

FIGURE 5.4   This multiplanar display is from a small-volume CBCT data set acquired with a 40 × 40 mm field of view. Counterclockwise from upper left show axial, coronal, and sagittal sections and a 3D emulation through tooth American Dental Association (ADA) #30. This tooth has a large periapical lesion (abscess), which has perforated through the buccal plate of the mandible. This cortical bone perforation would not have been detected with conventional periapical imaging and would not have been correctly diagnosed and treatment planned.

CBCT was introduced to dentistry at the end of the twentieth century, and its use has been steadily increasing since then. With a wide range of fields of view and multiplanar and various dental-specific software emulations, CBCT has become a powerful diagnostic and treatment planning tool, especially in areas of dental-implant treatment planning, three-dimensional positioning of impacted teeth relative to other teeth and critical structures, detailing bone lesions and defining root canal configurations and possible root fractures in endodontic therapy (root canal treatment). The range of doses are extremely wide, as the field of view and presets allow considerable variability in the anatomy covered and the voxel size. Unfortunately, due to its 3D capabilities and beautiful software emulations, as well as a widely held misconception that it is a fancy panoramic machine, CBCT has the capacity to be markedly overused. Figure 5.4 shows a typical multiplanar display from a small-volume CBCT acquisition.

5.1.1 Doses: Frequently Encountered Dentomaxillofacial Imaging Modalities

5.1.1.1 Intraoral Imaging

Doses from conventional intraoral imaging are extremely small compared to those from diagnostic imaging generally but are quite variable depending on collimation and receptor type. The most significant factors that control dose in intraoral imaging are: (a) rectangular collimation, (b) receptor speed, (c) digital device "presets," and (d) selection criteria. Other factors include thyroid shielding in children, lap/chest shielding in young women, and film chemistry for those still using film-based imaging. Any reduction in dose if applied nationally impacts 0.5–1 billion exposures annually.

Representative doses from intraoral imaging procedures are shown in Table 5.1.

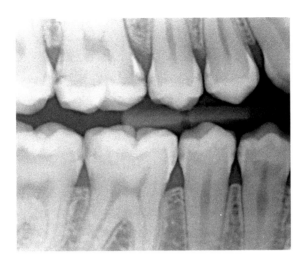

FIGURE 5.5 This right premolar bitewing demonstrates numerous carious lesions of mild-moderate severity. These appear as demarcated, relatively radiolucent areas, which interrupt the normal enamel and dentin densities. Mild lesions, primarily in approximal enamel, are seen on ADA teeth 4, 5, 28, and 29. Moderate lesions, well into dentin, are seen in ADA teeth 3, 30, and 31.

The bitewing radiograph, almost always used to image posterior teeth (premolars and molars), shows the contacting surfaces in both arches and allows inspection of these surfaces for the telltale loss of mineral density seen with dental caries. This technique is highly reliable for detecting cavitated lesions and 70%–80% reliable for detecting lesions confined to enamel (White 2013). Figure 5.5 shows typical bitewing images with enamel, dentinal, and very deep carious lesions.

5.2.1.2 Periodontal Disease

Periodontal disease (Newman 2014) is a complex, multifactorial inflammatory/infectious process, which affects the gingiva (gums) and alveolar bone that support the teeth. The patient's immune and inflammatory response to microbes, particularly subgingival microbes, is a key to the severity and progression of periodontal disease, and periodontal disease has been shown to have associations with the overall health of the patient. Inflammatory periodontal disease begins in the gingival tissues and spreads progressively into the tooth-supporting periodontal ligament and alveolar bone where it becomes the destructive form of periodontal disease, called periodontitis. This can ultimately lead to irreversible bone loss and subsequent tooth loss. Moderate periodontal diseases affect a majority of adults greater than 50 years or age, and severe periodontitis effects 5%–15% of adults (NIH 2017). Factors known to increase susceptibility to and severity of periodontal disease include smoking, diabetes mellitus, some anti-seizure medications, anti-cancer drugs, oral contraceptives, and some calcium channel blockers (NIH 2017).

In mild or moderate chronic periodontitis, periodontal disease is very treatable by conventional, non-invasive methods. Severe periodontitis requires more invasive treatments, including efforts to regenerate supporting bone, but will often result in tooth loss. Thus, as with dental caries, early detection is critical to maintaining the health of the teeth and their supporting structures.

Marginal periodontal structures and their relationships to the teeth are best imaged with bitewing projections in the posterior areas and periapical projections in the anterior areas. Isometric projection of these relationships is critically dependent on proper geometrical positioning of the x-ray tube and the receptor relative to the angulations of the teeth and the curvatures of the arches. It is essential to properly show the crest of the alveolar bone and its relationship to the cementoenamel junctions of the teeth. Panoramic images give an overall and gross view of marginal bone configurations and levels but

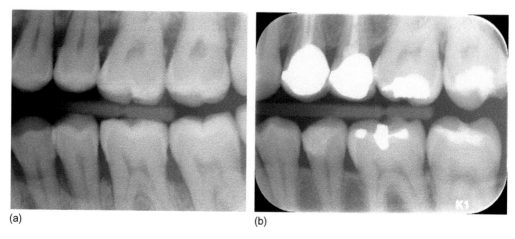

(a)   (b)

FIGURE 5.6 (a) Shows normal marginal periodontal structures. The alveolar crestal bone is sharply defined, corticated, and located 1–2 mm apical to the cementoenamel junctions. ADA teeth 15 and 20 have undergone end-odontic treatment (root canal treatment). Various types of restorations are seen on ADA teeth 14, 15, 18, 19, and 20. (b) Shows frank evidence of marginal periodontitis. Bone levels have resorbed markedly toward the root apices in a variety of patterns, and cortical bone at the alveolar crests has been lost. Abundant calculus deposits, seen as opacities projecting from the crowns and roots of most of the teeth, are present.

do not give the detail of the intraoral views. Treatment of marginal inflammatory periodontal disease is best accomplished by including home care with proper toothbrushing and flossing as well as periodic examinations and professional cleanings. In most cases, this will prevent the progression of inflammatory periodontal disease to the destructive form of periodontal disease. Treatment of destructive forms of periodontal disease include proper home care but may also include a variety of therapeutic interventions, including scaling and planning of the roots, local surgical procedures, and bone regeneration procedures.

In a healthy patient, free of periodontal disease and having no history of periodontal disease, the alveolar crestal bone should be corticated and be located 1–2 mm apical to the cementoenamel junctions of the teeth. The appearance of a normal, healthy marginal periodontium on imaging is shown is Figure 5.6a. The progression of periodontal disease into and through the alveolar bone results in loss of the periodontal ligament attachment between teeth and bone, and loss of bone in a variety of patterns; frank marginal bone loss is shown in Figure 5.6b.

5.2.1.3 Apical Periodontitis

Apical periodontitis (Hargreaves 2015) is usually the outcome of dental caries penetrating into and infecting the pulp. Following pulp infection, microbial products may leach into the periapical tissues via apical foramina, exert immunomodulatory effects on various periapical cells, and cause periapical inflammation (i.e., apical periodontitis). A dynamic encounter between microbial factors and host defenses at the interface between infected root canal and periodontal ligament results in local inflammation, resorption of hard tissues, and eventual formation of what are commonly referred to as periapical lesions. Periapical inflammatory diseases manifest themselves in a wide variety of clinical and radiographic features, which include apical osteolysis, which may be seen as a periapical radiolucent area in the radiograph, periapical pain, periapical abscess formation, swelling, and sinus tract ("fistula") formation. Unlike necrotic pulp, inflamed periapical tissues do not normally harbor bacteria, because any bacteria extruded into the periapical tissues is likely to be eliminated by host tissue immune mechanisms. However, transitory bacteria in periapical tissues may sometimes cause periapical infection, which, if not managed, may lead to severe and—at times—life-threatening perioral infections.

The treatment of periapical lesions is the realm of the endodontist, who has to decide if the tooth can be saved with root canal treatment or needs to be extracted. This decision is based on patient signs and symptoms, clinical findings, and imaging features. Endodontic treatment consists of cleaning, shaping, and disinfection of the root canals, followed by obturation of the canals, usually with gutta percha and paste sealer. Antibiotics are contraindicated, as there is generally not any infection in the osseous tissues. Endodontically treated teeth are usually then treated with a cast post-and-core restoration, which forms the base for a cast restoration of the crown. Traditional imaging has been with periapical radiographs, and panoramic images show overt periapical lesions; however, early apical periodontitis will often not be detected with any conventional imaging techniques. Several new modalities have appeared in recent years, including root apex locators (an electronic device which can identify the narrowest portion of the canal), microendoscopic canal evaluations, and CBCT that add to the strength of defining the extent of overt lesions and mapping canals, but still are not very reliable in detection of early disease.

The essence of the imaging diagnosis is bone change at and surrounding the root apex, and in the case of maxillary posterior teeth, changes in the cortical floor and mucoperiosteal lining of the maxillary sinus. The extent of bone demineralization and sclerosis, the presence of cortical perforation, and the possible presence of root fractures are also critical features to demonstrate as they effect treatment decisions. Additionally, it is important to develop a map of the root canal system so that all of the pulp can be removed and all of the canals can be obturated. Traditional periapical imaging, shown in Figure 5.7, tends to show the canal system of single-rooted teeth relatively well; however, multi-rooted teeth such as molars and some premolars can have accessory canals, extra canals, and require more advanced imaging techniques to properly demonstrate the canal system. CBCT has become the primary imaging modality for situations that demand detailed imaging of

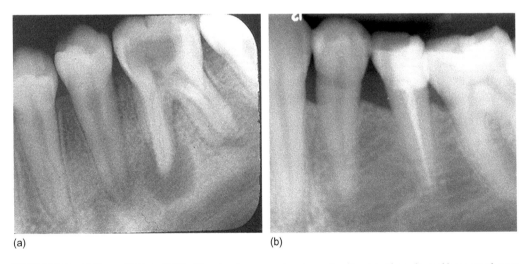

(a)   (b)

FIGURE 5.7   (a) Shows ADA tooth 19 with extensive coronal caries, extending into the pulp, and bone rarefaction at the root apices surrounded by marked bone sclerosis (apical periodontitis). (b) Shows ADA tooth 20 with obturation of the root canal by gutta percha.

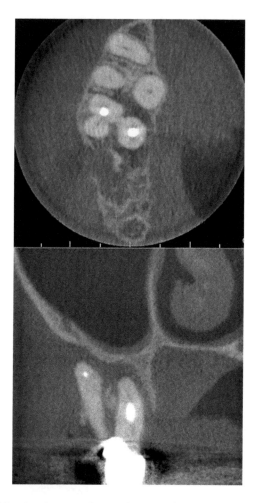

FIGURE 5.8 Shows axial (above) and corrected coronal sections from a small volume CBCT acquisition of end-odontically treated ADA tooth #2 using 60×60 mm FOV. The patient continued having pain and an active buccal fistula. The CBCT showed no obturation of the distobuccal canal, extensive periradicular and furcational bone rarefaction, loss of the buccal cortex, and mild mucoperiosteal swelling of the overlying maxillary sinus floor. The only abnormality evident on periapical imaging was persistence of a periapical radioluceny.

roots, canal morphology, extension of periapical lesions, and identifying possible causes of prior endodontic treatment failure (AAE and AAOMR 2015). Traditional images of a tooth with apical periodontitis and an endodontically treated tooth are shown in Figure 5.7, while more complex imaging of a tooth with a failed endodontic treatment, subsequently imaged with small-volume CBCT, is shown in Figure 5.8.

5.2.2 Evaluation of Growth and Development

Orthodontists, oral and maxillofacial surgeons, and pediatric dentists are frequently involved in treating patients with disorders of growth and development. Such disorders range from mild

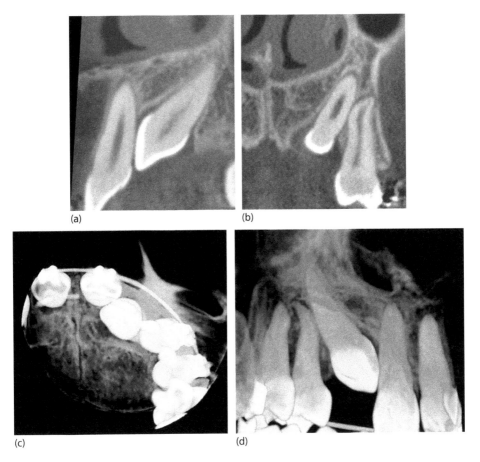

(a)   (b)

(c)   (d)

FIGURE 5.9 This young patient has an impacted maxillary canine, ADA tooth 11, in her palate. CBCT imaging was used to precisely image this tooth and define its relationships to the roots of adjacent teeth 9 and 12 (tooth 10 was congenitally absent). (a) A corrected sagittal section, shows the crown of impacted tooth 11 adjacent to, but not resorbing, the root of tooth 9. (b) A corrected coronal section, shows the root of impacted tooth 11 adjacent to the apical one-third of the root of tooth 12. There is a slight curvature of the apical one-third of the root of 12, likely developmental in response to the adjacent root of 11. (c) 3D maximum intensity rendering (MIP) of the acquired CBCT volume in the axial plane. Orthodontic brackets and wires are present on all of the teeth except impacted tooth 11. (d) 3D MIP of the acquired CBCT volume in reversed sagittal rendering—as if you are inside the mouth looking out. These types of views, taken together, are very helpful in planning the best route for moving the impacted canine into the proper position in the arch.

dental crowding to severe syndromes with palatoschesis, mandibular hypoplasia, macroglossia, airway difficulties, and other associated craniofacial and systemic problems. Diagnostic evaluation is complex, and may involve panoramic, cephalometric, CBCT, MDCT, and MRI imaging of the craniofacial and gnathic structures. Treatments are often multimodality, involving teams of health-care providers from a variety of specialties. CBCT imaging of a patient with an impacted maxillary canine is shown in Figure 5.9. CBCT imaging of a patient with Ehlers-Danlos syndrome is shown in Figure 5.10.

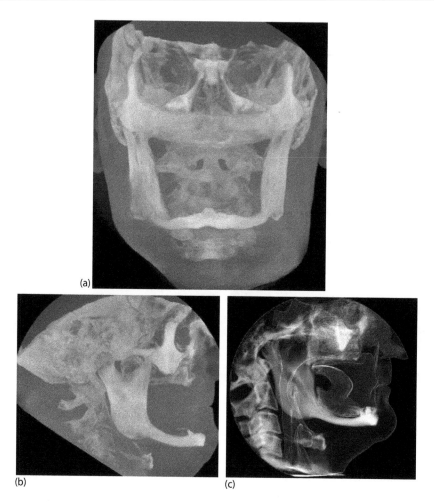

FIGURE 5.10 3D MIP, rendered from a large volume CBCT acquisition, of a 38-year-old male patient with Ehlers-Danlos syndrome. Facial bones are severely underdeveloped, and the patient is edentulous. The CBCT volume was needed to plan bone augmentation for dental implant placement to build a functional occlusion and an aesthetic mid-facial result. (a) Sagittal rendering, (b) coronal rendering, and (c) colorized, sagittal rendering of bone surfaces with superimposed soft-tissue renderings.

5.2.3 Delineation of Unusual and Advanced Pathology of the Gnathic Structures

Textbooks are filled with a vast array of odontogenic cysts, neoplasms, dental abnormalities, extensive infections, temporomandibular joint disorders, and salivary gland disorders, as well as trauma to the teeth, supporting structures, and facial bones. Extensive DMF diseases that involve salivary glands and fascial spaces are the domain of MDCT and MRI and will not be discussed here.

Bone lesions of the gnathic structures and temporomandibular joint (TMJ) disorders, more often than not, present with pain and/or swelling. Malignant lesions involving the gnathic bones are often accompanied by paresthesia. The imaging component of the diagnosis of such a lesion almost always begins with a panoramic image (Figure 5.11). This may be sufficient, but is often followed by advanced imaging, with CBCT being superior for bone lesions due to its extremely high spatial resolution. Where soft-tissue involvement is suspected, MDCT and/or MRI become the imaging of choice. CBCT is especially effective in defining the extent of intraosseous lesions and bone involvement by extraosseous

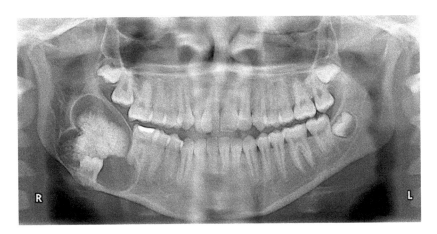

FIGURE 5.11 This panoramic image shows a patient with a large, odontogenic tumor—an ameloblastic fibro-odontoma. A moderately locally aggressive tumor, surgical treatment planning was complex, as it required a mandibular segmental resection. Further imaging was accomplished using MDCT.

lesions as well as mass effect and effects of lesions on teeth. All of these features are important in the final diagnosis of the problem. CBCT is the imaging of choice for defining the extent and severity of medication-related osteonecrosis of the jaws in patients receiving anti-bone-resorptive pharmacotherapy as part of treatment for malignant lesions, most often multiple myeloma and breast and prostate carcinomas. Osteoarthritic changes are frequently encountered in the TMJs. Evaluating TMJ area pain for osseous pathology is the realm of CBCT, which provides excellent detail of the configurations and osseous fine structures, enabling identification of the presence and severity of these changes (Figure 5.12).

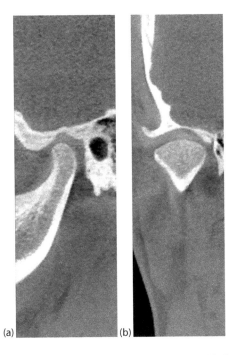

(a)   (b)

FIGURE 5.12 Small-volume CBCT imaging of this normal temporomandibular joint (TMJ) shows exquisite detail of the temporal and mandibular osseous elements of this complex and unique joint. (a) Sagittal section and (b) coronal section. Morphology and details of the cancellous and cortical components are clearly seen.

Even normal findings in the TMJ osseous components is diagnostically helpful as it can lead to further evaluation of the TMJ soft-tissue components, namely, of the articular disc and its osseous attachments as well as the presence or absence of inflammatory effusions within the joint spaces using MRI.

5.2.4 Treatment Planning Implant and Extensive Surgical Procedures

Replacement of missing teeth with dental implants is a widespread and rapidly growing procedure; over 5.5 million implants were placed by U.S. dentists in 2006 and increasing by some 500,000 per year. It is estimated that the U.S. market for implants and their restorations will reach $6.5 billion by 2018 (American Academy of Implant Dentistry 2017). CBCT imaging demonstrates bone quality, quantity, and location of critical structures in the projected implant sites with excellent spatial resolution; as such, it is the gold standard modality of choice for implant treatment planning (Tyndall 2012). An example of the use of corrected-angle cross-sectional images to plan a mandibular implant are shown in Figure 5.13.

Oral and maxillofacial surgery (OMFS) can be very complex and require precise advanced imaging for orthognathic surgical procedures. Most OMF surgeons will image their patients initially with panoramic acquisitions. Advanced imaging—MDCT, MRI, or CBCT—is dictated by historical, clinical, and panoramic findings. Typical imaging needs of OMFS include impacted third molars, bone lesions, infections, TMJ disorders, dental-implant treatment planning and growth, and development disorders. CBCT is especially useful in defining the relationships between third molar roots and the inferior alveolar canal, showing bone quality, dimensions, and locations of critical structures for implant treatment planning, defining extent of bone lesions and extensive infections and their effects on adjacent structures, and planning complex surgical procedures for patients with craniofacial deformities. Imaging needs tend to be more extensive in this domain than for other DMF diagnosis and treatment planning. An example of CBCT imaging for third molar evaluation is shown in Figure 5.14.

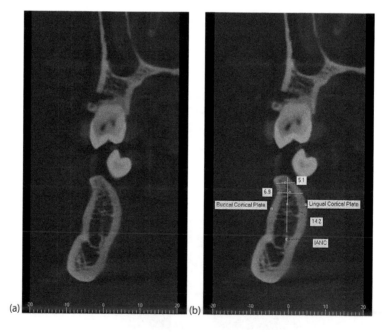

FIGURE 5.13 Adjusted cross-sectional images from a small-volume CBCT of a projected site for a dental implant fixture in the mandibular premolar region. (a) Shows the projected site of the implant with a positioning stent above the edentulous site. (b) Shows the site with critical anatomic features identified and measurements to help select and align the implant fixture prior to placement (IANC = inferior alveolar nerve canal).

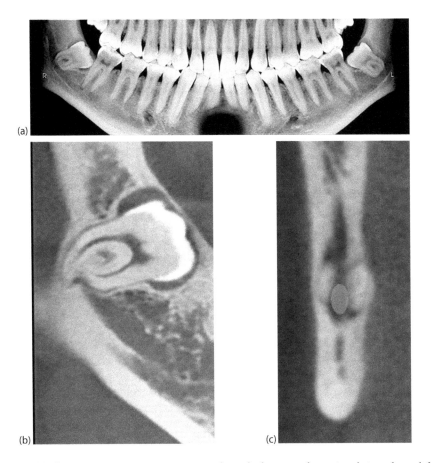

FIGURE 5.14 (a) Shows a panoramic reconstruction through the jaws of a patient being planned for third-molar extractions. ADA tooth #32 (mandibular right third molar) is impacted and the roots appear to be adjacent to the IANC. The canal is mapped in red. (b) Section in corrected sagittal plane from the multiplanar display, shows that the roots of this tooth are, in fact, severely curved (dilacerated) but do not clearly show the relationships between the roots and the IANC. (c) Section in corrected coronal plane with the IANC highlighted in red, shows that the roots flank the IANC. These images showed that there was a high likelihood of damage to the inferior alveolar nerve if the tooth was extracted in a conventional manner. The crown of the tooth was sectioned and removed, and the roots were left in ("coronectomy"). The patient did fine.

5.3 Risks of Dentomaxillofacial Imaging

In general, the risks from conventional DMF imaging, if any, are extremely small, perhaps negligibly so. Absorbed doses from conventional intraoral imaging, panoramic imaging, and cephalometric imaging are almost always in the low-uGy range. Thus, estimation of risk is based on studies of populations exposed to larger doses with extrapolations to very low doses. Layered on these extrapolations are the unknown and conflicting potential mechanistic modifiers of adaptive response and bystander effect. Thus, such calculations are fraught with uncertainties (NCRP 2012; UNSCEAR 2015), and assigning risk values to individuals undergoing DMF imaging is inappropriate. The linear non-threshold hypothesis for cancer induction at doses below 100 mGy has been a prudent and practical guideline for decades and continues to be so. The enormous numbers of DMF images exposed annually in the United States require careful consideration of risks to the population. CBCT is the fastest growing DMF imaging modality in the United States with over 5000 units in use and several million examinations conducted

TABLE 5.3   Terminology for Range of Absorbed Dose for
Low-LET (Linear Energy Transfer) Radiation

Dose Bands	Range of Absorbed dose for low-LET radiation
High	Greater than about 1 Gy
Moderate	About 100 mGy to about 1 Gy
Low	About 10 mGy to about 100 mGy

Source: UNSCEAR, Annex A: 2015, Table 1.

annually in 2014–2015 (Spelic and Farris 2015). Given that CBCT doses can equal or exceed those of MDCT for comparable examinations, risk to the individual, while small, is considerably greater than with conventional DMF imaging and needs to be considered even more carefully.

5.3.1 Intraoral Imaging

Doses from intraoral imaging have steadily decreased for decades, and with the widespread adoption of digital receptors are now so low they can be difficult to measure. Doses from a pair of properly exposed bitewings are 0.005–0.010 mGy, and from a full-mouth-series are 0.017–0.035 mGy, depending on the number of images. These doses fall into the low end of the dose band range as seen in Table 5.3 (UNSCEAR 2015). Intraoral images taken annually in the United States comprise the majority of dental imaging, with the majority of these employing digital technology, which generally reduces the dose versus film (Spelic and Farris 2015). The risk to the individual from a properly exposed intraoral examination is negligible. However, slower receptors, lack of rectangular collimation, and needless exposures, coupled with the enormous numbers of annual intraoral examinations, can significantly increase the collective dose, demanding that all such images be taken with optimal technique and appropriate selection criteria.

5.3.2 Panoramic Imaging

Doses from panoramic imaging have likewise experienced steady and substantial diminution during the last three decades, due initially to development of rare-earth film-screen combinations and more recently to the widespread adoption of digital technology. The range of effective doses reported in recent literature is 9–24 uSv (White 2013). This dose range is well within the very low category, and as such poses a minimal risk to the individual patient (UNSEAR 2015). While the numbers of panoramic machines and annual exposures are not presently available, the 1999 Nationwide Evaluation of X-Ray Trends (NEXT) survey found that over 28 million examinations were performed annually (CRCPD 2003) and that number is certainly substantially larger today. As with intraoral imaging, the numbers of panoramic images acquired annually, coupled with prudent practice, require optimal technique and appropriate selection criteria for panoramic imaging.

5.3.3 Cephalometric Imaging

Cephalometric imaging, primarily used by OMF surgery, orthodontics, and occasionally pediatric dentistry, results in doses in the range of a pair of bitewings (2–6 uSv). Far fewer cephalometric images are acquired annually than other DMF images. Thus, this type of imaging poses a relatively negligible risk to individual patients. However, the patient population is usually children and adolescents, and a growing number of practitioners are using CBCT to generate cephalometric images. Cephalometric images should be acquired with appropriate equipment, and with appropriate selection criteria.

5.3.4 Cone-Beam Computed Tomography Imaging

CBCT imaging has been used in dentistry for about 15 years and poses an entirely new risk consideration. The marked variations amongst manufacturers of FOV, voxel sizes, and machine presets present a wide range of doses and exposed tissues from these examinations. Also, the lack of standardized CBCT education in dental schools adds greatly to the variability in how these instruments are understood and used.

FOV varies greatly, from skull vertex to hyoid bone full coverage down to 40×40 mm, with many intermediate FOVs. Spatial resolution is controlled, in large part, by voxel size, which has varying levels of operator control amongst machines. The smaller the voxel size, the better the resolution and the higher the dose. Finally, most machines have presets with names such as high definition, high resolution, average definition, and standard resolution. There are dramatically different doses associated with the different presets, and many operators automatically choose the highest resolution because it generates the most aesthetically pleasing images. Unfortunately, the highest resolution presets also generate the highest patient dose, in some instances doses greater than those from MDCT. Thus, the risks to individual patients from CBCT examinations, while generally lower than those from MDCT, can be equal to or greater than those associated with MDCT examinations.

The ranges of CBCT doses were shown previously in Table 5.2. For small FOVs, doses are usually quite low, ranging from effective doses of 19–652 uSv. Medium-FOV effective doses range from 45 to 860 uSv. Finally, large-FOV effective doses range from 68 to 1073 uSv (White 2013). This latter range remains in the very low category (UNSCEAR 2015a) and thus carries small risk but is certainly greater than that of other DMF examinations. This makes it essential for the machine to be maintained in excellent working condition, and for the practitioner to be thoroughly educated and trained in the indications for CBCT as well as the operation of the machine.

5.4 Summary and Conclusions

Conventional DMF imaging, consisting of intraoral and panoramic images, is by far the most frequently employed imaging in dentistry. Modern technology, namely, digital imaging and intraoral rectangular collimation, has dramatically reduced the patient dose per image during the past three decades. The reduction in risk from such imaging is now so small that it is almost negligible for individual exposures (Pradhan 2013). Given the fact that over a billion intraoral images are acquired annually in the United States, there is certainly a risk consideration, albeit small, to the population.

CBCT is a rapidly growing imaging modality in dentistry, with powerful diagnostic applications in endodontic, impacted tooth assessment, and implant and orthognathic surgical treatment planning. As with MDCT, there are many acquisition settings and presets that profoundly affect the patient exposure. Thus, the smallest FOV and the largest voxel size compatible with the diagnostic task must always be used.

Finally, As Low As Reasonably Achievable (ALARA) considerations and uncertainties in carcinogenesis risk estimation at very low doses make it prudent and practical to continue to apply the optimization principle. The best way to minimize radiation dose and risk and maximize benefit to the patient is to have a strong, valid reason to employ x-ray-based imaging, that is, to use established, strong selection criteria when ordering and acquiring diagnostic images in dentistry, and acquire these images with the best techniques possible to yield the required diagnostic information.

References

AAE and AAOMR Joint Position Statement. 2015. Use of cone beam computed tomography in endodontics 2015 update. *Oral Surg Oral Med Oral Pathol Oral Radiol* 120(4):508–512.

American Academy of Implant Dentistry, www.aaid.com/about/press_room/dental)implants_faq.html, accessed August 10, 2017.

Conference of Radiation Control Program Directors. (CRCPD). 2003. Publication E-03-6, NEXT Tabulation and Graphical Summary of the 1999 Dental Radiography Survey.

Farman, A.G. (Ed.). 2009. Self-referral: An ethical concern with respect to multidimensional imaging in dentistry? *J Appl Oral Sci* 17(5). doi:10.1590/S1678-77572009000500001.

Fejerskov, O., B. Nyvad and E. Kidd. 2015. *Dental Caries: The Disease and its Clinical Management/ Edition 3*. Hoboken, NJ: Wiley-Blackwell.

Hargreaves, K.M. and L.H. Berman. 2015. *Cohen's Pathways of the Pulp/Eleventh Edition*. St. Louis, MO: Elsevier Mosby.

Kim, J.K., L.A. Baker, S. Davarian and E. Crimmins. 2013. Oral health problems and mortality. *J Dent Sci* 8(2):115–120.

NCRP. 2012. National council on radiation protection and measurements. Uncertainties in the estimation of radiation risks and probability of disease causation, NCRP Report No. 171. National Council on Radiation Protection and Measurements, Bethesda, MD.

Newman, M.G., H. Takei, P.R. Klokkevold and F.A. Carranza. 2014. *Carranza's Clinical Periodontology/ Twelfth Edition*. Philadelphia, PA: W.B. Saunders.

NIH. 2017. NIH Research Portfolio Online Reporting Tools (RePORT). Periodontal Diseases. Accessed July 7, 2017. https://report.nih.gov/NIHfactsheets/Pdfs/PeriodontalDiseases(NIDCR).pdf

Ozdemir, D. 2013. Dental caries: The most common disease worldwide and preventive strategies. *Int J Oral Biol* 5(4):55–61.

Pradhan, A.S. (Monitoring Editor) 2013. On the risk to low doses (<100 mSv) of ionizing radiation during medical imaging procedures – IOMP policy statement. J Med Phys 38(2):57-58.

Spelic, D. and K. Farris. 2015. Co-presentation: Nationwide evaluation of X-Ray trends: Highlights of the 2014–2015 NEXT dental survey. In: *Proceedings of 47th National Conference on Radiation Control*, August 2015. CRCPD Publication E-15-4. Conference of Radiation Control Program Directors, Frankfort, Kentucky.

Tyndall, D.A., J.B. Price, S. Tetradis et al. 2012. Position statement of the American Academy of Oral and Maxillofacial Radiology on selection criteria for the use of radiology in dental implantology with emphasis on cone beam computed tomography. *Oral Surg Oral Med Oral Pathol Oral Radiol* 113:817–826.

UNSCEAR. 2015. United Nations Scientific Committee on the Effects of Atomic Radiation. Sources and Effects of Ionizing Radiation, UNSCEAR 2012 Report to the General Assembly, with Scientific Annexes. Scientific Annex A, Attributing health effects to ionizing radiation exposure and inferring risks. Scientific Annex B, Uncertainties in risk estimates for radiation-induced cancer. United Nations, New York.

White, S.C. and M. J. Pharoah. 2014. *Oral Radiology: Principles and Interpretation, Edition 7*. St. Louis, MO: Elsevier Mosby, Table 3-3, Page 32.

6

Optimization and Dose Reduction in Mammographic Imaging

Lawrence N.
Rothenberg

6.1 Introduction

Mammography, or radiography of the breast, is one of the most technically challenging of all radiographic procedures. The mammographic image must have both high spatial resolution to visualize clusters of sub-millimeter calcifications and high contrast to demonstrate both subtle soft-tissue lesions and other clinically significant variations in breast structure. In addition, since much of mammographic imaging is performed to discover early breast cancer in large populations of asymptomatic women, the process of optimization of protection, where the radiation dose for the procedure must be kept low in order to maintain an acceptably high ratio of benefit to risk, applies. The benefit of mammography is early detection when the cancer may best be treated, while the risk being the low probability for induction of new breast cancers caused by the ionizing radiation from the procedure.

Historically, the image receptor used for mammography has been radiographic film or an intensifying screen and film combination. Currently, these film receptors have generally been replaced by several types of digital image detectors. In the United States more than 99% of mammography is performed with digital image receptors (FDA 2018). Therefore, film or screen-film mammography will not be discussed in this chapter. For a good discussion of screen-film mammography and proper techniques see NCRP Report 149, "A Guide to Mammography and Other Breast Imaging Procedures" (NCRP 2004).

In the United States, the federal government, through the Food and Drug Administration (FDA), must certify all mammography facilities and the personnel associated with the facility. These federal regulations were initially embodied in the Mammography Quality Standards Act of 1992. The law and associated regulations are in Title 21 of the Code of Federal Regulations, Part 900. The certification of the facilities includes their obtaining accreditation from recognized accrediting bodies such as the American College of Radiology Mammography Accreditation Program (ACR-MAP). In addition to the mammography facility having approved and properly functioning imaging equipment, it must also have specially trained and experienced radiologists, technologists, and medical physicists.

The FDA initially approved clinical use of full-field digital mammography (FFDM) units for patient examinations following a major comparison study: the American College of Radiology Imaging Network Digital Mammography Imaging STudy (ACRIN-DMIST) of digital versus screen-film mammography involving 50,000 women who had one breast imaged by both methods (Pisano et al. 2005). The ACRIN-DMIST study of 50,000 women with both screen-film and FFDM showed better sensitivity for digital over screen-film mammography for women under age 50 regardless of breast density, women of any age with dense breasts, and pre- and peri-menopausal women of any age. This represents about 65% of the U.S. screening population.

6.2 Fundamentals of Mammography

The components of the mammographic imaging system are mounted on a C-arm assembly that can be rotated by more than 90 degrees in either direction from the vertical to achieve the required mammography views (See Bushberg et al. 2012, Figure 8.3). An excellent source for the overall technical and clinical aspects of mammography is NCRP Report 149—"A Guide to Mammography and Other Breast Imaging Procedures" (NCRP 2004). Technical details of the mammography process are also comprehensively explained in Chapter 8 of *The Essential Physics of Medical Imaging* (Bushberg et al. 2012).

6.2.1 The Imaging Chain

6.2.1.1 X-ray Generator

The x-ray generator used for mammography is typically a low power, 3- to 10-kW, constant-potential generator.

6.2.1.2 X-ray Tube

X-ray tubes used for mammography are grounded anode, metal housing tubes with a very thin beryllium (Be) exit window. There are commonly two focal spots: a nominal 3.0 mm focal spot used for general breast imaging and a nominal 0.1 mm focal spot used for magnification studies. The tube and anode are angled so that the central ray of the x-ray beam points straight along the chest wall of the patient.

6.2.1.3 Targets and Filters

Several different target-filter combinations are used in digital mammography. The early digital units used the same target-filter combinations that had been used for screen-film imaging. These include a molybdenum (Mo) target with either molybdenum or rhodium (Rh) filters, and a rhodium target with rhodium filter. The Mo and Rh spectra with appropriate filtration have a strong characteristic x-ray peak in the mammography energy range, 17–19 keV for Mo and 19–22 keV for Rh. The x-ray spectra become increasingly shifted to higher energy for the same peak voltage setting with a shift of target-filter combinations from Mo-Mo to Mo-Rh to Rh-Rh. The resulting increased effective energy of the x-ray beam is necessary as the imaging task moves from thin more adipose breasts to thicker more glandular breasts.

More recently, optimization studies have shown that the mammography units designed for digital imaging should use tungsten (W) targets with various filters including aluminum (Al), molybdenum, rhodium, and silver (Ag) (Williams et al. 2008; Ranger et al. 2010). Note that any x-ray tube used for mammography must be specifically designed for breast imaging; general-purpose x-ray tubes with W targets may not be used.

6.2.1.4 Typical X-ray Spectra

Typical x-ray spectra from some of the various target-filter combinations are shown in Bushberg et al. 2012, Figures 8.9–8.11. The spectra from the Mo and Rh targets contain strong characteristic x-ray peaks in addition to broad bremsstrahlung distributions. A spectrum from a Mo target filtered by 0.1 mm Mo, has the characteristic peaks between 17 and 20 keV with a minimal background of bremsstrahlung x-rays, while the spectrum from a Rh target with a Rh filter has the characteristic peaks between 20 and 23 keV on the minimal bremsstrahlung background. When a Mo target is used with a Rh filter, an additional amount of bremsstrahlung radiation is included between 20 and 23 keV. The W target spectra contain only the bremsstrahlung distributions with a shift toward higher effective energy, or half-value layer (HVL), as one switches from Mo to Rh to Ag filter.

6.2.1.5 Collimation

The collimation on most mammography systems is set to cover the full image receptor and must extend slightly beyond the chest-wall edge of the image receptor, so that all of the breast tissue along the chest wall is imaged. There are visible light localizers on most of the collimation systems.

6.2.1.6 Compression Device

An important component of any mammography x-ray unit is a flat rigid sheet of radioluscent plastic, called a compression plate, which is pressed down on the breast before an x-ray exposure is made. The purpose of this compression device is to spread the breast tissue over a larger area with a reduction in the dimension in the direction of the x-ray beam.

Some of the benefits of using the compression device are: reduction in required radiation dose, since thinner tissue must be traversed by the x-rays; reduction in scattered radiation reaching the image receptor due to less thickness of scattering tissue; less motion blur due to almost no movement of the breast during the radiation exposure; less geometric blur (penumbra) due to reduced ratio of breast image distance to target image distance; spreading of the breast tissue to avoid overlapping structures in the image; and lower exit exposure dynamic range.

One major problem with applying firm compression is great discomfort for some patients. Therefore, it is necessary to have a compassionate technologist to gently position the patient and to explain the benefits that come from the discomfort. In some cases, digital mammography may be performed with somewhat reduced compression force.

6.2.1.7 Automatic Exposure Control Detectors

An important component of an optimized mammography system is automatic exposure control (AEC), sometimes called a "phototimer." A single detector, or a set of x-ray detectors, is placed between the breast and the image receptor to measure the amount of radiation reaching the image receptor. On some digital units, the image receptor itself may act as the AEC detector. Precalibration of the system with appropriate thicknesses of breast-equivalent materials is used so that the x-ray exposure will terminate when the x-ray flux striking the receptor is sufficient to obtain a high-quality image.

6.2.1.8 Grid

A low ratio anti-scatter grid is used for imaging of thicker breasts. The grid ratio is typically 5:1 with a Bucky Factor of 2.

6.2.1.9 Digital Image Receptors

There are three types of image receptors used in digital mammography. Two types are fixed in the mammography unit and the third type is the computed radiography (CR) cassettes. The two fixed receptors are referred to as direct and indirect image receptors (See Bushberg et al. 2012, Figures 8.26, 8.27).

The direct image receptors employ a layer of amorphous selenium over a thin-film transistor (TFT) array. The x-rays transmitted through the breast interact in the selenium layer, which acts as a photoconductor and delivers a charge distribution to the TFT layer where an image is formed.

The indirect image receptor contains a cesium iodide layer, which interacts with the x-rays and converts the x-ray image to an image of light photons, which is then captured in a TFT layer to produce an image.

The CR system for mammography utilizes light-tight plastic cassettes containing plastic sheets coated on both sides with a photostimulable phosphor. These cassettes are placed in image receptor slots on mammography units that were initially designed for screen-film mammography. Following exposure of the patient with the mammographic x-ray beam, the cassette is removed and inserted into a processing unit. In the processor, the photostimulable phosphor plate is removed from the cassette and read out with a scanning red helium neon laser beam to form a digital image suitable for processing. The CR system used for mammography differs from a conventional CR system used for general radiography in several ways. The CR plate is coated on both sides with the photostimulable phosphor layer for increased sensitivity. The laser beam reads out both sides of the sheet simultaneously. The laser spot is kept small enough, so that a detector element (del) size of 0.05 mm is maintained. All new mammography x-ray systems utilize the fixed-image receptors described above, so it is likely that CR mammography will gradually disappear.

The detector elements in the array are on the order of 0.1 mm for the indirect systems, 0.07 mm for the direct systems, and 0.05 mm for the CR systems. The size of these elements limits the high-contrast spatial resolution of the digital mammography systems to about 5, 7, and 10 line pairs (lp)/mm, respectively, significantly coarser than that of screen-film mammography, which has a limiting spatial resolution of 11–13 lp/mm or better. The lower resolution of the digital imaging systems is more than compensated for by other factors discussed below.

6.2.1.10 Processing of the Digital Mammographic Image

The raw digital images obtained from the mammography unit or the CR plate must be processed before being presented on the high-quality viewing monitor. First, corrections are applied for bad pixels and varying gain across the detector area. Next, linear contrast and brightness corrections are applied. Then, the final image has skin equalization and non-linear enhancements applied so that the final image closely approximates high-quality screen-film mammograms.

6.2.1.11 Advantages of Digital over Screen-Film Mammography

Despite its coarser spatial resolution, digital mammography has several technical advantages over film images of the breast. Screen-film systems have an exposure dynamic range of about 25:1, while a highly glandular breast can have a transmitted x-ray exposure range of more than 200:1. Digital receptors respond to an exposure range of 1000:1.

There is improved conspicuity of lesions allowed by image processing and display manipulation, including window/level control by the viewer.

There are also data showing that the typical FFDM techniques being utilized show a reduction in mean glandular dose to about 1.2 mGy compared to 1.8 mGy for screen-film techniques (Bushberg 2012, Page 276).

Another major advantage of digital mammography with the fixed image receptors is a reduction in examination time, since little or no time is required for processing, handling, and distribution. Furthermore, the digital mammograms are stored in a Picture Archive and Communication System

(PACS) from which they can be easily retrieved and also sent to other locations for remote viewing by radiologists. Digital mammographic images can also be directly entered into Computer-Assisted Detection, or Diagnosis, systems (CAD), which can aid the radiologist in identifying suspicious areas of the images (Bushberg 2012, Page 270).

These many advantages of digital mammography appear to have overcome the somewhat poorer resolution and higher equipment costs relative to film methods. As stated earlier, more than 99% of all mammography units in the U.S. are now digital units.

6.2.1.12 Typical Technique Setting for Mammography

The peak voltage settings for dedicated mammography units can typically be adjusted from 22 to 35 kVp. Units that will be used for contrast-enhanced digital mammography (CEDM) (See Section 6.4.2) can set tube voltages to as high as 49 kVp. The tube current settings for the standard focal spot are typically on the order of 100–200 mA, and the very small focal spot used for magnification views is used with a current of about 25 mA.

Exposure time settings are normally controlled by the AEC system and are ideally kept well below 2.0 seconds to minimize any blurring of the image due to possible patient motion.

6.2.1.13 Typical Patient Views for Mammography

The standard breast views for screening mammography are a craniocaudal (CC) view and a mediolateral oblique (MLO) view. Diagnostic mammography examinations on symptomatic patients or those with other family background or genetic indications of breast disease may include lateral and other views.

6.2.1.14 Viewing Monitors for Mammography

The digital mammographic images must be viewed on monitors approved by the FDA for mammography. Conventional computer monitors must not be used for mammographic diagnosis. These monitors must have at least 5 megapixels, have a contrast ratio of at least 350 to 1, and maximum luminance of at least 450 candela/m^2. They are much brighter and have much higher resolution than conventional computer monitors. In addition, the reading room in which the monitors are being used must have very low ambient lighting, less than 20 lux. No bright objects such as light from a doorway or lighted view boxes used to view previous film mammograms should be in a position to cause reflections on the monitors.

6.2.2 Patient Dose in Mammography

Over the years several different parameters have been used to describe radiation dose in mammography.

6.2.2.1 Skin Dose and Air Kerma

The easiest dose parameter to measure for a mammography exam is exposure in air, in units of roentgens (R or mR), at the position of the x-ray beam entrance surface of the breast. If System Internationale (SI) units are to be used, then air kerma, K_{air}, in Gy or mGy would be measured.

6.2.2.2 Breast Mid-Plane Dose

In the past, some investigators have used a measurement or calculation of breast mid-plane dose to obtain a better estimate of the meaningful dose from a mammographic procedure.

6.2.2.3 Mean Glandular Dose

The consensus at the present time is that the best estimator of radiation risk from mammography is the mean (average) dose to the glandular tissue of the breast, because it is assumed that the low risk of radiation-induced cancers in the breast will more likely arise in the glandular tissue.

Mean glandular dose (MGD) cannot be measured directly, but must be calculated from measurements of exposure or air kerma and beam HVL, combined with use of MGD per unit air kerma or

TABLE 6.1 Normalized Mean Glandular Dose vs. kVp and HVL

HVL (mm)	kV 25	26	32
0.25	0.140		
0.26	0.144	0.147	
0.27	0.149	0.151	
0.28	0.153	0.156	
0.29	0.159	0.161	
0.30	0.164	0.166	
0.31	0.168	0.171	
0.32	0.173	0.175	0.183
0.33	0.177	0.180	0.188
0.34	0.183	0.184	0.192
0.35	0.188	0.190	0.197
0.36	0.192	0.195	0.202
0.37		0.199	0.205
0.38			0.210
0.39			0.214
0.40			0.220

Source: Adapted from ACRQC Manual, (1999).

exposure factors that have been determined from breast tissue phantom measurements or Monte Carlo calculations. The MGD factor that is used for a specific calculation depends on many parameters such as kVp, beam quality, breast tissue composition (adipose-glandular tissue mix), and compressed breast thickness. Table 6.1 shows values of the MGD factors for specific technique settings and breast thicknesses (ACR 1999).

Typical values of MGD for a breast of average composition and thickness, which is often assumed to be about 50% glandular and 50% adipose and 4.5 cm thickness, range from about 1.0–2.0 mGy for single views currently being performed with FFDM units. Of course, thinner or thicker and more or less adipose breasts will have a lower or higher MGD, respectively. The FDA regulations specify that the dose to a standard phantom representing the average breast must not exceed 3 mGy. It should be noted that the common assumption of an average breast being 50% adipose and 50% glandular tissue has been questioned (Yaffe 2009).

6.2.2.4 Instruments for Measurement of Dose in Mammography

The x-ray exposure or kerma measurements are performed with a thin window ionization chamber specifically designed for low energy mammography measurements. Beam quality measurements HVL should employ a series of very thin filters of high purity aluminum, 0.1 mm thickness each.

If it is desired to measure skin dose directly, then properly calibrated thermoluminescent dosimeters (TLDs) or optically stimulated detectors (OSDs) might be placed directly on the breast, or on a suitable tissue substitute phantom, during an actual image capture.

6.3 Quality Assurance Programs for Mammography

Mammography Quality Standards Act (MQSA) requirements for mammography facilities specify a very strict quality assurance (QA) program, involving specified measurements with different frequencies: daily, weekly, monthly, and semi-annually tests to be performed by specially trained mammography QA technologists whose work must be overseen by the facility's medical physicist(s). In addition, the physicists must perform an additional series of QA tests annually or whenever major modifications are made to the mammography x-ray units and associated equipment.

A special mammography phantom containing test objects simulating fibrous structures, microcalcifications, and subtle solid masses was developed for image-quality tests as part of the specified QA program. These phantoms are available for a small number of manufacturers who assure consistent production of the phantoms.

Previously, in the case of screen-film mammography units, the tests were specified in a publication such as the ACR Mammography Quality Assurance Manual (ACR 1999) and in the FDA regulations.

When FFDM units were first introduced, each manufacturer provided a set of recommended quality-assurance test procedures for their particular units. In mid 2017, the ACR introduced a new test phantom and QA manual that can be used for testing of all FFDM units. Units that also perform digital breast tomosynthesis (DBT) or contrast-enhanced digital mammography (CEDM) (See Section 6.4 below) cannot be tested with this new protocol and are still subject to the manufacturer's testing recommendations.

6.4 Related Special Mammographic X-ray Procedures

6.4.1 Digital Breast Tomosynthesis

Digital breast tomosynthesis is performed on mammographic x-ray units specifically designed for this purpose. During the exposure, the x-ray tube will rotate through a series of angles above the breast, and numerous digital images acquired from these different angles will be stored in the unit's memory. These images can then be electronically moved relative to each other in the direction of rotation in order to obtain sharp images of various planes within the breast. Information from overlying and underlying tissues is thereby purposely blurred out. The number of images obtained varies from 9 to 25 while the total angle through which the x-ray tube rotates ranges from 15 to 50 degrees. The technique settings for a DBT examination are set low enough so that the breast dose from the full set of images is similar to that obtained from a conventional two-view mammography procedure (Vedantham et al. 2015). Table 6.2 adapted from (Vedantham et al. 2015) lists some of the features of several of the DBT units.

DBT is sometimes referred to as "CT Mammography," although in principle the information obtained is somewhat different from that in a conventional computed tomography (CT) image set. Conventional CT images do not contain information from tissue slices surrounding the slice being imaged, except for a small amount of scatter. DBT images do contain information from surrounding slices, although it is blurred across the image.

The individual "slices" show information in various planes of the breast that is not obscured by sharp images of the overlying and underlying structures.

Since radiologists viewing the series of slices imaged in a DBT "3D" examination would typically also like to see a 2D conventional mammographic image of the breast, the digital information from the series of DBT images can be electronically combined to give a synthetic 2D image, claimed to be equivalent to a conventional mammographic image. Use of the synthetic 2D images avoids the need to give the patient additional radiation to obtain the conventional image (Vedantham et al. 2015).

TABLE 6.2 Specifications of Clinical Digital Breast Tomosynthesis Systems

Manufacturer	General Electric[a]	Hologic	Internazionale Medico Scientifica	Siemens
Model/Platform	SenoClaire/ Senographe Essential	Selenia Dimensions	Glotto Tomo	MAMMOMAT Inspiration
Source to Detector distance (cm)	66	70	68	65.5
Source to center-of-rotation distance (cm)	62	70	66	60.8
Source to breast support distance (cm)	63.8	67.5	65.8	63.8
X-ray tube angular range	±12.5°	±7.5°	±20°	±25°
X-ray tube motion	Step-and-shoot	Continuous	Step-and-shoot	Continuous
Detector angular range	Stationary	±2.1°	Stationary	Stationary
X-ray tube target material(s)	Mo/Rh	W	W	W
X-ray filter material(s)	Mo/Rh	Al	Rh/Ag	Rh
No. of projections	9	15	13	25
Equiangular distribution of projections	Yes	Yes	No[b]	Yes
Scan time (sec)	Typically <10	3.7	12	25
Detector type	a-Si indirect conversion	a-Se direct conversion	a-Se direct conversion	a-Se direct conversion
Detector pixel size (μm)[c]	100	70 (2 × 2 binned)	85	85
Equal milliampere-second/projection	Yes	Yes	No	Yes
Reconstruction method	Iterative (ASIR-DBT)	FBP/Iterative contrast	Iterative	FBP/Section thickness filter

Source: Vedantham, S. et al., *Radiology*, 277, 663–684, 2015.
FBP = Filtered back-projection.
[a] System uses a 5:1 linear antiscatter grid with focal distance of 65 cm.
[b] Finer angular sampling near center (0°) projection.
[c] Pixel size prior to binning.

6.4.2 Contrast-Enhanced Digital Mammography

Contrast-enhanced digital mammography (CEDM) involves the production of high-kVp and low-kVp images of the breast following injection of an intravascular iodinated contrast material. The two images are then electronically subtracted from each other leaving an image of only the regions in which the iodine contrast material is present, greatly enhancing the appearance of blood vessels within the breast. Since the k-absorption edge of iodine is at 33.2 keV, the high-kVp image is usually obtained at about 49 kVp with a tungsten or rhodium target and a copper (Cu) filter about 0.3-mm thick to produce a spectrum with many x-rays above the iodine absorption edge, and the low-kVp image is obtained at about 25 kVp with a standard filter to produce x-rays below the iodine k-absorption edge. The typical radiation dose, MGD, for CEDM procedures is about 20% to 30% higher than for a single low-kVp image. CEDM can improve the sensitivity of digital mammography. CEDM is less sensitive but more specific than breast magnetic resonance imaging (MRI) (see Section 6.5.1 below). In addition, since CEDM is significantly less expensive than breast MRI, it could possibly be used for screening patients who are unable to undergo breast MRI (Jochelson 2014).

6.5 Other Breast Imaging Procedures

The diagnostic workup of a patient with suspected breast disease often includes acquiring images with other well-established imaging techniques that complement x-ray mammography.

6.5.1 Magnetic Resonance Imaging of the Breast

Magnetic resonance imaging of the breast is quite useful for finding malignant disease in a variety of high-risk patients. MR imaging also does not involve any exposure to ionizing radiation. The MRI examination is so successful in detecting abnormalities in some patients with family history or genetic factors for breast cancer that conventional mammographic screening may not add any additional information and can be avoided. Two major disadvantages of MRI of the breast relative to mammography screening is that small, subtle indications of breast cancer that are seen on mammograms will be missed and the very high cost and limited availability of MRI compared to conventional mammography.

6.5.2 Ultrasonic Imaging of the Breast

Breast ultrasound is another imaging procedure that does not involve exposure to ionizing radiation. Its greatest success in breast imaging is in differentiating cystic from solid masses. Ultrasound is also used for guidance of biopsy needles in the breast.

6.6 Conclusion Optimization of Mammography

The following strategies should be considered to optimize image quality with a low radiation dose in mammography:

1. The appropriately designed mammographic x-ray units must be subject to a rigorous QA program carried out by properly trained and well-experienced technologists and medical physicists. Also, the entire program must be overseen by well trained, experienced physicians, normally radiologists.
2. The technique settings, including target, filter, kVp, and possibly filter selection, position of the AEC detector, rotation of the x-ray unit to get the proper view, and any additional choices on the unit available to the technologists, must be properly chosen.
3. The patient must be properly positioned on the x-ray unit by a suitably trained and compassionate mammography technologist who can explain the necessity of the mild, or in some cases greater, discomfort from firm compression that the patient must endure in order that the best possible image can be obtained.
4. The captured digital image must be properly processed and presented to a well-trained and experienced diagnostic mammography radiologist.
5. The radiologist must view the digital mammogram on a suitable high-quality monitor in a room that is properly set up so that the image is seen in a setting with minimal ambient light and no bright objects behind the viewer that might cause confusing or distracting reflections.
6. In addition, when these proper techniques are employed the need for repeat examinations with additional radiation exposure to the patient is avoided.
7. When feasible, some of the alternate breast imaging techniques described in Section 6.6 should be employed to further minimize the need for the use of additional ionizing radiation exposures to the breast.

References

ACR Mammography Control Manual. 1999. *American College of Radiology.* Reston, VA.

Bushberg, J., Seibert, J., Leidholt, E., Boone, J. 2012. *The Essential Physics of Medical Imaging.* 3rd ed. Philadelphia, PA: Lippincott Williams & Wilkins.

Fda.gov/RadiationEmittingProducts/MammographyQualityStandardsActandProgram/FacilityScorecard January 2018.

Jochelson, M. 2014. Contrast-enhanced digital mammography. *Radiol. Clin. N. Am.* 52: 609–616.

National Council on Radiation Protection and Measurements, Report No. 149, A guide to mammography and other breast imaging procedures, Bethesda, MD, 2004.

Pisano, E.D., Gatsonis, C., Hendrick, E. et al. 2005. Diagnostic performance of digital versus film mammography for breast-cancer screening. *N. Engl. J. Med.* 353: 1773–1783.

Ranger, N., Lo, J., Samei, E. 2010. A technique optimization protocol and the potential for dose reduction in digital mammography. *Med. Phys.* 37: 962–969.

Vedantham, S., Karellas, A., Vijayaraghavan, G., Kopans, D. 2015. Digital breast tomosythesis: State of the art. *Radiology* 277: 663–684.

Williams, M., Raghunathan, P., Mitali, J. et al. 2008. Optimization of exposure parameters in full field digital mammography. *Med. Phys.* 35: 2414–2423.

Yaffe, M., Boone, J., Packard, N. et al. 2009. The myth of the 50–50 breast. *Med. Phys.* 36: 5437–5443.

7

Optimization and Dose Reduction in Fluoroscopy and Interventional Imaging

Dustin W. Lynch
and Daniel J. Long

7.1 Introduction

While there are multiple definitions for fluoroscopically guided procedures, International Commission on Radiological Protection (ICRP) Publication 85 defines these as "procedures comprising guided therapeutic and diagnostic interventions, by percutaneous or other access, usually performed under local anesthesia and/or sedation, with fluoroscopic imaging used to localize the lesion/treatment site, monitor the procedure, and control and document the therapy." This definition is echoed in the more recent National Commission on Radiological Protection (NCRP) Report 168.

Fluoroscopy as an imaging modality shares many of the same components that are used in radiography (Schueler 2000). However, it is the way in which the images are formed in the imaging chain and recorded that distinguishes fluoroscopy from its radiographic imaging counterpart. While radiography seeks to produce a high-quality 2D image of a given area, the primary purpose of fluoroscopy is to use the 2D imaging principles of radiography to provide "real time" imaging visualization. This can be achieved by using a much smaller exposure rate per x-ray pulse than radiography while using many more pulses to image the area of interest over time. This allows temporal changes in the area of interest to be observed as a given process proceeds. Fluoroscopy procedures include biopsies, drug delivery, stent placements, embolizations, tumor ablations, and many others procedures (Jones et al. 2014). Because of the unique properties of fluoroscopy and the many ways in which it can be utilized to observe and treat patients, each procedure presents particular challenges to recording, optimizing, and ultimately reducing the patient's radiation dose while maintaining the high image quality needed to perform the specific task. This chapter examines the use of fluoroscopy and its development over time as a broadly applicable tool to accomplish many different clinical tasks using the powerful imaging tool. It also addresses the radiation dose challenges facing the use of modern fluoroscopic imaging equipment for increasingly more complicated interventional procedures.

7.2 Fundamentals of Fluoroscopy

Fluoroscopy has been used since the early part of the 1930s. In its most basic form, a fluoroscopic system consisted of an x-ray tube and a fluorescent screen. Originally, the physician would position the patient between the x-ray and the fluorescent screen. This would allow for the x-rays to penetrate the patient and release visible light directly into the eyes of the physician. While lead glass may have been used to reduce the dose to the observing physician, there was still a significant dose delivered both to the patient and directly to the physician who would typically hold up the screen or place it in a holder in front of themselves. Along with the obvious problems of occupational and patient exposures using this system, the most notable problem was that the fluorescent screen could not produce images that were bright enough to practically evaluate the patient since the human eye is not sensitive enough at such low light levels (Schueler 2000). To remedy the obstacle of prohibitively poor sensitivity, spatial resolution, and contrast at low light levels, a device known as the image intensifier was created. This device would produce a large brightness gain so that the human eye could sensitively perceive differences in brightness across the image. The image intensifier would use a series of lenses and mirrors to magnify the output image. Originally, the viewing angle of the image was very narrow so the device needed to be repositioned frequently, and only one person could view the image at a time, making it difficult to describe changes in the image (Schueler 2000). These initial challenges were overcome by the addition of a video camera to view the output image. The image intensifier has continued to be improved upon through recent decades and now many modern units have replaced the image intensifier with a flat panel detector and the latest in high-resolution digital monitors to achieve excellent spatial resolution.

7.2.1 The Imaging Chain

7.2.1.1 X-ray Generator

The x-ray generator can impart energy to the fluoroscopy unit in two different modes, continuous and pulsed. The continuous method, similar to radiography, provides a steady tube current for the duration of activation (via foot pedal or hand switch) of the unit. Pulsed mode delivers exposure in short-interval sequences each typically 3–10 msec in duration (Schueler 2000). The pulse rate can often be adjusted on modern units from 1 to 30 pulses per second, depending on the model and/ or manufacturer. Pulsed fluoroscopy is useful for the operator not only because it reduces dose but also improves temporal resolution by shortening the pulse acquisition time (Jones et al. 2014). One of the most important aspects of the generator is the automatic exposure rate control (AERC) formerly known as automatic brightness control (ABC). This is a feedback mechanism used to maintain constant levels of x-ray flux at the entrance to the detector that change with the varying degrees of attenuation throughout the field of view. This is maintained by the generator, which automatically adjusts the technique factors like kVp and mA to maintain the brightness. This is especially important for specific interventional and cardiology applications where contrast is important. The way in which the AERC is configured to adjust the techniques or baseline brightness varies by mode and manufacturer settings.

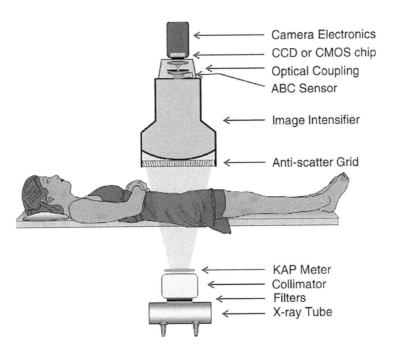

FIGURE 7.1   Typical fluoroscopic imaging chain shown with an image intensifier detector and KAP meter for measuring output from the x-ray tube. (Adapted from Bushberg, J.T. et al., *The Essential Physics of Medical Imaging*, Lippincott Williams & Wilkins, Philadelphia, PA, 3rd ed., 2012.)

7.2.1.2 X-ray Tube

X-rays are produced in a fluoroscopic system (shown in Figure 7.1) in the same manner as an x-ray tube in a radiographic system. Electrons are ejected from a filament across a voltage difference to a rotating (usually tungsten) anode, producing x-rays as the electrons are stopped in the anode. Electrons hit an area on the anode called the focal spot or target. It is common to also have multiple focal spots of varying dimensions, as in radiographic systems. This allows for normal fluoroscopy imaging as well as image recording, the latter requiring a higher x-ray flux and therefore a tube current (Nickoloff 2011). The anode for fluoroscopy units is typically configured as high-speed rotating anode to control the heat energy accumulated from long exposure times. Fluoroscopic systems also have different filters that are used in conjunction with the automatic brightness control to modulate the x-ray beam energy (i.e., eliminate more lower-energy photons) and increase penetration for larger patient thicknesses. The degree of filtration used will also vary based on manufacturer configuration.

7.2.1.3 Collimator

The collimator is an important tool in "shaping" the radiation beam. The collimator contains multiple sets of radiopaque shutter blades. These can be in the shape of a circular iris or a rectangle. Each blade can be manipulated manually to limit the radiation beam to the size of the relevant area of interest. Likewise, changing the magnification mode will automatically limit the x-rays to an area no larger than the new field of view. This limits the radiation exposure impacting the surrounding tissue volume and reduces the overall patient radiation dose. Modern interventional fluoroscopic units also contain wedge filters. These are used to further shape the x-ray beam and to reduce glare coming from unattenuated x-rays in areas just beyond the contour of the patient's body.

7.2.1.4 Image Intensifier System

The most significant developments in the advancement of fluoroscopic imaging have come in the form of better detectors and digital image processing. Originally, an image intensifier (II) was utilized in order to convert the incoming x-rays to usable light photons displayed on a television screen (Nickoloff 2011). The traditional II consisted of the input phosphor layer (usually made of cesium iodide [CsI]), photocathode, focusing electrodes, and an output phosphor contained in a large vacuum tube. X-rays enter the curved input surface and strike the input phosphor. This energy is converted to light and guided along the CsI crystals to be absorbed by the photocathode. The photocathode then emits electrons, following the photon absorption, that accelerate through the vacuum by a series of electrostatic focusing electrodes gaining substantial energy before striking the output phosphor. This is known as electronic gain. A portion of the energy gained is then converted back to light photons as the electrons hit the smaller output phosphor. This light at the output phosphor is much brighter than at the input phosphor due to both the electronic gain and the minification gain due to the large difference in the diameter of the input and output phosphors. Lastly, the light moves into a lens system (optical coupling) before striking the television camera, which converts the light into an electrical signal that is displayed on a television monitor (Nickoloff 2011).

7.2.1.5 Optical Coupling and the Television System for Image Intensifier System

For a traditional II, the detector would be optically coupled to a video camera and image recording devices. This would distribute the outgoing light produced from the II through a beam-splitting mirror to a video camera and an image recorder. The video camera would have an adjustable aperture size to control the amount of light entering the camera. When the aperture is set to a small setting, the automatic exposure control would have to increase the amount of radiation to maintain the same level of noise preset in the system. Likewise, the inverse is true when the aperture is opened to a wide setting (Schueler 2000).

The video camera converts the light entering it to a voltage signal that is displayed on a monitor system. The video camera consists of a vacuum tube cylinder with a photoconductive target and scanning electron beam. The optical coupling leaves a latent image on the camera that is then read out by the electron beam. The scanning is done in a series of horizontal raster lines that are then displayed on the television system. The scanning produces a series of continuous 2D images with varying voltages (Van Lysel 2000; Granfors et al. 2003).

Older vacuum tube video cameras were later replaced with charge coupled devices or CCDs. These solid-state cameras store images as pixel arrays until they are read out. These CCD cameras were smaller and required less power than traditional vacuum tube cameras (Van Lysel 2000; Granfors et al. 2003). It should be noted that the use of the flat-panel detector (FPD) subsequently has eliminated the need for optical coupling and the television system, but there are many II systems still in use.

7.2.1.6 Flat-Panel Detector System

Since the early 2000s, FPDs have become more common for use in interventional procedures (Jones et al. 2014). These detectors consist of thin-film transistor arrays that store charge both indirectly (using amorphous Silicon with a CsI scintillator) and directly (using amorphous selenium), depending on the configuration (Jones et al 2014). The size of the configuration depends on the manufacturer but can range from 25×25 to 40×40 cm^2 in area. Each detector is made up of multiple detector elements or "DELs." The DELS store charge (indirectly or directly) based on the flux of the photons incident of the surface of the DEL. Each DEL is then read row by row to form an electronic image displayed on a monitor of the x-ray distribution across the detector. This eliminates the need for a television camera system (Nickoloff 2011). The advancement in FPD detectors has increased the radiation detection sensitivity and decreased radiation dose while maintaining image quality.

7.2.1.7 Automatic Exposure Rate Control

Modern fluoroscopic units operate with AERC algorithms that continue to become more and more complex (Jones et al. 2014). As stated previously, these algorithms control the output of the machine while automatically changing the technical parameters such as kV, mA, pulse width, and filtration. These vary by manufacturer and detector type and depend on how the manufacturer chooses to maintain a preset output signal. Because of the variability among manufacturers and the variability within unit types, it becomes increasingly important for the individual operator to optimize the radiation dose to the patient and staff based on the capabilities of the fluoroscopy unit itself.

7.3 Fluoroscopy Rooms

The configuration of components comprising a fluoroscopy system can vary depending on the application, for example, with the x-ray tube fixed under the table for gastrointestinal (GI) and interventional procedures (Bushberg et al. 2012). Some of the more specialized systems can have a biplane configuration where two tubes can be operated simultaneously or independent of one another depending on the task. Overhead x-ray tube configurations are often used in urology. There are also mobile units for orthopaedic, pain management, and pediatric imaging purposes. The mobile units typically have the computer and video components separate from the x-ray tube and generator components.

7.3.1 Radiographic/Fluoroscopic Rooms

Some hospitals have a room for radiographic/fluoroscopic ("R and F"), shown in Figure 7.2a, that performs studies involving the GI systems. These rooms typically consist of a fluoroscopic unit with a large table that can be rotated from a horizontal to vertical position. The large detector is mounted over the table so that the x-ray tube is permanently positioned below the patient. The table can also be used for radiographic purposes with the radiographic x-ray tube mounted on a crane that allows the tube to be moved freely around the room (Bushberg et al. 2012).

7.3.2 Interventional Rooms

In the typical setup in an interventional radiology or cardiology suite, the fluoroscopic unit is mounted on a c-arm apparatus that allows the x-ray tube and detector to move in tandem around the patient. The detector is mounted above the patient but can be rotated in all directions to allow for postero-anterior and lateral projections. The table for the patient can "float" side to side to position the patient where needed (Bushberg et al. 2012). Depending on the purpose of the unit, the unit can be equipped in different configurations. Many angiography suites contain units with a single c-arm (Figure 7.2b). Many newer interventional cardiology suites have a unit with two mounted c-arms with smaller detectors known as a bi-plane unit (Figure 7.2c). This configuration supplies additional viewing coverage over the smaller anatomy of the heart. One other configuration contains permanent magnet attachment for electrophysiology procedures (Figure 7.2d).

7.3.3 Mobile C-arms

Mobile c-arms are popular for vascular imaging, pain management, and orthopedic surgical procedures (Schueler 2000). These units are compact and often consist of two separate parts, an x-ray generator with a c-arm x-ray tube and small II system attached and set on wheels and a display monitor that separately attaches to display the images. There are also smaller units that can make and display images on the same unit. These are typically used for extremity imaging. More recently, mini c-arms equipped FPDs and digital monitors have become available.

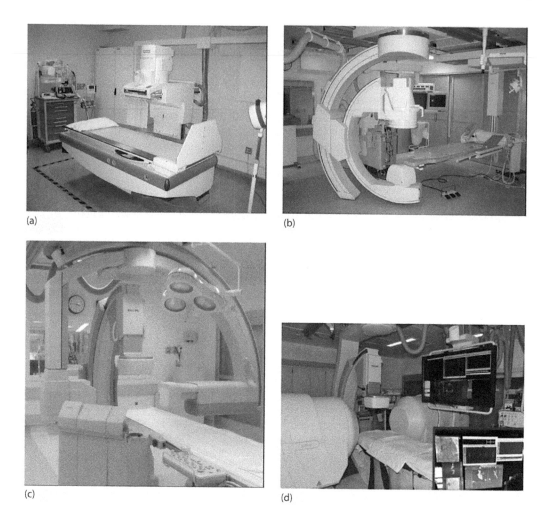

FIGURE 7.2 Various fluoroscopy room configurations: (a) Radiographic/Fluoroscopic "R and F" room typically used for GI/GU applications. (b) Typical interventional angiography setup with single c-arm and large II. (c) Bi-plane interventional cardiology suite with smaller digital flat panel detectors for greater coverage. (d) Setup shown for an electrophysiology lab. A single c-arm unit is positioned between a permanent magnet, which allows cardiac functions to be mapped remotely. (Adapted from Bushberg, J.T. et al., *The Essential Physics of Medical Imaging*, Lippincott Williams & Wilkins, Philadelphia, PA, 3rd ed., 2012.)

7.4 Patient Dose in Fluoroscopy and Interventional Imaging

Observable radiation effects from fluoroscopic x-rays have been reported for both workers and patients (Jones et al. 2014). Observed effects in workers are commonly stochastic, caused by long-term radiation exposure. Effects observed in patients are most commonly deterministic injuries typically caused by the delivery of high radiation doses over a short time period. These effects are most commonly seen for large patients and can range from short-term (weeks to months) effects such as injury to skin and underlying tissues to long-term (years to decades) effects such as cataractogenesis and osteonecrosis. Long-term stochastic risks are typically the primary concern for abdominal and thoracic procedures performed on pediatric patients, due to the increased radiosensitivity and longer potential lifespan of that population (Miller et al. 2010).

7.4.1 Fluoroscopy Radiation Dose Metrics

Four major metrics have been developed over the years for the purpose of patient radiation dose estimation in fluoroscopy: the peak skin dose, the reference point air kerma, the kerma-area product (also known as dose-area product), and fluoroscopy time. Kerma (an acronym for "kinetic energy released per unit mass") refers to the energy deposited from the x-ray beam per unit mass in a small volume of some specified material. Therefore, air kerma refers to the amount of energy delivered from an x-ray beam per unit mass of a small volume of air.

7.4.1.1 Peak Skin Dose

Peak skin dose (PSD), measured in grays (Gy), refers to the highest absorbed dose in any portion of the patient's skin during a procedure. This value includes contributions from the primary x-ray beam, tube leakage radiation, and scattered radiation from deeper in the patient's body that travels back to the skin (backscatter). This dose is the best predictor of potential deterministic effects to the patient's skin post-procedure. Estimation of this parameter can range from simple to complex depending upon the type and amount of dose information available. Methodologies to estimate the peak skin dose based on a number of scenarios involving the availability of manufacturer dose reports, Digital Imaging and Communications in Medicine (DICOM) image headers, machine-reported dose metrics, or simply patient images from the procedure have been discussed in the research literature and are beyond the scope of this chapter (Dauer et al. 2009; Jones and Pasciak 2011, 2012). While historically the estimation of this parameter was limited to post-procedure reconstruction using one of the several available methodologies, studies in the literature have begun to explore solutions for real-time estimation of the PSD during performance of the procedure (Bednarek et al. 2011; Johnson et al. 2011; Khodadadegan et al. 2013; Rana et al. 2014, 2016).

7.4.1.2 Reference Point Air Kerma

The reference point air kerma (also known as cumulative dose, cumulative air kerma, or reference air kerma) is the cumulative air kerma measured over the course of an entire procedure to a specific point in space (i.e., the patient entrance reference point) relative to the fluoroscopy gantry. The concept of the reference point air kerma first emerged in 2000 in the International Electrotechnical Commission (IEC) standard for interventional fluoroscopy equipment and was adopted by the U.S. Food and Drug Administration in 2005, with the second edition of the standard being published in 2010 (IEC 2000, 2010; FDA 2005). Currently, all fluoroscopes sold in the United States after June 2006 are required to measure or calculate and display this parameter at the working position of the interventionalist. The position of the reference point depends upon the type of fluoroscopy system; however, for isocentric c-arms used in most interventional procedures, the point lies on the central ray of the x-ray beam, 15 cm toward the x-ray tube from the isocenter of rotation (IEC 2010). The point is roughly located at skin entrance for a 30-cm-thick patient with isocenter positioned at patient midpoint (see Figure 7.3a); however, changing imaging geometry can strongly influence where this point lies in relation to the patient, as it is fixed with respect to the gantry but not to the patient's relative position (see Figure 7.3b). It is important to note that the point's location also does not change when the source-to-image distance (SID) and table height changes. Despite this variation in location with respect to the patient, the reference point usually is close to the patient's skin and the reference point air kerma can therefore be thought of as the total air kerma incident on the skin summed over all areas of skin irradiated during the procedure (NCRP 2010).

For procedures in which the x-ray beam is moved with respect to the patient, the reference point air kerma will provide an overestimate of the patient's peak skin dose and therefore overestimate the risk of deterministic effects (Miller et al. 2003; NCRP 2010). Conversely, as beam motion during a procedure decreases, the reference air kerma will better estimate the peak skin dose and associated deterministic risks (Miller et al. 2004). It is important to be aware that the radiation dose delivered to the patient's skin differs from the entrance air kerma due to both differences in energy absorption coefficients of air,

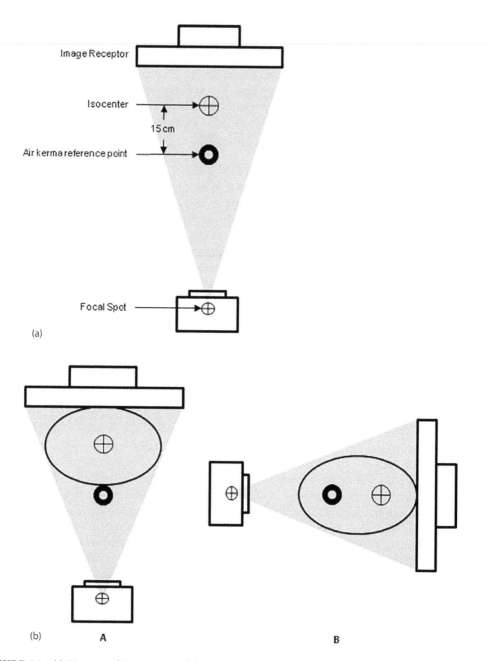

FIGURE 7.3   (a) Diagram of the location of the air kerma reference point for isocentric systems as defined by the IEC and U.S. Food and Drug Administration. The point moves with the gantry and is located 15 cm from the isocenter toward the x-ray tube. (b) Diagram illustrating how the air kerma reference point can move relative to the patient's skin depending on gantry positioning. In configuration A, the reference point is located on the skin surface while in configuration B the point has moved inside the patient.

tissue, and backscatter. As noted, backscatter refers to the photons that get scattered back toward the skin from deeper within the patient, which results in a patient skin kerma approximately 40% larger than the incident air kerma alone (Balter and Miller 2014). Additionally, for the same x-ray beam and entrance skin air kerma, the dose absorbed by bone is two to three times as high as that absorbed by soft tissue due to the increased absorption coefficient of calcium (Balter and Miller 2014). Therefore, though most efforts for patient dose management revolve around the skin, it is important to keep superficial bone structures (e.g., the ribs) in mind as well. Despite this variability, with appropriate implementation, the reference point air kerma can be used not only as a reliable metric for predicting the likelihood of deterministic effects but also overall procedure dose performance (Balter et al. 2011).

7.4.1.3 Kerma-Area or Dose-Area Product

The kerma- or dose-area product (KAP or DAP, respectively) represents an approximation of the total x-ray energy leaving the x-ray tube, which is typically measured by a flat, square ion chamber located near the collimator assembly (NCRP 2010). The units for KAP and DAP are typically gray-square centimeters (Gy cm^2), which represents the total dose measured within the collimated field multiplied by the field area. Since dose is inversely proportional to the square of the distance from the x-ray tube focal spot and field size is directly proportional to the same distance, KAP is independent of SID. So long as the x-ray field is confined to the patient, KAP can be used to provide an estimate for the total x-ray energy absorbed by the patient during the procedure and is therefore used to provide a measure of the stochastic risks of a procedure. However, KAP is a poor metric for estimating peak skin dose since, due to its formulation, delivering a very large radiation dose to a small area can result in an identical KAP as a small radiation dose delivered over a large area. In fact, estimation of absorbed skin dose from KAP data can have potential errors of at least 30%–40% (McParland 1998a, 1998b).

7.4.1.4 Fluoroscopy Time

Fluoroscopy time is the metric that has historically been in place longest on fluoroscopic equipment, though it has become largely outdated with the arrival of the previously described dose metrics. It is simply a measure of the "beam on" time for fluoroscopy during a procedure, with no additional dosimetric information. It is independent of the techniques (i.e., kVp, mA, pulse width, etc.) delivered by the system during a procedure and does not include the contributions of fluoroscopic images (e.g., digital subtraction angiography runs). Since it correlates poorly with other dose metrics, the NCRP recommends that it should only be used with respect to monitoring patient irradiation if no other metric is available on a system (Fletcher et al. 2002; Miller et al. 2004; NCRP 2010). With all newly manufactured equipment since 2006, newer metrics are required to be displayed, and fluoroscopy time will be abandoned as older units are replaced.

7.4.2 Accuracy of Dose and Risk Estimation

It is important to keep in mind that any estimation of patient dose and associated stochastic and deterministic risk comes with a degree of uncertainty. The displayed values of reference point air kerma and KAP are allowed an uncertainty of ±35%, and using these metrics to convert to patient dose introduces more uncertainty due to patient size and position relative to the x-ray beam during the procedure (Miller et al. 2004). Additionally, patient-to-patient biological variability with regards to radiation effects can make exact prediction of risk more uncertain (Balter et al. 2010). However, a qualified medical physicist can help quantify these uncertainties such that the most informed decisions about patient radiation dose management can be made. One always should remember that even if perfect knowledge of each patient's risk cannot be obtained, these metrics can be used to help achieve the goal of maximizing the margin of patient benefit over harm.

7.5 Methods for Dose Optimization

Dose optimization with the use of fluoroscopy can seem like a daunting task given the large number of operator and unit-specific variables affecting dose encountered over the course of a procedure. As with all other imaging modalities, the task of optimization relies on the balance between the need for diagnostically adequate image quality to accomplish the goal of the procedure and minimizing the dose to the patient (ICRP 2007). This is especially important in the case of fluoroscopy because the modality is often used as a tool to assist surgical procedures that can last hours at a time, and the patient dose is highly dependent on the continuous use of the operator. It is this prolonged exposure to the patient (and the staff receiving the prolonged scatter radiation) alike that warrants the need for optimization protocols for procedures involving fluoroscopy. The ICRP describes optimization as consideration of "the likelihood of incurring exposures, the number of people exposed, and the magnitude of their individual doses should all be kept as low as reasonably achievable, taking into account economic and societal factors" (ICRP 2007). To help one optimize the radiation dose to patients and staff, many considerations must being considered. This includes everything from the equipment selection and room layout itself to the dose-sparing techniques that can be used before, during, and after the procedure.

7.5.1 Prior to Procedure

7.5.1.1 Equipment Selection

One of the biggest advances in fluoroscopic imaging has been the development of the FPD, an alternative technology to the II. This critical advancement changed the way radiation could be utilized in the interventional setting and subsequently has allowed for better image quality for clinical tasks and the possibility of lower doses to the patient. The image quality has improved for multiple reasons. First, a modern flat panel does not suffer from the same distortions associated with an II such as the "pincushion" caused by the curvature of the input phosphor on the II, "vignetting" caused by uneven distribution of photons striking the flat output phosphor surface, and "S" distortions caused by magnetic fields interfering with the trajectory of electrons within the II. As discussed earlier, multiple DELs are fixed in place that allows for much more consistent uniformity across the detector. The decreasing size of electronics allows the FPD to be much less bulky than the II systems. This makes it easier to position it over multiple smaller areas of the patient throughout the procedure. The FPD also eliminated the need for a television camera to convert x-ray intensity into electronic signal, reducing electronic noise. This also makes for a more streamlined fluoroscopic unit and eliminates further variability in signal. Eliminating the analog signal conversion has also resulted in less noise being produced along the imaging chain.

Aspects to consider for FPD:

While FPDs have improved fluoroscopy imaging by eliminating II-related artifacts and relieved the user of the more cumbersome television system, there are trade-offs to consider when optimizing the dose to the patient. As noted above, there are multiple DELs that are read out to form the electronic signal that produces the image. The number of DELs and the dimensions of the DELs ultimately limit the spatial resolution that can be achieved by the unit. The distance from the center of one DEL to the adjacent DEL is known as the pitch. The maximum spatial resolution for a given FPD system is given by the following relationship:

$$\text{Maximum spatial resolution} = \frac{1}{2 * \text{pitch}}$$

Using this relationship, one can conclude that using the smallest pitch possible will optimize the spatial resolution. However, one must also consider that each DEL contains readout electronics on the circuit that take up some of the area within the DEL. The smaller the pitch, the greater the fractional area the electronics occupy on each DEL, decreasing the sensitive area for x-ray absorption (Nickoloff 2011). The fraction of the sensitive area relative to the area of the pitch is known as the fill factor:

$$\text{Fill factor} = \frac{\text{sensitive area of DEL}}{\text{pitch *pitch}}$$

Manufacturers must therefore balance the size of pitch to increase spatial resolution with the size of the sensitive area used for each DEL to not introduce excessive noise in the image (Nickoloff 2011).

Regardless, the presence of the electronic noise in the readout and sampling of the detector place a dose "floor," i.e., a minimum, on the input air kerma needed to maintain proper detector quantum efficiency at low input air kerma rates for FPDs (Jones et al. 2014). Because FPD systems do not have the same gain ability as an II system given that they use similar input phosphors (for indirect FPD), the dose "floor" for FPD is higher than that for II systems (Jones et al. 2014).

One other consideration related to the detector is the time it takes to read the data for given field of view: the larger the field of view, the longer the time it takes to read out the full signal given the size of the data sample. To reduce the data rate, manufacturers often group data from adjacent DELs together. This is a process known as binning (Nickoloff 2011). Binning reduces the data rate by reducing the number of DELs that need to be read and reduces noise by increasing the signal per grouped DEL. However, this also reduces the spatial resolution because the pitch is increased (refer to above equation). So, while the spatial resolution generally remains constant for FPDs and is better than for an II system, for larger fields of view where binning is used, the spatial resolution falls by approximately 50% and closely matches what would be achieved by an II system for the same large field of view (Nickoloff 2011). While the benefits of FPD still outweigh the limitations of FPD, these factors should be considered when attempting to optimize the dose during a given procedure.

7.5.1.2 Room Layout

An aspect to consider when analyzing overall dose management starts before the patient ever enters the room. Once the equipment is selected, the planning and construction of the fluoroscopy suite must take place. It is critical that the space that occupies the unit is designed in such a way that limits the amount of radiation to anyone outside the room and minimizes the dose to personnel in the room. This involves making sure the walls are properly shielded based on the workload of the room. Attention must be paid to the doors of fluoroscopy suites. Conventional radiographic units may have door interlocks that disable the x-ray when the door is opened. However, this cannot be the case due to the dynamic nature of the fluoroscopic procedure and can put patient safety at risk. Instead, a manual switch, in the form of a button can be placed in the room to disable the x-ray tube, is necessary. All corridor doors should remain closed during the procedure. Should a door be left open (e.g., the control room door) while the procedure is underway, there should not exist any line of sight between the iso-center of the unit and an occupied area. Lastly, doors should be placed such that no person must travel through an unshielded area to enter or exit the corridor or control room (Jones et al. 2014). Within the room, dose reduction to the staff performing the procedure includes the use of tableside lead drapes to lower immediate scatter levels from underneath the patient and use of various versions of leaded glass, which are typically suspended from the ceiling or are on wheels and moved throughout the room. It is imperative to use these items as much as possible during a procedure to keep the staff exposure as low as possible.

7.5.1.3 Pre-procedure

A key part to dose optimization is a pre-procedure risk-benefit analysis. While the benefits for patients undergoing fluoroscopic procedures typically outweigh the potential risks, factors such as the patient's previous interventional procedure history, current weight, pregnancy status, current medications, and any previous medical problems should be considered when planning the use of radiation with possible deterministic effects (Miller et al. 2010).

It is important that the benefits and the risks associated with procedure be explained in the informed consent presented to the patient. The discussion of risk should include potential skin burn effects, fetal risk for pregnant patients, and cancer risks associated with the use of radiation.

7.5.1.4 Time-Out

Once the procedure has been deemed appropriate and the patient has been prepared for the procedure, it is important that the patient's identity be verified and the fluoroscopic unit set up for the appropriate procedure via the system procedure-specific protocols. The pre-procedure time-out should also include verification that all staff present are wearing their dosimetry badges and protective lead garments, including lead aprons, glasses, and gloves when necessary.

7.5.2 Optimization of Dose During a Procedure

The operator of fluoroscopic and interventional imaging equipment is ultimately responsible for optimizing the dose delivered to the patient during a procedure—that is, using the least amount of dose required to provide sufficient image quality for successful clinical performance. In addition to benefiting the patient, the operator's dose will also be minimized if this optimization is performed. Several national and international organizations have provided publications describing recommendations for such optimization, including the following references: ICRP 2000, 2010; Hirshfeld et al. 2004; NCI 2005; NCRP 2010; ACR 2018. Table 7.1, adapted from Bushberg et al. 2012, also provides a succinct overview of dose and image quality trade-offs that are described in greater detail in the following sections.

TABLE 7.1 Summary of Operational Factors That Affect Image Quality and Radiation Dose to the Patient and Staff

Operational Change	Effect on Image Quality and Radiation Dose		
	Image Quality	Radiation Dose to the Patient	Radiation Dose to the Staff
Increase in patient size	Worse (increased scatter fraction)	Higher	Higher
Increase in tube current (mA) with constant kV (i.e., AERC off)	Better (lower image noise)	Higher	Higher
[a]Increase in tube potential (kV) with AERC active	Soft tissue: Better (lower noise) Bone and contrast material: Worse (decreased subject contrast)	Lower	Lower
Increase in tube filtration with AERC active	Little change	Lower	Lower
Increase in source to skin distance	Slightly better	Lower	Little change
[b]Increase in skin to image receptor distance	Slightly better (less scatter)	Higher	Higher
Increase in magnification factor	Better (improved spatial resolution)	Higher	Higher
Increase in collimator opening	Worse (increased scatter fraction)	Little change (however higher integral and effective dose)	Higher
Increase beam on time	No effect	Higher	Higher
Increase in pulsed fluoroscopy frame rate	Better (improved temporal resolution)	Higher	Higher
Grid is used	Better (decreased scatter fraction)	Higher	Higher
Image recording modes (cine, digital subtraction angiography (DSA), radiographic)	Better (lower noise, higher resolution)	Higher	Higher

Source: Bushberg, J.T. et al., *The Essential Physics of Medical Imaging*, 3rd, ed., Lippincott Williams & Wilkins, Philadelphia, PA, 2012.

[a] When kV is increased with AERC activated, the system decreases the mA to maintain a constant signal level at the image receptor.

[b] Fixed source to skin distance.

7.5.2.1 Techniques to Minimize Total Dose

There are several equipment and technical parameters under the direct control of the operator that can heavily influence the total dose delivered to the patient during fluoroscopic and interventional procedures (Wagner 2007). Understanding their influence on dose and making proper adjustments when applicable can help minimize dose while maintaining clinical performance of a procedure.

Positioning of the x-ray tube, image receptor, and patient can have a major impact on the dose. Since dose is inversely proportional to the square of the distance from the x-ray tube focal spot, the distance between the patient and the x-ray tube should be maximized in any positioning setup. Additionally, the relative positioning of the x-ray tube, patient, and image receptor has significant impact on the function of the automatic exposure control integrated into all modern fluoroscopic equipment. This control system automatically adjusts the radiation output delivered based on the thickness and composition of the body part of the patient being imaged as well as the distance of the image receptor from the x-ray tube in order to maintain a constant and acceptable image quality. A thicker portion of the body as "seen" by the beam will be more difficult to penetrate and thus require higher radiation output as compared to a thinner portion. Owing to this, steep oblique orientations of the x-ray tube increase the path length the radiation travels through the body as compared to a postero-anterior orientation (see Figure 7.4), and will thus cause an increase in radiation output up to a factor of 10 or more (Miller et al. 2010). The steep angles also tend to reduce the distance between the x-ray tube and skin surface, also increasing dose.

The image receptor should be positioned in any orientation as close to the patient as possible for multiple reasons. First, maintaining proximity to the image receptor will reduce the presence of focal spot blurring and magnification effects in the patient images. Second, this positioning will generally reduce the dose for both fixed (mobile c-arm) and variable (angiography, interventional, etc.) focal spot-to-image receptor distance units. In fixed units, positioning the patient as close to the image receptor will maximize the x-ray source to patient skin distance and will therefore minimize dose. For variable-distance units, for a given source-to-skin distance, keeping the receptor close to the patient will reduce the total focal spot-to-image -receptor distance. The farther away the image receptor is from the patient,

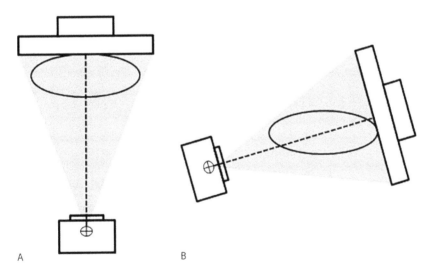

A   B

FIGURE 7.4 Diagram illustrating the effect of gantry positioning on the path length the x-ray beam travels through the patient. The beam travels a longer path length through the patient in the steep oblique positioning in configuration B as compared to the postero-anterior positioning in configuration A. Therefore, in order to maintain a sufficient signal to the detector, the AERC system will increase the output for configuration B as compared to A, which will result in a higher dose to the patient.

the higher the output the AERC system will prescribe to ensure sufficient image quality level, so minimizing that distance for a given source-to-skin distance will minimize dose. Therefore, for variable distance units, first position to maximize the source-to-skin distance and then position the image receptor as close to the patient as possible.

An additional consideration is to utilize positioning and beam collimation to reduce to the presence of bone in the beam. Since bone is more attenuating soft tissue, the AERC system will increase x-ray output to compensate, thus increasing dose. Superimposed bone also makes visualization of medical devices such as catheters difficult, and thus can necessitate additional procedure time (Miller et al. 2010). Related to this, the arms should be positioned out of the field of view unless they are to be imaged as part of the procedure (Wagner 2007). The arms introduce additional body thickness and bony structure in the path of the beam, so removing them from the field of view will ensure no unnecessary increase of radiation output by the AERC system.

Aside from positioning, there are multiple operational parameters that can be chosen during the procedure that can help minimize dose. Beam collimation should always be employed to limit the x-ray field size to the smallest area appropriate for the imaging task, which not only reduces the area of skin being irradiated but reduces the amount of radiation scattering back to the personnel in the procedure room. The use of magnification and high-dose-rate modes should be limited only to clinical necessity. Low-dose-rate AERC modes should be used when clinically acceptable. The use of pulsed fluoroscopy can also have a major impact on dose minimization (Wagner et al. 2000). However, it is important to note that each manufacturer implements pulsed fluoroscopy modes in their own manner, with some modes yielding the same or higher dose rate than conventional, continuous fluoroscopy (Miller et al. 2002). Therefore, it is good practice to consult with the manufacturer of the equipment or with a qualified medical physicist in order to determine the actual dose rate for each pulsed fluoroscopy mode before implementation (Miller et al. 2010). Judicious use of these operational parameters is crucial, as their effects on patient dose are multiplicative and can result in differences in patient skin doses of up to 8 Gy for long procedures (Wagner et al. 2000).

The use of last-image hold, fluoroscopy loops, and "virtual" collimation, when available and carefully used, can lead to reductions in fluoroscopy time and patient dose. Fluoroscopy is only necessary for guidance of devices within the body or to observe motion. For intraprocedural review, last image hold and fluoroscopy loops are sufficient while not requiring any additional radiation dose to the patient (Miller et al. 2010). If last-image hold and fluoroscopy loops provide adequate information for diagnosis or documentation for a particular procedure, there is no need to perform higher-dose fluororadiography or digital subtraction angiography runs, which can save a substantial amount of dose. For cases in which images or runs are clinically necessary during a procedure, the number of images obtained can be minimized by using preprogrammed acquisition sequences with lower frame rates that reduce the total number of images taken without the loss of important information.

Adjustments to technical parameters, such as AERC calibration, pulse rate, pulse width, beam filtration, imaging sequences, image processing, and video frame averaging (for image noise reduction), should be performed with the assistance of a qualified medical physicist and field service engineer to ensure their proper use (Miller et al. 2010). It is of the utmost importance to remember that dose reduction should never impair the image quality necessary to successfully perform clinical procedures, and any implementation of the aforementioned strategies should always be centered in the justification of providing the patient more benefit than harm.

7.5.2.2 Techniques to Minimize Peak Skin Dose and Eye Dose

In general, utilizing the previously described techniques to minimize total dose will also reduce patient peak skin dose and eye dose; however, there are specific techniques that can be employed to further reduce these doses.

For peak skin dose, dose spreading and collimation are two methods for dose minimization (Miller et al. 2002). The purpose of using these methods is to reduce the skin dose at any single location on the patient skin surface by irradiating different portions of the skin over the course of the procedure. Dose spreading refers to the practice of changing the position of the x-ray field on the patient's skin by using table movement and/or gantry angulation (Mahesh 2001; Norbash et al. 1996). By altering the x-ray location on the patient skin surface, both the peak skin dose and the size of skin area subjected to the peak skin dose are reduced. Collimating the x-ray beam to only the anatomy of clinical interest reduces the area of skin irradiated as well as the overlapping of x-rays fields while using dose-spreading techniques. Since these overlapping areas produce higher dose, tight collimation can help improve the effectiveness of dose spreading at reducing peak skin dose and the areas subjected to the peak dose (Miller et al. 2002).

To minimize patient eye dose when allowed by the clinical requirements of the procedure, there are two methods that can be employed. First, when imaging the head and face the beam should be positioned such that it enters the head in the posterior-anterior direction. In this configuration, the anatomy posterior to the eyes act to shield the eyes and reduce their radiation dose. Second, as with peak skin dose reduction, beam collimation can be employed in order to exclude the eyes from the field when clinically permissible and thus reduce eye dose.

7.5.2.3 Monitoring Dose During Procedures

Multiple guidelines recommend that radiation dose should be monitored during a procedure (Stecker et al. 2009; NCRP 2010). Monitoring ensures that the operator of the fluoroscopic equipment is aware of the amount of radiation being delivered, such that radiation dose is included the ongoing risk-benefit balance used to determine the value of continuing a procedure (Miller et al. 2010). Since it is routine for the operator to lose awareness of the radiation dose while concentrating on the clinical performance of the procedure, it is recommended that another individual (technologist, nurse, or other) be designated to monitor dose and keep the operator informed.

Dose monitoring involves the tracking of one of the four dose metrics previously described in the chapter. Since real-time monitoring of peak skin dose and maps of patient skin dose distribution is generally not widely available as of 2017, most institutions will perform monitoring using reference point air kerma, KAP, or fluoroscopy time. As was previously described, fluoroscopy time correlates poorly with patient dose metrics and should be used with caution and only when other dose metrics are unavailable (Fletcher et al. 2002; Miller et al. 2004; NCRP 2010). Monitoring should involve the operator being notified when the patient's radiation dose exceeds specified values and at regular intervals thereafter (Stecker et al. 2009). First presented in Stecker et al. in 2009 and subsequently adopted in the recommendations of NCRP Publication 168, the notification levels shown in Table 7.2 were selected such that three notifications during a procedure result in a *significant radiation dose*, a radiation dose that might produce an injury in an average patient and would therefore prompt a post-procedure patient follow-up (Stecker et al. 2009; NCRP 2010). These recommended significant radiation dose levels can be found in Table 7.3, though lower values can be used at the discretion of an institution, especially for cases in which previously irradiated skin is involved (NCI 2005; ACR 2018). It should be noted that for procedures utilizing a biplane fluoroscopic unit, dose metrics from each plane should be considered

TABLE 7.2 Summary of Radiation Monitoring Dose Notification Thresholds

Parameter	First Notification	Subsequent Notifications
Peak skin dose	2,000 mGy	500 mGy
Reference point air kerma	3,000 mGy	1,000 mGy
Kerma-area-product	300 Gy cm^{2a}	100 Gy cm^{2a}
Fluoroscopy time	30 min	15 min

Source: Stecker, M.S. et al., *J. Vasc. Interv. Radiol.*, 20, S263–S273, 2009.
[a] Assuming a 100-cm^2 field at the patient's skin. The value should be adjusted to the actual procedural field size.

TABLE 7.3   Thresholds for Patient Follow-up

Parameter	Threshold
Peak skin dose	3,000 mGy
Reference point air kerma	5,000 mGy
Kerma-area-product	500 Gy cm^2
Fluoroscopy time	60 min

Source: Stecker, M.S. et al., *J. Vasc. Interv. Radiol.*, 20, S263–S273, 2009.

independently if the fields do not overlap, and the doses added for cases in which the fields overlap or if it is uncertain if they overlap (Miller et al. 2010).

A procedure should never be stopped solely due to the fact that the significant radiation level has been exceeded, especially considering that the clinical benefit to the patient from a successful procedure almost always outweighs any potential harm as the result of radiation exposure (Balter and Moses 2007). Additionally, halting the procedure on this basis alone without attaining the desired clinical result will have subjected the patient to radiation risks without the associated clinical benefit. Even with judicious implementation of dose optimization techniques, it is not always possible to prevent the peak skin dose from reaching a threshold for patient skin effects. A variety of factors including patient anatomic variation, radiation doses from previous procedures, disease and lesion complexity, type of procedure, and the clinical indication for the procedure may combine such that a prolonged procedure with a high radiation is unavoidable and not indicative of poor operator technique (Miller et al. 2010).

7.5.3 Post Procedure

It is important to record the patient dose following a patient procedure using fluoroscopy. As stated earlier, time alone is not an indicator of patient dose. All available metrics, such as peak skin dose, reference dose, and KAP should be recorded in conjunction with fluoroscopy time. Most modern equipment has the ability to record these following the procedure. If the patient is pregnant, these metrics should be used to evaluate the dose to the fetus and recorded.

Should a significant patient dose be recorded or trigger levels activated during the procedure, the operator should respond in accordance to the hospital's protocol. This may include noting in the patient's record that a significant amount of radiation has been delivered along with stating a reason. If necessary, this should also include consultation with a qualified medical physicist for further evaluation of the patient dose. If it is determined that the patient received a significant enough amount of radiation to cause a deterministic effect, follow up with the physician should occur at intervals of two weeks and one month following the procedure. It is at this point that a patient can report any skin changes following the procedure (Miller et al. 2010). If a follow-up is found to be necessary or precautionary due to previous interventional procedures, it should be scheduled prior to the patient leaving the hospital and the patient given written follow-up instructions along with their other discharge information.

7.5.4 Future of Fluoroscopic Dose Optimization

In order to address the increasingly complex issue of patient and staff dose, some novel approaches have been introduced in order to raise awareness and further control radiation dose to all involved. In a recent paper, Balter et al. (2016) explored the concept of eye tracking software for interventional radiology. The study used a new technology called eye-controlled region of interest. Eye-tracking software monitors the movement of the physician's gaze to automatically adjust the aperture of the fluoroscope to the clinical region of interest. With this software engaged, a dynamic collimator moves with the operator's eye movement. The collimator attenuates a significant portion of the field of view outside a smaller diameter linked to where the operator is looking. Image processing is then added to bring the brightness and contrast of the highly attenuated surroundings to match the fully irradiated area. In doing this, general workflow is not disturbed and the KAP to the patient and operator irradiation were found to be drastically reduced.

While considerable effort is expended to reduce the patient dose, there has also been a concerted effort to enhance the training of the fluoroscope operators with regard to scatter radiation awareness. To highlight the areas in the clinical setting that receive the highest amount of scatter radiation, Rodas and Padoy (2015) used a series of digital cameras to photograph a typical clinical set up for an interventional procedure. A 3D model was constructed of the setup and various Monte Carlo simulations were run with measurement verification from real-time electronic dosimeters. The resulting simulation was overlaid on the 3D model of the setup to show the highest areas of radiation scatter. This then gives physicians a real-time scatter dose map simulation that engenders a heightened awareness to the users for future cases. This kind of visualization is crucial in the continuing education of fluoroscopy operators as it shows the highest and lowest areas of scatter radiation and how this changes with distance and patient positioning.

7.6 Conclusion

The use of fluoroscopic imaging has increased over the past several decades as the modality has become an integral part in increasingly complex procedures across multiple disciplines. While several advances in detector technology, dose-sparing modes, and advanced image processing have made significantly lowering the patient dose possible, the responsibility still remains with the operator to ensure that these technologies and techniques are utilized to optimize the patient dose for a given procedure. Variability in vendor design and varying degrees of user training make it essential that the operator become familiar with the unique functions and modes of their fluoroscopy unit and also employ basic dose optimization practices within the context of the patient procedure. While this is an ongoing challenge, the strategies presented in this text in conjunction with the highlighted ongoing research show that dose optimization within fluoroscopy is achievable in concert with continually improving technology to ultimately benefit both the patient and staff involved.

References

American College of Radiology. (ACR). 2018. ACR-AAPM technical standard for management of the use of radiation in fluoroscopic procedures. *ACR Practice Guidelines and Technical* Standards. Available at https://www.acr.org/-/media/ACR/Files/Practice-Parameters/MgmtFluoroProc.pdf. accessed January 30, 2018.

Balter, S. & Miller, D. 2014. Patient skin reactions from interventional fluoroscopy procedures. *Am J Roentgenol* 202: W335–W342.

Balter, S. & Moses, J. 2007. Managing patient dose in interventional cardiology. *Catheter Cardiovasc Interv* 70(2): 244–249.

Balter, S., Hopewell, J.W., Miller D.L., Wagner, L.K. & Zelefsky, M.J. 2010. Fluoroscopically guided interventional procedures: A review of radiation effects on patients' skin and hair. *Radiology* 254(2): 326–341.

Balter, S., Rosenstein, M., Miller, D., Schueler, B. & Spelic, D. 2011. Patient radiation dose audits for fluoroscopically guided interventional procedures. *Med Phys* 38(3): 1611–1618.

Balter, S., Simon, D., Itkin, M. et al. 2016. Significant radiation reduction in interventional fluoroscopy using a novel eye controlled movable region of interest. *Med Phys* 43(3): 1531–1538.

Bednarek, D.R., Barbarits, J., Rana, V.K., Nagaraja, S.P., Josan, M.S. & Rudin, S. 2011. Verification of the performance accuracy of a real-time skin-dose tracking system for interventional fluoroscopic procedures. *Proc SPIE Int Soc Opt Eng* 7961: 796127.

Bushberg, J.T., Seibert, J.A., Leidholdt, E.M. & Boone, J.M. *The Essential Physics of Medical Imaging.* Philadelphia, PA, Lippincott Williams & Wilkins, 3rd ed., 2012.

Dauer, L.T., Thornton, R., Erdi, Y., Ching, H., Hamacher, K., Boylan, D.C., Williamson, M.J., Balter, S. & St. Germain, J. 2009. Estimating radiation doses to the skin from interventional radiology procedures for a patient population with cancer. *J Vasc Interv Radiol* 20(6): 782–788.

Fletcher, D.W., Miller, D.L., Balter, S. & Taylor, M.A. 2002. Comparison of four techniques to estimate radiation dose to skin during angiographic and interventional radiology procedures. *J Vasc Interv Radiol* 13(4): 391–397.

Food and Drug Administration (FDA). 2005. Electronic products; performance standard for diagnostic x-ray systems and their major components. Final rule. *Fed Regist* 70(111): 33997–34042.

Granfors, P.R., Aufrichtig, R., Possin, G.E. et al. 2003. Performance of a 41 × 41 cm² amorphous silicon flat panel x-ray detector designed for angiography and R&F imaging applications. *Med Phys* 30(10): 2715–2726.

Hirshfeld, J.W. Jr., Balter, S., Brinker, J.A. et al. 2004. ACCF/AHA/HRS/SCAI clinical competence statement on physician knowledge to optimize patient safety and image quality in fluoroscopically guided invasive cardiovascular procedures. A report of the American College of Cardiology Foundation/American Heart Association/American College of Physicians Task Force on Clinical Competence and Training. *J Am Coll Cardiol* 44(11): 2259–2282.

International Commission on Radiological Protection (ICRP). 2000. Avoidance of radiation injuries from medical interventional procedures. ICRP publication 85. *Ann ICRP* 30(2): 7–67.

International Commission on Radiological Protection (ICRP). 2010. Radiological protection in fluoroscopically guided procedures outside the imaging department. ICRP publication 117, *Ann. ICRP* 40(6)."

International Commission on Radiological Protection. 2007. The 2007 recommendations of the international commission on radiological protection. ICRP publication 103. *Ann ICRP* 37(2–4): 1–332.

International Electrotechnical Commission (IEC). 2000. Medical electrical equipment: Part 2-43-particular requirements for the safety of x-ray equipment for interventional procedures. IEC Report 60601. Geneva, Switzerland: International Electrotechnical Commission.

International Electrotechnical Commission (IEC). 2010. Medical electrical equipment: Part 2-43-particular requirements for the safety of x-ray equipment for interventional procedures. IEC Report 60601 2nd ed. Geneva, Switzerland: International Electrotechnical Commission.

Johnson, P.B., Borrego, D., Balter, S., Johnson, K., Siragusa, D. & Bolch, W.E. 2011. Skin dose mapping for fluoroscopically guided interventions. *Med Phys* 38(10): 5490–5499.

Jones, A.K. & Pasciak, A. S. 2011. Calculating the peak skin dose resulting from fluoroscopically-guided interventions. Part I: Methods. *J Appl Clin Med Phys* 12: 231–244.

Jones, A.K. & Pasciak, A.S. 2012. Calculating the peak skin dose resulting from fluoroscopically-guided interventions. Part II: Case studies. *J Appl Clin Med Phys* 13: 174–186.

Jones, A.K., Balter, S., Rauch, P. & Wagner, L. 2014. Medical imaging using ionizing radiation: Optimization of dose and image quality in fluoroscopy. *Med Phys* 42(1): 014301-1-26.

Khodadadegan, Y., Zhang, M., Pavlicek, W. et al. 2013. Validation and initial clinical use of automatic peak skin dose localization with fluoroscopic and interventional procedures. *Radiology* 266(1): 246–255.

Mahesh, M. 2001. Fluoroscopy: Patient radiation exposure issues. *RadioGraphics* 21(4): 1033–1045.

McParland, B.J. 1998a. A study of patient radiation doses in interventional radiological procedures. *Br J Radiol* 71(842): 175–185.

McParland, B.J. 1998b. Entrance skin dose estimates derived from dose-area product measurements in interventional radiological procedures. *Br J Radiol* 71(852): 1288–1295.

Miller, D. Balter, S. Schueler, B. Wagner, L. Strauss & K. Vañó, E. 2010. Clinical radiation management for fluoroscopically guided interventional procedures. *Radiology* 257(2): 321–332.

Miller, D.L., Balter, S., Cole, P.E. et al. 2003. Radiation doses in interventional radiology procedures: The RAD-IR study: part II: skin dose. *J Vasc Interv Radiol* 14(8): 977–990.

Miller, D.L., Balter, S., Noonan, P.T. & Georgia, J.D. 2002. Minimizing radiation-induced skin injury in interventional radiology procedures. *Radiology* 225(2): 329–336.

Miller, D.L., Balter, S., Wagner, L.K. et al. 2004. Quality improvement guidelines for recording patient radiation dose in the medical record. *J Vasc Interv Radiol* 15: 423–429.

National Cancer Institute (NCI). 2005. Interventional fluoroscopy: Reducing radiation risks for patients and staff. NIH Publication No. 05-5286. Bethesda, MD: National Cancer Institute.

National Council on Radiation Protection and Measurements (NCRP). 2010. Radiation Dose Management for Fluoroscopically-Guided Interventional Medical Procedures. NCRP Report No. 168. Bethesda, MD: National Council on Radiation Protection and Measurements.

Nickoloff, E.L. 2011. Physics of flat panel fluoroscopy systems. *RadioGraphics* 31(2): 591–602.

Norbash, A.M., Busick, D. & Marks, M.P. 1996. Techniques for reducing interventional neuroradiologic skin dose: Tube position rotation and supplemental beam filtration. *AJNR Am J Neuroradiol* 17(1): 41–49.

Rana, V.K., Rudin, S. & Bednarek, D.R. 2014. Improved-resolution, real-time skin-dose mapping for interventional fluoroscopic procedures. *Proc SPIE* 9033: 903340.

Rana, V.K., Rudin, S. & Bednarek, D.R. 2016. A tracking system to calculate patient skin dose in real-time during neurointerventional procedures using a biplane x-ray imaging system. *Med Phys* 43(9): 5131–5144.

Rodas, N.L., Padoy, N. 2015. Seeing is believing: Increasing intraoperative awareness to scattered radiation in interventional procedures by combining augmented reality, Monte Carlo simulations and wireless dosimeters. *Int J. CARS* 10: 1181–1191.

Schueler, B.A. 2000. The AAPM/RSNA physics tutorial for residents' general overview of fluoroscopic imaging. *RadioGraphics* 20(4): 1115–1126.

Stecker, M.S., Balter, S., Towbin, R.B. et al. 2009. Guidelines for patient radiation dose management. *J Vasc Interv Radiol* 20(7 suppl): S263–S273.

Van Lysel, M.S. 2000. The AAPM/RSNA physics tutorial for residents: Fluoroscopy—Optical coupling and the video system. *RadioGraphics* 20(6): 1769–1786.

Wagner, L.K. 2007. Radiation injury is a potentially serious complication to fluoroscopically-guided complex interventions. *Biomed Imaging Interv J* 3(2): e22.

Wagner, L.K., Archer, B.R. & Cohen, A.M. 2000. Management of patient skin dose in fluoroscopically guided interventional procedures. *J Vasc Interv Radiol* 11(1): 25–33.

8

Optimization and Dose Reduction in CT Imaging

Usman Mahmood
and Yusuf E. Erdi

8.1 Introduction

The task of optimizing radiation dose and image quality for computed tomography (CT) imaging systems is one of the most challenging tasks faced by the imaging community. With the constant evolution of technology outpacing the knowledge of users, technological innovations aimed at reducing radiation dose while improving, or maintaining, image quality are often left underutilized and in some instances are improperly applied. Within this section, we aim to review methods available to optimize radiation dose from CT imaging systems.

8.2 Radiation Dose in CT

Since its invention in 1973, CT has transformed medical care.[1] The technology has replaced exploratory surgeries, improved treatment options, and has contributed to the improvement of quality of life and prolonged life expectancy for millions of patients.[2-4] However, cancer risks associated with the use of ionizing radiation in CT has driven regulators, the media, and several national and international organizations to call for standards designed to limit and reduce the amount of radiation delivered per procedure.[5-13]

As with any medical procedure, there are risks and benefits associated with CT. For CT, there may be a small, statistical risk of radiation-associated cancer induction through exposure.[5,14] Because this risk is not zero, it is prudent, particularly in pediatric patient populations, to perform a critical analysis of imaging parameters with the goal of optimizing and/or reducing the amount of ionizing radiation used in CT procedures.

The first step in the optimization frame work is to determine if the ordered procedures are justified for the diagnostic task and the individual patient.[15,16] Both the prescribing physician and radiologist share the responsibility of determining the appropriate imaging procedure. The American College of Radiology (ACR) has outlined appropriateness criteria that can be used by all physicians when making such decisions.[17] The overall intent is to ensure that the benefits of the imaging examination will outweigh any associated risks.

In a clinical setting, the focus for optimization is balancing the image quality and radiation dose in the context of the diagnostic task. With any reduction in radiation dose, the primary detriment to image quality is the increase in image noise, which could compromise relevant diagnostic information. Although there is no one strategy that fits all optimization frameworks, primary consideration must be placed on acceptability of the image quality to the radiologist interpreting the CT scan. Determination of this threshold of acceptability is a difficult balancing act that may require changes to imaging protocols. Regardless of the strategy chosen, manufacturers now offer several tools to help facilities achieve a radiation dose that is as low as diagnostically achievable (ALADA) for a specific diagnostic task.[18]

An initial step towards the optimization process is to thoroughly understand the image quality requirements for each particular diagnostic task. For example, lung cancer screening studies primarily involve the detection of soft-tissue nodules in a background consisting of air-filled lung parenchyma.[19-21] In this context, the contrast difference between the background and nodule is inherently high, allowing for lower radiation dose and more image noise without sacrificing diagnostic confidence.[19-21] However, with liver CT examinations, the diagnostic task involves the detection of more subtle, low-contrast lesions, which tolerate less image noise.[22-24] Adding to the complexity of balancing radiation dose and image quality are differences in the perception of medical images by various radiologists, where some may tolerate more noise than others.[25] Prior to attempting to reduce radiation dose in CT, a team consisting of radiologists, technologists, and physicists actively involved in CT applications is critical.[25] Radiologists are able to direct the effort by informing participating colleagues about protocols producing poor image quality or requiring investigation. Technologists contribute expertise in terms of the departmental workflow and are critical to the successful implementation of protocols. Physicists are able to guide the team about risk from radiation exposure, experiments to assess image quality with phantoms, monitoring of protocol changes, and novel technologies that can help in the dose optimization process. A collaborative team effort is essential to the success of any such optimization effort.

In this section, we aim to provide an overview of the fundamentals of CT imaging physics, with an emphasis on methods one can use to optimize imaging parameters. In addition, we include a checklist to aid CT dose optimization efforts (see Table 8.1).

TABLE 8.1 Methods to Optimize CT Protocol

Options	Recommendation	References for Guidance
Size-based CT Protocols	- Develop technique charts according to patient size. Size is best determined from the patient scout or localizer exam. - Patient weight and BMI have also been used, but such parameters are not as informative as patient size. - Pediatric protocols should especially be size based.	26–30
Automated Tube Current Modulation (ATCM)	- The setting of the ATCM options should be applied according to the characteristics of the scanner, image quality requirements of the physician, and radiation dose for the imaging task. - Selection of the image quality reference parameter should be based on acceptable image quality.	26,27,30–40
Iterative Reconstruction	- Apply in stages. Seek out manufacturer and CT optimization team support. - Carefully consider the imaging task at hand. Some studies have noted a loss in low contrast resolution with a reduction in radiation dose and application of iterative reconstruction.	GE: 42–48 Siemens: 19–54 Phillips: 55–57 Toshiba: 58–61
Peak Tube Voltage Selection (kVp)	- Obese patients who exceed a certain size threshold or if the mAs is maxed out, a higher kVp setting should be used. - A reduced kVp should be considered for slim or pediatric patients. Designation of a slim patient should be based on physical size measurements and image quality requirements. - A reduced kVp could be used for CT angiography with iodine studies. - Consider implementing manufacturer-specific automated tube potential selection. However, prior to implementing, a thorough understanding of how the automated software works should be required.	62–71
Dual-Energy CT (DECT)	- Consider applying to studies where contrast enhancement is desirable. - Due to the lower energy spectrum, excessive noise may be problematic in larger patients. An upper size or weight threshold may need to be applied.	72–76

8.3 Methods to Optimize CT Radiation Dose

Within any optimization framework, the primary contributing factor to non-diagnostic image quality is excessive noise. There is an inverse relationship between the amount of radiation dose and noise present in a CT image. If not careful, excessive reduction in radiation dose can increase noise to a point that diagnostic information is lost, as seen in Figure 8.1.

The concept of noise associated with radiation dose originates from two primary sources: the first arising from the number of X-ray photons incident on the detector, and subsequently detected, (i.e., quantum mottle) and the second concerning the electronic components of the CT system, otherwise known as electronic noise.[78,79] A third form of noise is referred to as anatomical noise, which arises from variations in tissue textures and is described as the presence of anatomy that is not relevant to the diagnostic task.[78] A complete understanding of the sources of noise from the entire imaging chain will assist any optimization

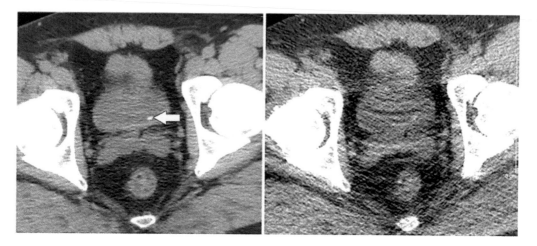

FIGURE 8.1   Left, Standard-dose CT image shows 4.5-mm calculus (arrow) at left ureterovesical junction. Right, Axial low-dose CT image at pelvic level shows that multiple streak artifacts reduce quality of interpretation. No ureteral calculus has been reported at low-dose CT analysis. (Reprinted with permission Poletti, P.A. et al., *Am. J. Roentgenol.*, 188(4), 927–933, 2007.)

framework. The type of detector, X-ray tube, image processing features, and other hardware and software components play a critical role in the final image formation process and noise content.

8.3.1 Detectors

The CT detector acts as a means to transfer energy from X-rays to electrical signals that can then be reconstructed into an image. If the conversion process is inefficient, some of the X-ray energy may not be converted into a useful electronic signal. A detector that has an inefficient detection and conversion process will result in an increase in image noise, which will require longer scan times and higher radiation dose to reduce. Prior to purchasing imaging equipment, medical physicists can help in determining and comparing the efficiency of various detectors that come with CT systems. Another consideration is electronic noise, which arises from random fluctuations of electronic signals that are not associated with the original X-ray energy deposition. CT examination, at routine doses, may not be impacted by electronic noise, but low-dose examinations, and for large patients, electronic noise may produce undesirable artifacts.[79] Duan et al.[80] demonstrated in phantoms and patients how differences in detector electronic systems result in excessive noise and artifacts at low radiation dose levels.

8.3.2 Bowtie Filter

The bowtie filter, or beam shaper, is a filter that is designed to shape the X-ray beam for the body habitus under evaluation. These filters play an important role in image quality and how radiation dose is distributed in the body. Most people are approximately oval-shaped, where the path length of radiation traverses through is shorter towards the periphery of the patient and longer towards the center. For this reason, these filters are shaped like a bowtie, where they are thicker at the periphery and thinner towards the center.[78] By reducing the beam intensity profile to match the greater tissue attenuation at the center of patients and the lesser attenuation at the periphery, the dose in the outer portions of the body is reduced. Because the dose to the center of the body is primarily a result of scatter radiation from the periphery, the dose at the center of the patient also decreases. The final result is a more homogenous distribution of radiation dose throughout the patient, which improves the uniformity of noise in the reconstructed image.

In practice, the shape and composition of bowtie filters vary among manufacturers. For example, GE has designed multiple bowtie filters that can be selected by the user.[31] During protocol development, facilities should ensure that the appropriate filter is matched to the particular body region. Consultation with the manufacturer and medical physicist will help determine the optimal filter for any imaging exam.

8.3.3 Scan Range

Scan range is an important clinical parameter that is directly controlled by the technologist. Limiting the scan range to the area of interest reduces unnecessary radiation exposure, particularly for pediatric patients. Although not possible for all situations, literature reports have demonstrated certain applications where scan range can be limited. One example is from a study by Patel et al.[81] where they showed a reduction in radiation dose of 48% when scan range for CT pulmonary angiographic studies was limited from the top of the aortic arch to the bottom of the heart, without any loss of sensitivity. Maintaining up-to-date knowledge, developing standards of practice or benchmarks for each protocol, and periodic review of scan ranges for exams performed by technologists will aid in the optimization of CT radiation dose. An exception to the above is helical CT acquisitions, which require additional data at the end of the scan length, outside of the originally prescribed scan range. The additional dose from this extension during helical acquisitions increases with higher pitches. However, modern CT scanners are typically equipped with dynamic z-axis tracking or collimation. This dynamic collimation allows the CT system to shape the X-ray beam so that any portion not used at the beginning or end of helical scans for the final image will be attenuated.[78] The amount of dose saving varies by manufacturer and is dependent on the CT pitch, scan length, and total collimation for a specific exam prescribed by the user. Again, medical physicists may assist by determining the available technical features during the purchasing of CT systems.

8.3.4 Automatic Tube Current Modulation or Automatic Exposure Control

Modulation of tube current according to the size (i.e., thickness) of the body part being imaged, attenuation characteristics, and acceptable amount of noise has been demonstrated as an effective means of reducing radiation dose while maintaining adequate image quality. In fact, all modern CT systems offer software that automatically modulate the tube current in the x-y (transverse) plane, z-axis (longitudinal) plane, or both, depending on the attenuation, size of the area being imaged, and the image quality reference parameter.[26,27,32–40] Tube current adjustment relative to the x-y plane, also known as angular modulation, accounts for the fact that patients are not circular in cross-section (Figure 8.2).

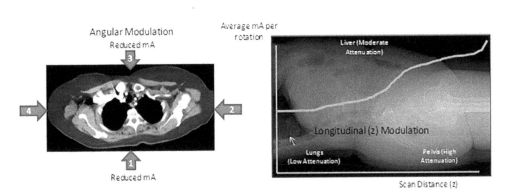

FIGURE 8.2 Left, Angular tube current modulation takes into account that patients are not round. As compared to the direction of 4–2, less radiation is needed in the direction of 1–3. Right, Longitudinal or z-axis modulation adjusts tube current for the differences in thickness and attenuation along the patient's body.

Since all CT manufacturers have different methods to implement automatic tube current modulation (ATCM), it is essential to understand the principles by which each manufacturer's tube current modulation software version operates.[38] For example, GE uses the noise index (NI), which is a user-selectable parameter that represents the acceptable noise level at the center of patient images.[34,35] To maintain the acceptable noise level, the system will adjust the mA by taking into account the NI, scan technical parameters, patient size/attenuation, and the desired slice thickness for the first reconstruction.[35] Other manufacturers, such as Siemens and Phillips, use real-time mA modulation with an online feedback system.[36] Regardless of the mechanism, the goal is to deliver the minimum amount of radiation to yield a diagnostic-quality image for a specific task. However, a common and critical component to all tube current modulation schemes is the selection of an image quality reference parameter. The underlying intent of the image quality reference parameter is to allow users to predefine the amount of noise radiologists are willing to accept in their final reconstructed images.

Once this parameter has been established, the scanner will automatically modulate the tube current to compensate for the thickness and attenuation of the relevant body part.

Due to the complexities associated with each manufacturer's automatic exposure control (AEC) system, prior to introducing tube current modulation into the clinic, all staff should undergo intensive training to understand the mechanisms by which the modulation system works.

AEC systems determine the patient size and attenuation information from the scout (a.k.a. localizer or topogram) scan, and/or via an online feedback system. During the scout acquisition, lead aprons, or heavily attenuating material, should not be placed within the field of view, as they will unnecessarily increase the radiation dose.

For all ATCM systems, the patient should always be properly centered. If the patient is off-center and positioned closer to the X-ray tube, they will appear larger in size.[37] In such a case, the tube current will increase, and the patient will receive more radiation than necessary (Figure 8.3).[37] Similarly, if the patient is too far from the tube during the scout, then the patient will appear smaller. In this situation, the system will reduce the mA and produce excessive noise in the reconstructed images. Patient centering is therefore essential to the appropriate application of AEC in CT.

Positioning of the X-ray tube Anterior-Posterior versus Posterior-Anterior for the scout acquisition could also increase radiation dose when AEC is used. Studies have demonstrated that when the scout is acquired with the tube posterior to the patient, with the patient in a supine position, the dose is increased by 36%–60%. Similarly, when the tube was in the lateral position, the dose increased by 15%. Hence, it is preferred that when the patient is in the supine position, the X-ray tube during a scout exam should be anterior to the patient.[38,39]

Finally, the user is responsible for determining the image quality reference parameter. This parameter controls amount of noise one is willing to accept in the final reconstructed image and selects the tube current to be used for each slice in the scan. Hence, when less noise is desired, the image quality reference parameter can be set up to deliver more tube current and radiation dose. When more noise

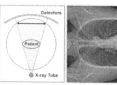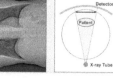

FIGURE 8.3 Left, Patient is properly centered. Middle, Patient is off-center and closer to the x-ray tube, making the patient appear wider than they actually are. This may result in more dose to the patient. Right, The patient is positioned too far from the x-ray tube, making them appear smaller than they actually are. This may result in underdosing the patient. (Reprinted with permission Bruesewitz, M.R. et al., Smart mA—automatic exposure control (AEC): Physics principles and practical hints, in *Annual Meeting of the Radiological Society of North America*, Chicago, IL, 2008, May.)

is acceptable, the image quality reference parameter can be adjusted so that the tube current will be reduced throughout the scan. The definition of this parameter varies across manufacturers, and since it has a significant impact on all tube current modulation systems, sites should consult with the manufacturer and the CT protocol optimization team when setting it up.

8.3.5 Peak Tube Voltage Selection

Peak tube voltage, or kVp, is a parameter that has traditionally not changed for the majority of imaging protocols. However, several studies have demonstrated improvement in image quality, primarily improved contrast discrimination, and/or reduction in radiation dose when appropriately adjusting tube potential for patient size and the imaging task.[62-70]

The benefit of optimizing kVp is primarily related to the k-edge of the materials being imaged. For example, iodine, which is a typical contrast media used in CT imaging, has a k-edge energy of 33 keV. The closer the average X-ray energy is to the iodine k-edge, the greater the number of photoelectric interactions that will occur. The attenuation coefficient of iodine increases as the photon energy approaches the k-edge of iodine. Hence, using a lower kVp will enhance visualization of the iodine distribution throughout the body and improve conspicuity between hyper- or hypovascular structures. However, reducing kVp also results in lower energy X-rays that are more readily absorbed by the patient, which could result in an undesirable increase in image noise if attention is not paid to the tube current or AEC settings. For example, if the mA and all other factors are kept constant, a reduction in kVp from 140 to 120 may result in an approximate dose reduction of 30%. In this case, image noise may be expected to increase by about 20%. However, with the application of AEC, lowering kVp will automatically result in an increase in mA to produce the specified noise level throughout the body of the patient.

To aid users in applying the optimal kVp for specific imaging needs, several manufacturers have introduced software that automatically selects the kVp for a particular diagnostic task.[71] These systems operate similarly to the AEC software, where patient size and attenuation information is extracted from the scout or localizer image. This patient-specific information, combined with the user-defined image quality reference parameter, is then used by the system to select an optimal kVp for the task at hand.[71] It is important to realize that each CT manufacturer's implementation of automated tube potential selection works in different ways. Prior to implementation it is critical to understand the complexities of the software being used.

8.3.6 Iterative Reconstruction

Iterative reconstruction is an image reconstruction technique designed to reduce statistical noise in diagnostic images while preserving structural details.[41-61] All major CT manufacturers have introduced their own iterative techniques for CT image reconstruction (Table 8.2). Although the implementation

TABLE 8.2 Currently Available Iterative Reconstruction Algorithms from Major CT Manufacturers

Vendor	Name
GE	Adaptive Statistical Iterative Reconstruction (ASiR)
GE	Model-Based Iterative Reconstruction (VEO)
Siemens	Sinogram Affirmed Iterative Reconstruction (SAFIRE)
Siemens	Image Reconstruction Iterative Reconstruction (IRIS)
Philips	iDose
Toshiba	Adaptive Iterative Dose Reduction (AIDR)

of iterative reconstruction for each manufacturer differs, the underlying principle has been to use statistical noise models and improved modeling of the system optics to reduce image noise seen with dose reduction.[78]

Several investigations using phantoms and patients have demonstrated that application of iterative reconstruction produced acceptable diagnostic image quality while reducing radiation dose.[41–61] However, several factors need to be considered when incorporating iterative reconstruction as a dose optimization strategy in the clinical setting.

The first consideration is that the application of iterative reconstruction does not automatically result in radiation dose reduction.[41] The iterative reconstruction algorithm only serves to reduce image noise. Second, several manufacturers offer different implementation of iterative reconstruction. Each method is proprietary and consists of iterative reconstruction "strengths" that vary amongst manufacturers. Typically, the lower strength results in less noise reduction. Figures 8.4 and 8.5 are examples of images generated with different strengths of GE's adaptive statistical iterative reconstruction (ASiR) technology. Although the higher strength allows for more noise reduction and the potential to decrease dose further, changes in the image texture may be undesirable for some radiologists and diagnostic purposes (see Figure 8.4).

Although iterative reconstruction allows for the reduction of image noise, which then enables one to reduce radiation dose, the arbitrary reduction of radiation dose should be discouraged since important clinical information may become obscured. For example, diagnostic tasks involving the visualization or assessment of high-contrast objects, such as delineation of large vessels with iodinated contrast, one may be able to substantially reduce dose without compromising the information content of the image. However, for imaging tasks where the visualization of low-contrast lesions is necessary, excessive dose reduction reduces contrast differences between the lesions and background, regardless of the strength of

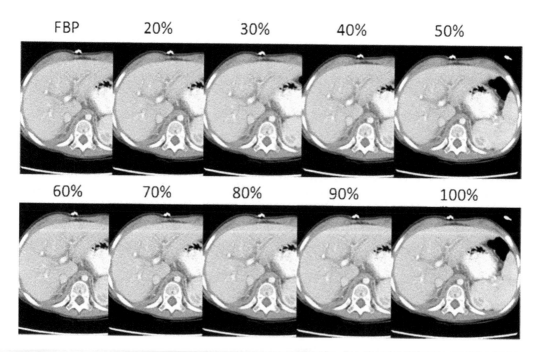

FIGURE 8.4 GE offers 10 strengths of their iterative reconstruction technology, known as ASiR. With each strength of ASiR, the appearance of the grainy image noise reduces.

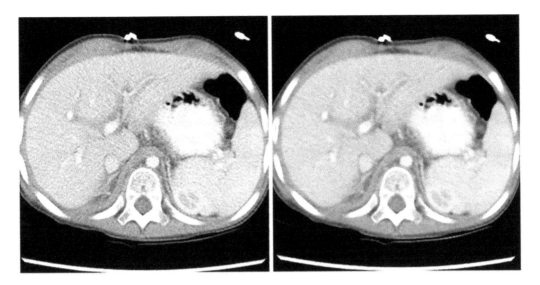

FIGURE 8.5 Both images are from abdominal CT scan. Left, Abdominal CT scan reconstructed with filtered back projection. Right, The same scan reconstructed with 100% ASiR.

IR applied.[59,82] The use of iterative reconstruction (IR) techniques to compensate for the increase in noise resulting from a decreased radiation dose must be implemented with care for examinations requiring low-contrast resolution.

8.3.7 Dual-Energy CT

Dual-Energy CT (DECT) is currently performed via two methods: (1) fast-switch kVp with one X-ray source and (2) with two X-ray sources. Although both methods are capable of generating material-specific images (e.g., images showing only iodine distribution),[72,73] the radiation doses from both methods are comparable to those of traditional single-energy CT applications.[74] For both methods, the true dose reduction potential has been the generation of virtual non-contrast images, which could allow users to omit pre-contrast scanning.[75] However, some studies have suggested image quality is inferior with virtual non-contrast images.[76] Table 8.1 summarizes important considerations when introducing DECT into the clinic.

8.3.8 Diagnostic Reference Levels

Diagnostic reference levels (DRLs) are defined as the third quartile (75%) of collected dose data from national, or international, dose databases. DRLs are a great resource for comparing radiation dose metrics and identifying high-dose protocols that may need to be optimized.[83] However, it is critical to understand that DRLs are often intended for specific patient populations. They may not apply to pediatric patients, large patients, or anyone who is not of the average normal size. They should not be used as dose thresholds.[84] The primary intent of a DRL is as a reference dose, a quantity above which users should investigate the potential for dose reduction measures.[15] Moreover, they are sometimes derived from phantoms and have no bearing on acceptable image quality.[18] As noted in Table 8.3, due to the difference in imaging tasks and patient body habitus, DRLs vary from region.

TABLE 8.3   Diagnostic Reference Levels of Computed Tomography Examinations in Adults

Exam	Dose Metric	ACR	NCRP	UK	Europe	IAEA	Korea	Taiwan	CRCPD/NEXT
Head	CTDI (mGy)	75	75	100	60	47	60	72	80
	DLP (mGy –cm)			930	1050	527	1000	850	1145
Chest	CTDI (mGy)			14	30	9.5	15		
	DLP (mGy –cm)			580	650	447	550		
Abdominal	CTDI (mGy)	25	25	15	35	10.9	20	31	24
	DLP (mGy –cm)			980	780	696	700	680	904

ACR—American College of Radiology. NCRP—National Council on Radiation Protection. IAEA—International Atomic Energy Agency. CRCPD—Conference of Radiation Control Program Directors. NEXT—Nationwide Evaluation of X-ray Trends.

8.4 Concluding Remarks

Emerging technological advances have resulted in several strategies to manage radiation dose and image quality. The best strategy will depend on the specific diagnostic task, equipment options and end users. However, for any clinically justified examination, scanning parameters should be managed according to the guiding principle of keeping radiation dose as low as diagnostically achievable (ALADA).

References

1. Hounsfield, G.N. (1973). Computerized transverse axial scanning (tomography). 1. Description of system. *The British Journal of Radiology*, 46(552), 1016–1022.
2. Alexander, R. E., & Gunderman, R. B. (2010). EMI and the first CT scanner. *Journal of the American College of Radiology*, 7(10), 778–781.
3. Zanzonico, P., & Stabin, M. G. (2014, May). Quantitative benefit-risk analysis of medical radiation exposures. *Seminars in Nuclear Medicine*, 44(3), 210–214.
4. Hendee, W. R. (2013). Policy statement of the international organization for medical physics. *Radiology*, 267(2), 326–327.
5. National Research Council. (2006). *Health Risks from Exposure to Low Levels of Ionizing Radiation: BEIR VII, Phase 2*. Washington, DC: National Academies.
6. Brenner, D. J., & Hall, E. J. (2007). Computed tomography—An increasing source of radiation exposure. *New England Journal of Medicine*, 357(22), 2277–2284.
7. Brenner, D. J. (2004). Radiation risks potentially associated with low-dose CT screening of adult smokers for lung cancer 1. *Radiology*, 231(2), 440–445.
8. Bogdanich, W. A. L. T. (2009). Radiation overdoses point up dangers of CT scans. *The New York Times*, 15.
9. Bogdanich, W. (2010). After stroke scans, patients face serious health risks. *The New York Times*, 31.
10. Boodman, S. (2016). Should you worry about the radiation from CT scans? *The Washington Post*. [Retrieved from https://www.washingtonpost.com/national/health-science/how-much-to-worry-about-the-radiation-from-ct-scans/2016/01/04/8dfb80cc-8a30-11e5-be39-0034bb576eee_story.html]. Accessed January 12, 2017.
11. Pearce, M. S., Salotti, J. A., Little, M. P., McHugh, K., Lee, C., Kim, K. P., & Parker, L. (2012). Radiation exposure from CT scans in childhood and subsequent risk of leukaemia and brain tumours: A retrospective cohort study. *The Lancet*, 380(9840), 499–505.

12. Wintermark, M., & Lev, M. H. (2010). FDA investigates the safety of brain perfusion CT. *American Journal of Neuroradiology*, *31*(1), 2–3.

13. Burke, L. M., Semelka, R. C., & Smith-Bindman, R. (2015). Trends of CT utilization in North America over the last decade. *Current Radiology Reports*, *3*(1), 1–5.

14. AAPM Policy Statement. (2011). AAPM Position Statement on Radiation Risks from Medical Imaging Procedures. [Retrieved from https://www.aapm.org/org/policies/details. asp?id=318&type=PP].

15. ICRP. (2007). The 2007 recommendations of the International Commission on Radiological Protection. ICRP publication 103. *Ann ICRP*, *37*(2–4), 1–332.

16. Sodhi, K. S., Krishna, S., Saxena, A. K., Sinha, A., Khandelwal, N., & Lee, E. Y. (2015). Clinical application of "Justification" and "Optimization" principle of ALARA in pediatric CT imaging: "How many children can be protected from unnecessary radiation?" *European Journal of Radiology*, *84*(9), 1752–1757.

17. American College of Radiology. ACR Appropriateness Criteria. [Retrieved from http://www.acr. org/Quality-Safety/Appropriateness-Criteria/New-and-Revised]. Accessed August 5, 2016.

18. Rehani, M. M. (2014). Limitations of diagnostic reference level (DRL) and introduction of acceptable quality dose (AQD). *The British Journal of Radiology*, *88*(1045), 20140344.

19. Marshall, H. M., Bowman, R. V., Yang, I. A., Fong, K. M., & Berg, C. D. (2013). Screening for lung cancer with low-dose computed tomography: A review of current status. *Journal of Thoracic Disease*, *5*(5), S524–S539.

20. National Lung Screening Trial Research Team. (2011). Reduced lung-cancer mortality with low-dose computed tomographic screening. *The New England Journal of Medicine*, *2011*(365), 395–409.

21. Saghir, Z., Dirksen, A., Ashraf, H., Bach, K. S., Brodersen, J., Clementsen, P. F., Døssing, M. et al. (2012). CT screening for lung cancer brings forward early disease. The randomised Danish lung cancer screening trial: Status after five annual screening rounds with low-dose CT. *Thorax*, *67*(4), 296–301.

22. Kanal, K. M., Chung, J. H., Wang, J., Bhargava, P., Kohr, J. R., Shuman, W. P., & Stewart, B. K. (2011). Image noise and liver lesion detection with MDCT: A phantom study. *American Journal of Roentgenology*, *197*(2), 437–441.

23. Schindera, S. T., Torrente, J. C., Ruder, T. D., Hoppe, H., Marin, D., Nelson, R. C., & Szucs-Farkas, Z. (2011). Decreased detection of hypovascular liver tumors with MDCT in obese patients: A phantom study. *American Journal of Roentgenology*, *196*(6), W772–W776.

24. Schindera, S. T., Odedra, D., Raza, S. A., Kim, T. K., Jang, H. J., Szucs-Farkas, Z., & Rogalla, P. (2013). Iterative reconstruction algorithm for CT: Can radiation dose be decreased while low-contrast detectability is preserved? *Radiology*, *269*(2), 511–518.

25. Hara, A. K., Wellnitz, C. V., Paden, R. G., Pavlicek, W., & Sahani, D. V. (2013). Reducing body CT radiation dose: Beyond just changing the numbers. *American Journal of Roentgenology*, *201*(1), 33–40.

26. Gies, M., Kalender, W. A., Wolf, H., Suess, C., & Madsen, M. T. (1999). Dose reduction in CT by anatomically adapted tube current modulation. I. Simulation studies. *Medical Physics*, *26*(11), 2235–2247.

27. Kalender, W. A., Wolf, H., & Suess, C. (1999). Dose reduction in CT by anatomically adapted tube current modulation. II. Phantom measurements. *Medical Physics*, *26*(11), 2248–2253.

28. Singh, S., Kalra, M. K., Moore, M. A., Shailam, R., Liu, B., Toth, T. L., Grant, E., & Westra, S. J. (2009). Dose reduction and compliance with pediatric CT protocols adapted to patient size, clinical indication, and number of prior studies 1. *Radiology*, *252*(1), 200–208.

29. Strauss, K. J., Goske, M. J., Kaste, S. C., Bulas, D., Frush, D.P., Butler, P., Morrison, G., Callahan, M. J., & Applegate, K. E. (2010). Image gently: Ten steps you can take to optimize image quality and lower CT dose for pediatric patients. *American Journal of Roentgenology*, *194*, 868–873.

30. Strauss, K. J. (2014). Developing patient specific dose protocols for a CT scanner and exam using diagnostic reference levels. *Pediatric Radiology, 44,* 450–459.
31. Toth, T., Ge, Z., & Daly, M. P. (2007). The influence of patient centering on CT dose and image noise. *Medical Physics, 34*(7), 3093–3101.
32. Lee, C. H., Goo, J. M., Ye, H. J., Ye, S. J., Park, C. M., Chun, E. J., & Im, J. G. (2008). Radiation dose modulation techniques in the multidetector CT era: From basics to practice 1. *Radiographics, 28*(5), 1451–1459.
33. Kalra, M. K., Sodickson, A. D., & Mayo-Smith, W. W. (2015). CT Radiation: Key concepts for gentle and wise use. *RadioGraphics, 35*(6), 1706–1721.
34. Kanal, K. M., Stewart, B. K., Kolokythas, O., & Shuman, W. P. (2007). Impact of operator-selected image noise index and reconstruction slice thickness on patient radiation dose in 64-MDCT. *American Journal of Roentgenology, 189*(1), 219–225.
35. McCollough, C. H., Bruesewitz, M. R., & Kofler Jr, J. M. (2006). CT dose reduction and dose management tools: Overview of available options 1. *Radiographics, 26*(2), 503–512.
36. Tack, D., De Maertelaer, V., & Gevenois, P. A. (2003). Dose reduction in multidetector CT using attenuation-based online tube current modulation. *American Journal of Roentgenology, 181*(2), 331–334.
37. Bruesewitz, M. R., Yu, L., Vrieze, T. J., Kofler, J. M., & McCollough, C. H. (2008, May). Smart mA—automatic exposure control (AEC): Physics principles and practical hints. In *Annual Meeting of the Radiological Society of North America.* Chicago, IL.
38. McKenney, S. E., Seibert, J. A., Lamba, R., & Boone, J. M. (2014). Methods for CT automatic exposure control protocol translation between scanner platforms. *Journal of the American College of Radiology, 11*(3), 285–291.
39. Keat N. (2005). MHRA Report 05016: CT scanner automatic exposure control systems. London, UK: Medicines and Healthcare Products Regulatory Agency.
40. Kalra, M. K., Maher, M. M., Toth, T. L., Schmidt, B., Westerman, B. L., Morgan, H. T., & Saini, S. (2004). Techniques and applications of automatic tube current modulation for CT 1. *Radiology, 233*(3), 649–657.
41. Hara, A. K., Paden, R. G., Silva, A. C., Kujak, J. L., Lawder, H. J., & Pavlicek, W. (2009). Iterative reconstruction technique for reducing body radiation dose at CT: Feasibility study. *American Journal of Roentgenology, 193*(3), 764–771.
42. Silva, A. C., Lawder, H. J., Hara, A., Kujak, J., & Pavlicek, W. (2010). Innovations in CT dose reduction strategy: Application of the adaptive statistical iterative reconstruction algorithm. *American Journal of Roentgenology, 194*(1), 191–199.
43. Marin, D., Nelson, R. C., Schindera, S. T., Richard, S., Youngblood, R. S., Yoshizumi, T. T., & Samei, E. (2009). Low-tube-voltage, high-tube-current multidetector abdominal CT: Improved image quality and decreased radiation dose with adaptive statistical iterative reconstruction algorithm—Initial clinical experience 1. *Radiology, 254*(1), 145–153.
44. Prakash, P., Kalra, M. K., Kambadakone, A. K., Pien, H., Hsieh, J., Blake, M. A., & Sahani, D. V. (2010). Reducing abdominal CT radiation dose with adaptive statistical iterative reconstruction technique. *Investigative Radiology, 45*(4), 202–210.
45. Sagara, Y., Hara, A. K., Pavlicek, W., Silva, A. C., Paden, R. G., & Wu, Q. (2010). Abdominal CT: Comparison of low-dose CT with adaptive statistical iterative reconstruction and routine-dose CT with filtered back projection in 53 patients. *American Journal of Roentgenology, 195*(3), 713–719.
46. Leipsic, J., LaBounty, T. M., Heilbron, B., Min, J. K., Mancini, G. J., Lin, F. Y., Taylor, C., Dunning, A., & Earls, J. P. (2010). Adaptive statistical iterative reconstruction: Assessment of image noise and image quality in coronary CT angiography. *American Journal of Roentgenology, 195*(3), 649–654.
47. Leipsic, J., LaBounty, T. M., Heilbron, B., Min, J. K., Mancini, G. J., Lin, F. Y., Taylor, C., Dunning, A., & Earls, J. P. (2010). Estimated radiation dose reduction using adaptive statistical iterative reconstruction in coronary CT angiography: The ERASIR study. *American Journal of Roentgenology, 195*(3), 655–660.

48. Singh, S., Kalra, M. K., Hsieh, J., Licato, P. E., Do, S., Pien, H. H., & Blake, M. A. (2010). Abdominal CT: Comparison of adaptive statistical iterative and filtered back projection reconstruction techniques 1. *Radiology, 257*(2), 373–383.
49. Yang, W. J., Yan, F. H., Liu, B., Pang, L. F., Hou, L., Zhang, H., Pan, Z. L., & Chen, K. M. (2013). Can sinogram-affirmed iterative (SAFIRE) reconstruction improve imaging quality on low-dose lung CT screening compared with traditional filtered back projection (FBP) reconstruction? *Journal of Computer Assisted Tomography, 37*(2), 301–305.
50. Schulz, B., Beeres, M., Bodelle, B., Bauer, R., Al-Butmeh, F., Thalhammer, A., Vogl, T. J., & Kerl, J. M. (2013). Performance of iterative image reconstruction in CT of the paranasal sinuses: A phantom study. *American Journal of Neuroradiology, 34*(5), 1072–1076.
51. Han, B. K., Grant, K. L., Garberich, R., Sedlmair, M., Lindberg, J., & Lesser, J. R. (2012). Assessment of an iterative reconstruction algorithm (SAFIRE) on image quality in pediatric cardiac CT datasets. *Journal of Cardiovascular Computed Tomography, 6*(3), 200–204.
52. Kalra, M. K., Woisetschläger, M., Dahlström, N., Singh, S., Lindblom, M., Choy, G. et al. (2012). Radiation dose reduction with sinogram affirmed iterative reconstruction technique for abdominal computed tomography. *Journal of Computer Assisted Tomography, 36*(3), 339–346.
53. Fletcher, J. G., Grant, K. L., Fidler, J. L., Shiung, M., Yu, L., Wang, J., Schmidt, B., Allmendinger, T., & McCollough, C. H. (2012). Validation of dual-source single-tube reconstruction as a method to obtain half-dose images to evaluate radiation dose and noise reduction: phantom and human assessment using CT colonography and sinogram-affirmed iterative reconstruction (SAFIRE). *Journal of Computer Assisted Tomography, 36*(5), 560–569.
54. Moscariello, A., Takx, R. A., Schoepf, U. J., Renker, M., Zwerner, P. L., O'Brien, T. X. et al. (2011). Coronary CT angiography: Image quality, diagnostic accuracy, and potential for radiation dose reduction using a novel iterative image reconstruction technique—Comparison with traditional filtered back projection. *European Radiology, 21*(10), 2130–2138.
55. Noël, P. B., Fingerle, A. A., Renger, B., Münzel, D., Rummeny, E. J., & Dobritz, M. (2011). Initial performance characterization of a clinical noise–suppressing reconstruction algorithm for MDCT. *American Journal of Roentgenology, 197*(6), 1404–1409.
56. Hu, X. H., Ding, X. F., Wu, R. Z., & Zhang, M. M. (2011). Radiation dose of non-enhanced chest CT can be reduced 40% by using iterative reconstruction in image space. *Clinical Radiology, 66*(11), 1023–1029.
57. Renker, M., Ramachandra, A., Schoepf, U. J., Raupach, R., Apfaltrer, P., Rowe, G. W., Vogt S. et al. (2011). Iterative image reconstruction techniques: Applications for cardiac CT. *Journal of Cardiovascular Computed Tomography, 5*(4), 225–230.
58. Gervaise, A., Osemont, B., Lecocq, S., Noel, A., Micard, E., Felblinger, J., & Blum, A. (2012). CT image quality improvement using adaptive iterative dose reduction with wide-volume acquisition on 320-detector CT. *European Radiology, 22*(2), 295–301.
59. Willemink, M. J., Leiner, T., de Jong, P. A., de Heer, L. M., Nievelstein, R. A., Schilham, A. M., & Budde, R. P. (2013). Iterative reconstruction techniques for computed tomography part 2: Initial results in dose reduction and image quality. *European Radiology, 23*(6), 1632–1642.
60. Schindera, S. T., Odedra, D., Raza, S. A., Kim, T. K., Jang, H. J., Szucs-Farkas, Z., & Rogalla, P. (2013). Iterative reconstruction algorithm for CT: Can radiation dose be decreased while low-contrast detectability is preserved? *Radiology, 269*(2), 511–518.
61. Yamada, Y., Jinzaki, M., Hosokawa, T., Tanami, Y., Sugiura, H., Abe, T., & Kuribayashi, S. (2012). Dose reduction in chest CT: Comparison of the adaptive iterative dose reduction 3D, adaptive iterative dose reduction, and filtered back projection reconstruction techniques. *European Journal of Radiology, 81*(12), 4185–4195.
62. Kalva, S. P., Sahani, D. V., Hahn, P. F., & Saini, S. (2006). Using the K-edge to improve contrast conspicuity and to lower radiation dose with a 16-MDCT: A phantom and human study. *Journal of Computer Assisted Tomography, 30*:391–397.

63. Yu, L., Li, H., Fletcher, J. G., & McCollough, C. H. (2010). Automatic selection of tube potential for radiation dose reduction in CT: A general strategy. *Medical Physics, 37*(1), 234–243.

64. Winklehner, A., Goetti, R., Baumueller, S., Karlo, C., Schmidt, B., Raupach, R., Flohr, R., Frauenfelder, T., & Alkadhi, H. (2011). Automated attenuation-based tube potential selection for thoracoabdominal computed tomography angiography: Improved dose effectiveness. *Investigative Radiology, 46*(12), 767–773.

65. Gnannt, R., Winklehner, A., Eberli, D., Knuth, A., Frauenfelder, T., & Alkadhi, H. (2012). Automated tube potential selection for standard chest and abdominal CT in follow-up patients with testicular cancer: Comparison with fixed tube potential. *European Radiology, 22*(9), 1937–1945.

66. Park, Y. J., Kim, Y. J., Lee, J. W., Kim, H. Y., Hong, Y. J., Lee, H. J., Hur, J., Nam, J.E., & Choi, B. W. (2012). Automatic tube potential selection with tube current modulation (APSCM) in coronary CT angiography: Comparison of image quality and radiation dose with conventional body mass index-based protocol. *Journal of Cardiovascular Computed Tomography, 6*(3), 184–190.

67. Yu, L., Bruesewitz, M. R., Thomas, K. B., Fletcher, J. G., Kofler, J. M., & McCollough, C. H. (2011). Optimal tube potential for radiation dose reduction in pediatric CT: Principles, clinical implementations, and pitfalls. *Radiographics, 31*(3), 835–848.

68. Schwarz, F., Grandl, K., Arnoldi, A., Kirchin, M. A., Bamberg, F., Reiser, M. F., & Becker, C. R. (2013). Lowering radiation exposure in CT angiography using automated tube potential selection and optimized iodine delivery rate. *American Journal of Roentgenology, 200*(6), W628–W634.

69. Lee, K. H., Lee, J. M., Moon, S. K., Baek, J. H., Park, J. H., Flohr, T. G., Kim, K. W., Kim, S. J., Han, J, K, & Choi, B. I. (2012). Attenuation based automatic tube voltage selection and tube current modulation for dose reduction at contrast-enhanced liver CT. *Radiology, 265*(2), 437–447.

70. Bischoff, B., Hein, F., Meyer, T., Hadamitzky, M., Martinoff, S., Schömig, A., & Hausleiter, J. (2009). Impact of a reduced tube voltage on CT angiography and radiation dose: Results of the PROTECTION I study. *JACC: Cardiovascular Imaging, 2*(8), 940–946.

71. Hu, L., Wang, Y., Hou, H., Wei, F., Yang, G., & Chen, Y. (2014). Radiation dose and image quality with abdominal computed tomography with automated dose-optimized tube voltage selection. *Journal of International Medical Research, 42*(4), 1011–1017.

72. Johnson, T. R., Krauss, B., Sedlmair, M., Grasruck, M., Bruder, H., Morhard, D., Fink, C. et al. (2007). Material differentiation by dual energy CT: Initial experience. *European Radiology, 17*(6), 1510–1517.

73. Goodsitt, M. M., Christodoulou, E. G., & Larson, S. C. (2011). Accuracies of the synthesized monochromatic CT numbers and effective atomic numbers obtained with a rapid kVp switching dual energy CT scanner. *Medical Physics, 38*(4), 2222–2232.

74. Schenzle, J. C., Sommer, W. H., Neumaier, K., Michalski, G., Lechel, U., Nikolaou, K., Becker, C.R. et al. (2010). Dual energy CT of the chest: How about the dose? *Investigative Radiology, 45*(6), 347–353.

75. Graser, A., Johnson, T. R., Hecht, E. M., Becker, C. R., Leidecker, C., Staehler, M., Stief, C. G. et al. (2009). Dual-energy CT in patients suspected of having renal masses: Can virtual nonenhanced images replace true nonenhanced images? 1. *Radiology, 252*(2), 433–440.

76. Lundin, M., Lidén, M., Magnuson, A., Mohammed, A. A., Geijer, H., Andersson, T., & Persson, A. (2012). Virtual non-contrast dual-energy CT compared to single-energy CT of the urinary tract: A prospective study. *Acta Radiologica, 53*(6), 689–694.

77. Poletti, P. A., Platon, A., Rutschmann, O. T., Schmidlin, F. R., Iselin, C. E., & Becker, C. D. (2007). Low-dose versus standard-dose CT protocol in patients with clinically suspected renal colic. *American Journal of Roentgenology, 188*(4), 927–933.

78. Bushberg, J. T., & Boone, J. M. (2011). *The Essential Physics of Medical Imaging.* Philadelphia, PA: Lippincott Williams & Wilkins.

79. Hsieh, J. (2006). *Computed Tomography: Principles, Design, Artifacts, and Recent Advances.* Bellingham, WA: SPIE Press.

80. Duan, X., Wang, J., Leng, S., Schmidt, B., Allmendinger, T., Grant, K., Flohr T., & McCollough, C. H. (2013). Electronic noise in CT detectors: Impact on image noise and artifacts. *American Journal of Roentgenology*, *201*(4), W626–W632.

81. Patel, H., Coughlin, B., LaFrance, T., et al., Eds. (2007). Comparison of full chest CTA with limited CTA and triple rule-out CTA for PE detection and effective dose implications. Chicago, IL: Radiological Society of North America.

82. McCollough, C. H., Yu, L., Kofler, J. M., Leng, S., Zhang, Y., Li, Z., & Carter, R. E. (2015). Degradation of CT low-contrast spatial resolution due to the use of iterative reconstruction and reduced dose levels. *Radiology*, *276*(2), 499–506.

83. Tsapaki, V., Aldrich, J. E., Sharma, R., Staniszewska, M. A., Krisanachinda, A., Rehani, M., Hufton, A. et al. (2006). Dose reduction in CT while maintaining diagnostic confidence: Diagnostic reference levels at routine head, chest, and abdominal CT—IAEA-coordinated research project 1. *Radiology*, *240*(3), 828–834.

84. Rehani, M. M. (2014). Limitations of diagnostic reference level (DRL) and introduction of acceptable quality dose (AQD). *The British Journal of Radiology*, *88*(1045), 20140344.

9

Optimization and Dose Reduction in Nuclear Medicine

Adam Kesner

9.1 Principles of Optimization in Nuclear Medicine

X-ray imaging has been around since the early twentieth century, since scientists first realized that high energy photons (e.g., x-rays) can transverse human tissue and be used to provide information, or an image, detailing the internal structure of a patient.

9.1.1 Transmission Imaging

Radiographic imaging, often referred to as x-ray imaging, is a technology that is still heavily utilized in modern clinical practice. In radiographic imaging, a patient, or the portion of a patient being imaged, is placed in between an x-ray generation tube and a detector. High-energy photons are generated and directed towards the patient, and a radiographic image is produced by detecting the variation in attenuation of the photons along the different paths through the body. Because the detected x-rays that are used to generate an image are transmitted through the body, this type of imaging is called transmission imaging. Transmission imaging is used to image patient anatomy.

9.1.2 Emission Imaging

Nuclear medicine imaging, also called nuclear imaging, is based on a similar principle to radiographic imaging, in that it uses the information carried by photons sent through the body to create an image. What makes nuclear imaging unique is the source of the photons in the system. In nuclear imaging, a radioactive radiotracer is introduced into the body, and is distributed according to the body's biological response for that tracer. As the tracer distributes in the body, it will emit photons from radioactive decay events. These high-energy photons will leave the body and can be measured with a detector, which will indicate presence of the tracer. Because the detected x-rays that are used to generate an image are emitted from inside the body, this type of imaging is called emission imaging. Emission imaging is used to image function, that is, the interaction and distribution of a tracer inside a biological system.

9.1.3 Optimization Principles in Nuclear Imaging

Transmission and emission imaging provide different types of information, but both rely on the same principle, generating images by detecting high-energy photons that have traveled through the body. Better images can be created when more photons are used in the system because more photons in turn provide more information. However, when photons traverse the body, there is a certain probability that they will interact with the body tissue. The energy of photons that are attenuated or completely stopped in the body is dissipated into the site of interaction. This energy is referred to as 'radiation dose' and can initiate negative biological effects.

Nuclear imaging provides invaluable information that is fundamental to creating images but the trade-off is the deposited dose. In any procedure, increasing the number of x-rays will improve the images (see Figure 9.1), and decreasing the number of x-rays will improve, therefore, reducing the radiation dose.

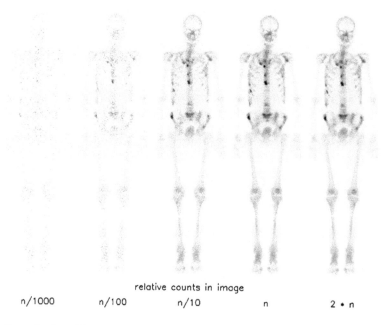

relative counts in image

n/1000   n/100   n/10   n   2•n

FIGURE 9.1 Example clinical bone scan (gamma camera) image generated using differing amounts of count statistics, with statistics increasing from the left to right. The figure demonstrates the two main principles of optimization (1) having more counts in an image with result in a better image, and (2) a highest-quality image may not be required for an accurate diagnosis, in this case determining the presence of bone metastasis can be made using the n/10 counts image—it does not require the highest count image.

The use of x-ray imaging is justified because there is a favorable risk/benefit ratio—the benefits of having clinical information are thought to outweigh the risks associated with the image. The term "optimization" encompasses the concept of finding the ideal balance between the amount of source radiation that creates the x-rays that can be used so that resulting images are good enough to provide the necessary clinical benefit, while limiting the source radiation to minimize the negative effects of radiation dose.

In nuclear imaging practice, clinical protocols are used to provide the best balance between risk and benefit for the patient being imaged. There are, however, some limits to how much can be optimized because the risks (radiation dose) and benefits (clinical image quality) for any study are impacted by many factors. In this chapter, we will discuss these different factors, current clinical protocols, and opportunities there may be for improved optimization.

9.2 Optimization Principles Relating to the Imaging Hardware

Hardware plays a fundamental role in nuclear imaging. The physical components of a system determine how the system will detect the signal from the photons emitted by a decaying radiotracer within a patient, and thus provides the source signal for image generation. Hardware relates to optimization because better, more efficient hardware can translate into more efficient information collection per unit of radiation emission.

9.2.1 Photon Detection

In nuclear imaging, patients are administered a radiotracer and placed in front of a camera. Radiotracers used in nuclear imaging contain an isotope that will decay to a more stable atomic configuration and, in the process, will emit high-energy photons. Most of the emitted photons will leave the body and can be detected by radiation detectors placed near the patient.

Photon detection is a fundament aspect of nuclear medicine. There are several types of radiation detectors that are commonly found in nuclear medicine clinics.

9.2.1.1 Gas-Type Detectors

Gas-type detectors are used to measure radiation and are based on the effect that high-energy photons or charged particles, when traversing a volume of gas, will interact with the gas molecules to generate positively and negatively charged ions. If the ions are generated within an electric field, they can be detected and converted to an electronic signal. The rate of ionic interactions in a gas detector can thus be inferred through a measurement of the resulting electric current. This measurement will be indicative of the amount of radiation reaching the detector. Examples of gas-type detectors found in nuclear medicine clinics include ion chambers, Geiger-Müller counters, and dose calibrators. Gas-type detectors are relatively inexpensive (compared to other types of radiation detectors) and can be uses to measure relatively large amounts of radiation. Gas-type detectors are used in the clinic to check radiation rates at the location of the detector, for dose preparation and safety purposes. Gas-type detectors are not used for image generation.

9.2.1.2 Scintillation-Type Detectors

Scintillation-type detectors are based on an effect when radiation interacts with matter, ionization events will occur in the exposed material. After initial excitation, ionized molecules will recombine and in the process release energy in the form of heat or vibrations. In some crystal materials, a portion of the energy released during recombination will be in the form of optical photons that can traverse the crystal structure. These crystals are called *scintillation crystals* and are utilized for this radiation-sensing behavior in scintillation detectors. A scintillation detector, generally speaking, is a scintillation crystal that is coupled with an optical photon detector. When high-energy photons, like those emitted from nuclear medicine radiotracers, encounter a scintillation crystal, there is a probability that the photon will interact in that crystal and initiate the creation of light photons, which can then be detected and converted into an electronic signal via a photomultiplier tube. Examples of scintillation-type detectors found in nuclear medicine clinics include well counters, thyroid probes, gamma cameras, and Positron Emission Tomography (PET) cameras. Scintillation-type detectors are denser than gas-type detectors and therefore stop more of the incident radiation. They also have superior energy discrimination and locational measurement capacity and are used almost ubiquitously in nuclear imaging systems.

9.2.1.3 Solid State Detectors

Solid state detectors utilize properties of semiconductor materials to detect and amplify the ionization events that occur after an interaction of incident radiation with the detector. Conceptually speaking, solid state detectors are similar to gas-type ion chambers, a solid material analog. However, because solid state detector materials are several thousands of times denser than gas, they are much more efficient at stopping and detecting incident radiation. Furthermore, the signal resulting from an incident particle on a solid state detector is capture in greater detail, and solid state detectors have very good energy resolution—that is, detector systems can determine the energy of incident photons with high precision, which can help optimize the selection of detected photons used in an image. Solid state detectors come in many forms, for example germanium (Ge) and cadmium zinc telluride (CZT). However, they have negative features that they are expensive, difficult to produce, and can have behavior that is highly temperature dependent. Traditionally solid state detectors are not used in nuclear medicine clinics. In recent years, however, vendors are bringing gamma cameras and PET systems to the market that have integrated solid state detector components. This is, in part, due to the fact that solid state detectors can be used to replace photomultiplier tubes and enable the combination of PET and Magnetic Resonance (MR) technologies.

9.2.2 Nuclear Medicine Camera Systems

The cameras used in nuclear imaging are similar in principle to common optical cameras used for photography. The difference is that nuclear imaging cameras are designed to detect high-energy photons, as opposed to photons that have energy in the optical light spectrum. The energy of photons used in nuclear medicine are in the range of 50–500 KeV. Isotopes that have photon emissions at these energies are deliberately utilized in nuclear imaging because these energies are high enough to traverse through the body, yet low enough to be stopped and measured efficiently by a detector. Nuclear medicine cameras utilize detectors of the type described in Section 9.2.1 and are designed to incorporate spatial information of the detected events to generate 2D or 3D images. The two main cameras commonly found in nuclear medicine clinics are scintillation crystal-based gamma cameras and PET cameras. Specific design components of the cameras can impact image quality, and therefore should be considered when discussing optimization.

9.2.2.1 Gamma Camera

Originally called an Anger camera, named after inventor Hal Anger, a gamma camera is a workhorse instrument for most nuclear medicine departments. Traditionally, a gamma camera consists of a large, flat scintillation crystal, coupled with an array of photomultiplier tubes and processing electronics. When used clinically, exchangeable collimators are placed at the front of the crystal to ensure the detected incident photons are directionally aligned—a feature required for proper event positioning. When radioactive sources, including patients injected with radiotracers, are placed in front of a gamma camera, a two-dimensional image of the source can be generated from the detected photons. If a gamma camera is rotated around a patient, a three-dimensional emission image can be created and is known as Single Photon Emission Computed Tomography (SPECT).

Gamma cameras can be designed with different specifications, which are relevant when considering optimization. For example, the thickness of the scintillation crystal component will affect the sensitivity of a gamma camera system. Thicker crystals will stop more of the incident photons on a system because of the increased probability of photon interaction in the detector, and thus reduced amounts of radiation will be required to obtain an image with an acceptable level of counts. However, the choice of crystal thickness of a camera is balanced with other considerations: thicker crystal systems have worse spatial resolution, energy resolution, dead time affects, and higher production costs. Typical current general-use systems use ≈ 1 cm crystal thickness.

Other camera components that affect the images and dose include detector surface size, collimator, processing electronics. Gamma cameras should always be tuned, through vendor recommended daily tuning, to ensure optimal image quality.

9.2.2.2 PET

PET cameras detect high-energy photons emitted from a radiotracer in a similar manner to gamma cameras, using coupled scintillation crystal/photomultiplier techniques. However, PET cameras usually come in a ring formation and use crystals that efficiently detect the 511 KeV photons that are produced during positron annihilations. The principle of PET is to use positron-emitting radiotracers for their unique positron annihilation characteristics. Positrons are the antiparticle of the electron and exist with opposite electric charge. When positrons are emitted from a radiotracer, they travel a short distance and then interact/combine with an oppositely charged local electron, in a process known as annihilation. The masses of the two charged particles will be transferred to energy and dissipate via two photons emitted at approximately 180° in opposite directions, each one having an energy of approximately 511 KeV.

PET scanners are more efficient for generating images than gamma cameras because they use electronic collimation instead of physical hardware. However, PET machines can only be used for positron emitting radiotracers. Positron-emitting isotopes generally deliver more radiation dose to a patient than isotopes used in gamma imaging because of the dose deposition characteristics of positrons and high-energy photons.

PET imaging is 3D by nature. Image quality is generally better with PET than with SPECT. However, the specific modality used in any specific nuclear medicine examination will depend on the purpose of the examination and the available radiotracers. PET and SPECT use similar activities of radiotracers (5–10 mCi). Positron-emitting isotopes generally deposit more dose per emission, but have shorter half-lives and thus dissipate from the body quicker. PET is more expensive to purchase and maintain. PET isotopes are also relatively expensive.

9.2.2.3 SPECT/CT, PET/CT, and PET/MR

SPECT and PET machines can be integrated with CT or MR machines to enable simultaneous acquisition protocols. Such acquisitions allow registration of functional and anatomical image data. Furthermore, CT or MR images can be used to correct the nuclear images for attenuation—important for quantitative imaging applications. Today, SPECT machines are sold with an optional CT component, and PET is only sold in hybrid PET/CT or PET/MR units. PET/CT is much more common, as PET/MR remains primarily a research tool.

Dual modality imaging, utilizing PET/CT and SPECT/CT, is very common in nuclear medicine clinics. More information on hybrid imaging: PET/CT and SPECT/CT can be found in Chapter 10. Because CT scans adds clinical benefit and radiation dose to a procedure, its use must be considered with respect to optimization. CT acquisition parameters are usually specified in nuclear imaging protocols. However, CT is a unique modality and its consideration with respect to optimization is often treated separately. In this book, more information on CT optimization can be found in Chapter 8.

More detailed information on radiation detectors and camera systems can be found in dedicated references (Knoll 2011).

9.3 Optimization Principles Relating to the Patient

The concept of optimization is often considered relative to technologies or populations. However, characteristics specific to individual patients being imaged will affect dose, dose concerns, image quality, and image quality concerns for that patient.

9.3.1 Patient Body Type

Patients come in all shape and sizes. Since nuclear images are generated from the photons that reach the detectors, any attenuating material, including the patients themselves, will degrade image quality. Since photon emissions need to transvers a greater amount of attenuating material in larger patients, these patients may need greater amounts of activity, or longer scanning times, to generate equivalent images to those of slimmer patients. In some cases, for larger patients, it may not be possible to generate high-quality images.

9.3.2 Task-Specific Demands

Nuclear imaging provides functional imaging and is used to support a variety of clinical tests. For example, nuclear imaging may be used to measure cardiac or lung function, tumor detection, or to facilitate oncology treatment planning. The image quality, which relates to radioactivity levels, required by physicians to confidently perform their diagnostic interpretations varies by test. For example, a lung perfusion scan can be suitably acquired with 5 mCi of the Tc-99m isotope and provide an accurate diagnosis, whereas a bone scan may require a higher dose of 30 mCi of the Tc-99 m isotope to provide suitable images.

9.3.3 Radiation Risk

The risk associated with radiation exposure at the levels typical of nuclear imaging studies, usually 2–20 mSv, remains uncertain. However, it is well recognized that we do not have any rigorous literature that is able to correlate risk with such low doses of exposure. The current policy statement issued by the American Association of Physicists in Medicine states in part: "Risks of medical imaging at effective doses below 50 mSv for single procedures or 100 mSv for multiple procedures over short time periods are too low to be detectable and may be nonexistent (AAPM Policy Statement (2011))."

Because the risks from low doses of radiation exposure are not precisely known, it is not straightforward to define the risk/benefit for any given nuclear medicine examination. The principle of optimization requires that, in the application of radiation, all risks be minimized, while benefits maximized. Therefore, we do not need to define the exact risk in our endeavor to reduce it, and we proceed with the principle that all potential risk should be justified with benefit.

At low doses of radiation, like the doses in nuclear imaging, we expect any potential risks to result from stochastic effects. Stochastic effects refer to probabilities that individual cells will be altered in a way that will affect the system, possibly inducing cancer for example. Stochastic effects are probabilistic in nature. More details about the effects of radiation in biological systems can be found in dedicated references (Hall and Giacca 2006).

9.3.4 Demographic Risk Considerations

Because we know that it can take decades for a cancer to develop from a radiation damaged cell, we can segregate the importance of protective actions and concerns to different age groups and/or diseased populations, as they likely have different risks and benefits associated with a nuclear imaging test. For example, it is appropriate to give more concern of radiation risk to younger or middle-aged populations, than to the elderly population. Also, the concept of optimization includes consideration of both the potential risks and clinical benefits, and populations utilizing imaging for cancer-related diagnostics for example, should appropriately place a higher priority on imminent clinical necessities rather than potential long-term risks.

In nuclear medicine, consideration of radiation exposure and the subsequent risks must also be considered for a patient's family. In nuclear imaging procedures, the patient becomes a source of radiation following a procedure and may expose others around them. If a patient will be in the vicinity of radiation-sensitive children, or is pregnant, the radiation dose to the children, or the fetus must also be considered when assessing the risks. If a patient is breastfeeding, literature should be consulted to determine if the radiotracer can inadvertently be transferred to the child via breast milk—in many cases a mother can refrain from breastfeeding following an exam to ensure minimal radiation exposure to their child. Because a patient who has received a radiotracer administration for an imaging procedure will themselves become a source of radiation, patients are often advised to maintain a physical distance between themselves and at-risk populations, for example, children. It should be remembered, however, that most imaging isotopes used in nuclear imaging have been selected because of their favorable half-lives and emission characteristics, and concern of secondary radiation dose/risk is usually minimal.

9.4 Optimization Principles Relating to Imaging Protocols

Nuclear medicine procedures are multi-staged by nature and are performed in standardized ways by adhering to study protocols. The protocols are designed to provide the optimal clinical benefit but also consider other practical factors, such as patient dose, patient experience, scanner throughput, and overall efficiency of the department.

9.4.1 Tracer and Isotope Selection

Different radiotracers have been developed for nuclear imaging protocols to evaluate different body functions. The function of the radiotracer is usually dictated by the pharmacological properties of the tracer. The radiation dosimetry and imaging capacity of the tracer, however, relate to the isotope being employed. Different isotopes have different radioactive decay properties. Relevant properties include half-life, decay mode, energy of emissions, and abundance of emissions. Isotopes are also subject to constraints of availability and cost. Common isotopes used in nuclear imaging include Technetium-99m (Tc-99m) for gamma imaging and Fluorine-18 (F-18) for PET.

Over the years the field of nuclear medicine has developed different pharmacological tracer and isotope combinations that can be used favorably for clinical tests and are approved by regulating agencies. Thus, utilization of radiotracers in a clinical setting is performed according to convention relative to the study, and there is little opportunity for optimization through isotope selection. In the research and development of new radiotracers, selection of pharmaceuticals and isotopes remains relevant.

9.4.2 Administration Activity

In nuclear imaging, actual patient dose may depend on many factors. One of these factors is the administration activity—the amount of source radiation in the nuclear medicine patient system. Usually, the radiation dose to a patient closely correlates with administration activity. Efforts to reduce dose at any point in the imaging protocol often begin with a focus on administration activity.

9.4.3 Uptake Time

In most nuclear imaging procedures, there is a deliberate delay between the time when a radiotracer is administered and the time when the patient is put on the scanner for imaging. This *uptake time* allows for physiologic processing and distribution of the tracer in the body to take place and is a fundamentally important aspect of nuclear medicine. Shorter uptake times may not properly provide an accurate picture of the targeted function because the tracer may not have fully been processed yet. Longer uptake times require an increased administration activity to ensure that enough radioactive emissions are coming from the patient at the time of imaging. Specific uptake times are usually standardized within a clinic and documented in the study protocol.

9.4.4 Image Acquisition Time

The amount of time a patient being imaged by a camera will have a large impact on the amount of signal available for image formation, and thus will impact image quality. Generally speaking, the longer a patient is on the bed in front of a camera, the more counts that will be acquired, and having more counts produces better quality images. Thus, it is possible to reduce dose without affecting image quality by increasing the acquisition time. However, there are competing influences that make increasing acquisition time unfavorable. Increased bed time may add discomfort to a patient's exam; it will increase the likelihood of motion artifacts that can degrade an image, and it will slow throughput and thus increase costs for the clinic. Furthermore, many protocols are acquired with multiple bed positions, and when measured over long acquisition times, images can be degraded by effects of physiological tracer redistribution. In this respect, it is favorable for all bed positions to be captured quickly. However, protocols cannot be set up to acquire images too quickly or else the images will not have enough detected counts and be of poor diagnostic quality. Increasing the administration activity could increase signal and theoretically allow for a faster imaging time, but this would cause the patient to receive increased radiation dose, the detectors would be subject to worsened performance due to detector dead time, and worsened scatter effects.

In practice, most protocols are built to operate with a reasonable balance between the benefits and concerns of raising or lowering dose. It is common, however, for departments to adjust protocols to meet local needs and physician preferences.

9.5 Optimization Principles Relating to the Acquisition Software

Once acquisition data is collected in a nuclear imaging examination, it usually needs to be processed to generate a clinically usable image. This processing will include image corrections and calibrations, and in the case of 3D imaging, image reconstruction.

9.5.1 Image Reconstruction

As the field is striving to reduce radiation burden to patients, there is active research into developing reconstruction algorithms that can produce betting images with less detected counts. Much of the clinical dose reduction innovation in tomographic nuclear imaging in the last decade has been enabled through advancements in data processing and image generation. For example, image reconstruction has progressed from filtered back projection techniques to now common iterative ones. Data processing and image generation remains an active area of development and can support a "do more with less" strategy to improve optimization.

9.6 Current Optimization Standards

Equipment manufacturers are continually striving to provide more powerful equipment to empower lower dosing and improved image quality. Since clinics are generally working with contemporary technologies, optimization in clinical nuclear imaging procedures is primarily addressed in the design of clinical protocols, as discussed in 9.4.

Protocols are usually designed around recommendations from the leading societies in the field, namely the Society of Nuclear Medicine and Molecular Imaging procedure standards, or the European counterpart the European Society of Nuclear Medicine procedure guideline documents. Both can be accessed from their respective organization websites (www.snmmi.org and www.eanm.org, respectively). Over time, protocol recommendations can change, and accessing them through the society websites will assure review of the most up-to-date guidance documents.

9.7 Conclusion

In nuclear imaging, optimization is perpetually at the forefront of consideration, as it balances the important concerns of patient safety and clinical care. As discussed in this chapter, optimization comes from balancing many considerations. The procedures used in nuclear medicine clinics should be revisited from time to time to ensure that they are appropriately incorporating modern technology, hospital constraints, and field standards.

References

AAPM Policy Statement. 2011. *AAPM Position Statement on Radiation Risks from Medical Imaging Procedures.* https://www.aapm.org/org/policies/details.asp?id=318&type=PP.

Hall, E. J. and Giacca, A. J. 2006. *Radiobiology for the Radiologist*, 6th ed. Philadelphia, PA: Lippincott Williams & Wilkins.

Knoll, G. F. 2011. *Radiation Detection and Measurement*, 4th ed. Hoboken, NJ: John Wiley & Sons.

10

Optimization and Dose Reduction in Hybrid Imaging: PET/ CT and SPECT/CT

Adam M. Alessio
and Frederic
H. Fahey

10.1 Introduction

The tomographic modalities of single-photon emission computed tomography (SPECT), positron emission tomography (PET), and computed tomography (CT) have been applied in clinical practice for over 40 years. As tomographic methods, these modalities require views from multiple angles in order to reconstruct internal structure. Both SPECT and PET are emission modalities that rely on the source of radiation in the patient and a scanner that detects photons emitted from within the body.[1] In contrast, CT is a transmission method in which an x-ray source is external to the patient and the scanner detects photons that originate outside and traverse the body.[2] Both SPECT and PET are functional modalities, while CT is generally employed for anatomic assessment. In many functional diagnostic tasks, correlative anatomic imaging provides added value. Basically, the localization of the SPECT or PET physiologic signal to anatomic information can improve the specificity and sensitivity of the study.

Hybrid PET/CT and SPECT/CT scanners, through shared mechanical components, provide a straightforward hardware solution for the alignment of functional and anatomic images. One of the original combined devices was a CT and SPECT system developed by Lang et al. in 1992.[3] The first PET/ CT scanner was introduced in 1998 through a collaboration of the National Cancer Institute, CTI PET Systems (Knoxville, TN), and the University of Pittsburgh.[4] This scanner was constructed from independent, previously developed CT and PET scanners. The combination of independent components

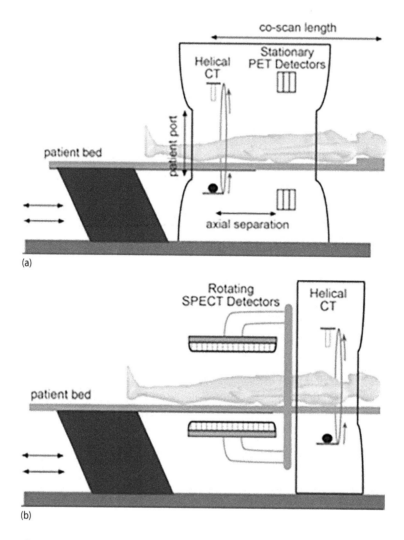

FIGURE 10.1 Illustration of the major PET/CT (a) and SPECT/CT and (b) scanner components for common designs.

remains the standard for scanner design today. Figure 10.1 illustrates the major components of common PET/CT and SPECT/CT systems. In brief, these systems offer sequential acquisition of the functional images (PET or SPECT) and anatomic images (CT).

While the primary benefit of these hybrid scanners is aligned functional and anatomic images, another key benefit is improved attenuation correction. The goal of SPECT and PET imaging is to form quantitative images of the distribution of a radiopharmaceutical in the body. Numerous sources of error interfere with this goal. The largest error for both modalities is photon attenuation, which includes any interaction that causes an emitted photon to deviate from its original linear trajectory. For both PET and SPECT, attenuation can cause substantial (1–200 times) reductions in the measured photons, and these reductions vary depending on the path through the body. Historically, stand-alone PET systems offered transmission scanning using an external radioactive source. This acquisition was relatively long (on the same order of time as the PET acquisition) and suffered from low-resolution and high noise.[5] Conventional SPECT systems offered approximate attenuation correction methods based on assumptions of the body habitus.[6,7] With hybrid systems, the CT image provides a fast, low-noise alternative, thereby significantly reducing overall scan time, noise contributions from the transmission scan in the case of PET, and errors from approximate methods in the case of SPECT.[8]

SPECT and PET attenuation correction with CT is now routinely employed, although there are several potential challenges with this approach that can lead to artifacts.[9] These challenges include: (a) conversion of attenuation maps from x-ray energies (~30–140 keV) to SPECT (energy depends on the radioisotope) and PET (511 keV) energies; (b) potential misalignment of functional and attenuation images due to respiratory or patient motion; and (c) errors in the CT image (from truncation, metallic components, photon starvation, etc.) propagating into functional images.

Along with the benefits of aligned functional plus CT images and attenuation correction, hybrid devices offer other practical advantages, including the ability to use the system for multiple purposes. For example, many clinics will purchase hybrid devices and use the CT component for conventional diagnostic CT exams for some studies and hybrid exams for others. Considering the substantial added value of the CT component, at present, all clinical PET systems are now only available as PET/CT systems.[10] SPECT/CT systems have not reached this level of prevalence in the market, although their availability has substantially increased in recent years.[11]

10.2 Dosimetry

In order to evaluate the most appropriate approaches to dose optimization for a particular imaging modality, it is essential to understand the basics of radiation dosimetry and the factors that affect this dosimetry. This section will provide discussion of radiation dosimetry as it applies to both the administration of radiopharmaceuticals for SPECT and PET and the practice of CT.

10.2.1 Dose from Radiopharmaceuticals

The dosimetry associated with the administration of radiopharmaceuticals is summarized by the basic equation developed by the Medical Internal Radiation Dosimetry (MIRD) Committee of the Society of Nuclear Medicine and Molecular Imaging.[12]

$$D(r_T) = \Sigma_S \tilde{A}(r_S) S(r_T \leftarrow r_S) \qquad (10.1)$$

In this equation, $D(r_T)$ represents the radiation dose to a particular target organ (r_T), $\tilde{A}(r_S)$ is the time-integrated activity in a selected source organ (r_S), and $S(r_T \leftarrow r_S)$ is the radionuclide-specific quantity representing the mean dose to the target organ per unit time-integrated activity present in the source organ. The symbol Σ_S indicates that the $\tilde{A}(r_S) S(r_T \leftarrow r_S)$ product is summed over all possible source organs. For example, since ^{18}F 2-fluoro-2-deoxy-D-gluocose (FDG) distributes to a number of organs, such as the brain, heart, liver, and skeletal muscle, as well as being cleared by the kidneys through the urinary bladder, all of these (and perhaps others) would be considered source organs when calculating the radiation dose from the administration of ^{18}F FDG to a selected target organ, for example, the ovaries. For any systemically administered radiopharmaceutical, some radioactivity will be deposited in all tissues of the body and all tissues are therefore source regions. The dose contributions of low-activity organ, however, are often incorporated into a "rest-of-body" source term.

The time-integrated activity, $\tilde{A}(r_S)$, depends on the amount of radioactivity administered to the patient, the fraction of that administration that went to the specific source organ, and its biokinetics, that is, its rate of uptake and clearance in the source organ. Basically, $\tilde{A}(r_S)$, with units of activity-time (e.g., MBq-hr), represents how much of the administered activity reached a particular source organ and how long it resided there. In other words, it signifies the total number of radioactive decays occurring in the source organ. In many instances, the uptake is considered to be very rapid, and the rate of clearance is assumed to be a combination of biological clearance and radioactive decay. If the biological clearance can be assumed to be exponential in nature, then the overall clearance can be characterized by the effective half-life, T_{eff}, a combination of the half-lives associated with radiation decay (T_{phys}), and biological clearance (T_{bio}).

$$T_{eff} = \frac{\left(T_{phys}\, T_{bio}\right)}{\left(T_{phys} + T_{bio}\right)} \tag{10.2}$$

Given this equation, the shorter of the two half-lives dominates the expression, that is, if T_{phys} is substantially shorter than T_{bio}, then T_{eff} is slightly shorter than T_{phys} and *vice versa*.

The S value in the MIRD equation is given by

$$S\left(r_T \leftarrow r_S\right) = \frac{\Sigma_i \Delta_i \varphi_i}{M_T} \tag{10.3}$$

In this equation, the subscript i represents the ith type of radiation emitted as the radionuclide in the radiopharmaceutical decays. For example, ^{18}F emits positrons leading to annihilation photons as well as a variety of other x-rays and Auger electrons, and all of these must be considered and summed (Σ_i) in the calculation of dose. The mean energy per nuclear decay for the ith radiation emitted by the radiopharmaceutical is given by Δ_i, which specifically depends on the radionuclide associated with the radiopharmaceutical (e.g., ^{18}F, ^{99m}Tc, or ^{123}I). Thus, the product of $\tilde{A}(r_S)$ and Δ_i is the total amount of energy emitted from the source organ via the ith radiation of the radionuclide of interest.

The fraction of energy emitted by the ith radiation from the source organ r_S that is absorbed by the target organ r_T is given by the absorbed fraction, $\varphi_i(r_T \leftarrow r_S)$. The absorbed fraction depends not only on the radionuclide of interest but also on the model of the patient including the size, shape, and orientation of the source and target organs and the spatial relationship between the two. Anatomic models exist for patients of various ages and sex.[13–15] There are also models for pregnant women at various stages of pregnancy.[16] The product of $\tilde{A}(r_S)$, Δ_i, and φ_i represents the total energy absorbed by the target organ from the ith radiation of the radionuclide, which is summed over i to yield the total absorbed energy in the target organ. Normalizing this quantity by the mass of the target organ, M_T (also model based), yields the radiation dose to the target organ from the specified source organ. As discussed above, the total dose to the target organ is determined by summing the dose from all source organs (including the rest of body) associated with the radiopharmaceutical of interest.

The radiation dose to the patient depends on the radiopharmaceutical (and its associated radionuclide) and the amount and route of administration of the radiopharmaceutical, as well as the physiology and the anatomy of the patient. It has been estimated that the uncertainty of dosimetric estimates using the MIRD formalism may be on the range of 50%–100% with most of this uncertainly being associated with the unknown physiology of an individual patient, which may vary significantly from what is assumed in commonly used patient models.[17] Within this dosimetric approach, the parameter that is most easily controlled is the activity administered to the patient, and this may be considered the dose index for the administration of radiopharmaceuticals. Other physiologic factors may also impact the actual doses, such as the patient's level of hydration, but these would be considered secondary. Thus, if one knows the amount and route of the administered activity and the size of the patient, a reasonable estimate of the radiation dose to a similar class of patients can be determined.

10.2.2 Dose from CT

In CT, the patient is exposed to x-rays emitted from an external x-ray source during the acquisition of the study. The number of x-rays emitted from the x-ray source and the duration of the scan, which both affect the radiation dose to the patient, can be controlled by adjusting several CT acquisition parameters.

The tube voltage (in kVp) controls the energy of the electrons impinging on the x-ray target material, which affects both the energy spectrum and the number of x-rays emitted. In general, the number of x-rays generated varies with roughly the square of the tube voltage. The tube current (in milliamperes, mA) determines the number of electrons that strike the target, which, in turn, directly affects the number of x-rays emitted. Since the duration of the scan is also linearly related to the number of x-rays emitted, the current (mA) and the duration (in seconds, s) are often combined and presented as the current-duration product (in mAs). In helical CT, where the patient table in translated through the imaging gantry as the x-ray tube and detector rotate about the patient, the degree to which the acquisition helix oversamples the axial z-direction of the scan is referred to as the pitch, which usually varies from about 0.5:1 (significantly oversampled) to 1.5:1 (slightly undersampled). Since the higher the pitch, the faster the patient bed will travel to image the same volume, the radiation dose is inversely proportional to the pitch. The current-duration product normalized by the pitch is referred to as the effective mAs.

$$mAs_{eff} = \frac{mAs}{pitch} \qquad (10.4)$$

Thus, the radiation dose should vary linearly as a function of effective mAs. The extent (i.e., length) of the scan and the portion of the body being scanned also affects the radiation dose to the patient. Clearly, a head and an abdominal CT will deliver different radiation doses to the patient. On the other hand, if only a small portion of the patient is scanned instead of a large axial extent, the amount of radiation delivered will be different.

The CT dose index (CTDI in mGy) refers to the radiation dose delivered by CT to defined locations within standard, cylindrical plastic phantoms (16 and 32 cm diameter for the head and the whole-body phantom, respectively). If CTDI is averaged over specified locations within the phantom and normalized by the pitch, it is referred to as $CTDI_{vol}$. The dose-length product (DLP in units of mGy-cm) is the product of the $CTDI_{vol}$ and the axial length of the CT acquisition. Values of $CTDI_{vol}$ and DLP are routinely displayed on the CT operator's console during an acquisition and are commonly used as dose indices for CT.

10.2.3 Effective Dose

Since the portion of the body that may be irradiated by a particular endeavor involving ionizing radiation may vary, it is often difficult to compare the risks of radiation doses of different exposures. For this reason, the concept of effective dose was developed by the International Commission on Radiological Protection (ICRP).[18] The effective dose (in mSv) is defined as a weighted sum of the organ doses associated with a particular exposure where each organ is weighted by its potential for a stochastic deleterious health effect resulting from the exposure. The values of these organ weights have been reevaluated several times by the ICRP, most recently in ICRP Report 103.[19] It must be kept in mind that these weights are broadly based on risk estimates across all ages and both genders, and thereby do not pertain to the risk of any individual patient. Therefore, effective dose should not be applied to a particular patient but characterizes the radiation risk across all patients. In addition, effective dose must be considered with caution when applied to a pediatric population, as the organ weights generally reflect risk applied to an adult population.[20]

10.3 CT Optimization for SPECT/CT and PET/CT

The CT acquisition during PET/CT or SPECT/CT imaging can be performed for a variety of purposes. In whole-body imaging, the CT portion can be used for:

1. Diagnosis,
2. Anatomic localization of PET or SPECT images, and/or
3. Attenuation correction of the PET or SPECT images.

The CT acquisition techniques should be tailored for their intended purpose. As a brief summary, CT imaging prescribed for diagnostic assessment (with contrast enhancement or otherwise) typically requires techniques delivering more radiation dose than the other two types of studies. If the CT scan is performed for anatomic localization of the functional image only, the acquisition technique can be adjusted to reduce patient doses substantially from those of diagnostic levels, often by 50%–80%.[21-23] Furthermore, if the CT study is only necessary for attenuation correction of the PET image, the technique can be adjusted to reduce dose even further, leading to a 10- to 100-fold reduction overall from those of diagnostic CT.[24,25] Some practitioners question this categorization of different techniques based on the argument that all three of these CT purposes are diagnostic in nature, in that they provide critical information for the clinical evaluation of a patient. In this work, we define "diagnostic CT" as an acquisition technique that may be performed on its own, without additional functional information from the PET or SPECT examination, for clinical assessment. In contrast, the localization and attenuation correction CT techniques discussed here are considered low-flux and not sufficient as a stand-alone examination for clinical evaluation.

Table 10.1 summarizes the typical ranges of techniques and dosimetry for the three categories of CT imaging for adult whole-body ^{18}F-FDG PET/CT for a scan range from the level of the eyes to mid-thigh. The absorbed dose from localization and attenuation correction examinations can be reduced compared to diagnostic-quality exams using standard methods for dose reduction in CT. In short, radiation dose from CT can be decreased by reducing the accumulated x-ray flux that contributes to images. This is commonly achieved with one or a combination of the following: (1) reducing tube current, (2) reducing tube voltage, (3) increasing pitch, (4) increasing rotation speed, and/or (5) increasing multi-slice collimation.

For the purposes of PET or SPECT attenuation correction, the CT image is used to generate a low-resolution attenuation map. Figure 10.2 presents the workflow for CT-derived attenuation correction. CT images used solely for attenuation correction can be typically substantially noisier than diagnostic-quality images because they will be smoothed to match the PET or SPECT spatial resolution prior to generation of attenuation correction factors. This smoothing is necessary and possible for two primary reasons. First, the spatial resolution of PET (4–8 mm) and SPECT (6–12 mm) is not as fine as that of CT (0.5–2 mm), mainly because the nuclear medicine modalities compared to CT (a) are imaging higher-energy photons that are more difficult to spatially position and (b) detect orders of magnitude fewer photons, resulting in lower signal, for each examination. Second, the influence of attenuation does not vary as much between different media for higher-energy photons. That is, PET and SPECT attenuation maps have less contrast resolution and a narrower range of attenuation values than CT attenuation maps. This allows for more aggressive smoothing operations without substantially biasing the attenuation correction factors.

When diagnostic CT studies are prescribed in addition to a PET or SPECT exam, the diagnostic CT region of interest may differ from the functional scan range. For example, in conventional whole-body

TABLE 10.1 Typical Acquisition Techniques Used for Whole-Body ^{18}F-FDG PET/CT Imaging

Study	Injected Activity	Effective Dose Estimate
PET[26,27]	[5–15] mCi 18F-FDG injected (185–555 MBq)	3.5–10.5 mSv
CT for diagnostic purposes[28]	[110–200] mAs[21] $CTDI_{vol} = [8-14]$ mGy	11–20 mSv
CT for anatomic localization[28]	[30–60] mAs[23] $CTDI_{vol} = [2-4]$ mGy	3–6 mSv
CT for attenuation correction only[28]	[5–10] mAs[24] $CTDI_{vol} = [0.3-1.0]$ mGy	0.5–1.0 mSv

Note: For ease of comparison, all CT studies presented are performed with 120 kVp, pitch 1.375, 40 mm collimation, 900 mm scan range, average tube current-time product.

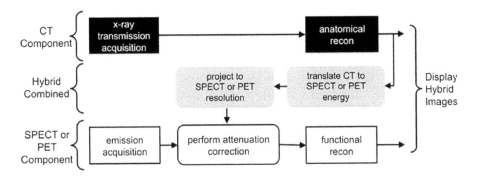

FIGURE 10.2 Workflow for standard SPECT/CT and PET/CT acquisition demonstrating the use of CT for display and attenuation correction.

FDG PET studies, the scan range is often skull base to mid-thigh, while diagnostic CT acquisitions rarely need to cover such a large range. Often the diagnostic CT study requires a limited field of view (FOV) (e.g., chest or pelvis only), which is a subset of the functional study. In these situations, the higher-dose CT study could be performed over only the diagnostic region of interest, and then a low-dose attenuation correction study could be performed over the remaining regions to provide attenuation correction throughout the functional study. Some PET/CT and SPECT/CT scanners offer this flexibility for axially varied low-high-low dose CT acquisitions to lower exposure where appropriate. This approach has recently been discussed in the context of pediatric PET/CT.[29-31]

Contrast-enhanced CT can lead to errors in attenuation correction for the functional images. This is a result of the quantitative limitations of CT and the multi-linear scaling method for deriving attenuation maps from CT images.[32] For the purpose of detection of lesions, these errors are arguably minor.[33,34] When quantitation is necessary, particularly for therapy assessment and clinical trials, correction steps are needed to account for the effects of contrast.[35] Many commercial systems offer corrections to reduce these errors, although they can lead to increased errors in other highly attenuating regions, such as bone, and at lower kVp settings.[36]

When the CT is acquired for the sole purpose of attenuation correction, the technique can be such that very low dose is delivered, as presented in Table 10.1. Often, these studies can be performed with the lowest-dose settings available on the CT system. It should be stressed that these images have a very low signal-to-noise ratio. While these images may not be visually interpretable, they provide more than sufficient information for attenuation correction of the PET and SPECT images. Individual sites should test these low-dose settings to ensure that other CT processing corrections do not fail at these settings. For example, some systems provide truncation correction of the CT image to effectively enlarge the image to match the PET FOV (i.e., the 50-cm CT FOV is enlarged to the 60–90 cm PET FOV). These truncation algorithms can fail in the case of extremely low-dose CT technique settings. Future development of PET/CT systems will lead to CT data processing methods designed specifically for low-dose attenuation correction acquisitions.[25]

10.4 Dose Optimization in Hybrid SPECT

10.4.1 Clinical Applications of SPECT/CT

Clinical SPECT/CT devices were first developed in the 1990s specifically to provide attenuation correction. These early devices were provided with a low-power, low-resolution CT that was low-cost with limited image quality. Since that time, SPECT/CT devices have been developed and marketed with a variety of CT capabilities from low-cost, low-dose units using a flat-panel plus cone-beam approach to state-of-the-art, high-quality helical CT. The use of hybrid SPECT/CT with either a low-dose or high-quality CT has demonstrated significant clinical value. This is particularly true in certain specific applications; although there are a number of routine cases where SPECT/CT may provide only incremental value. The areas where SPECT/CT has demonstrated its worth include cardiology and oncology, as well as the imaging of infection and the skeleton where the inclusion of CT-based attenuation correction and anatomical correlation has improved sensitivity in some cases but more commonly improved specificity. For certain clinical applications, SPECT/CT should be routinely applied in all cases. These include myocardial perfusion imaging and localization of parathyroid adenomas. However, in many other applications, SPECT/CT may only be useful in certain cases where anatomical correlation is considered essential or particularly helpful.

One of the initial motivations for the development of SPECT/CT was to improve non-uniform attenuation correction as it applies to myocardial perfusion imaging. Attenuation correction of thoracic SPECT has always been a challenge, and thus the potential to apply a very low-noise CT-based correction has been of substantial interest for almost 20 years. When applied accurately, CT-based attenuation correction can improve accuracy and reader confidence by, for example, reducing the cases with apparent inferior wall hypo-perfusion due to attenuation from overlapping tissue.[37] In addition, some groups have looked at combining the myocardial perfusion results with calcium scoring within the coronary arteries to yield a more complete view of the patient's cardiac health.[38]

In endocrinology, SPECT/CT has demonstrated considerable utility in the localization of parathyroid adenomas, particularly in cases where the ultrasound findings are inconclusive. Many sites utilize SPECT/CT routinely for parathyroid localization. In these cases, the inclusion of anatomical imaging has substantially improved the specificity of the results, further assisting the surgeon in knowing exactly where a potential adenoma may be located, including within the mediastinum.[39] With respect to thyroid disease, SPECT/CT has been shown to be more accurate in distinguishing between benign and malignant tissue for both diagnosis and staging of differentiated thyroid cancer.[40]

SPECT/CT has also shown promise in musculoskeletal imaging, although the evidence is limited. With respect to evaluating known or suspected malignant disease, SPECT/CT demonstrated improved diagnostic confidence and the ability to identify additional lesions, including metastases that may affect staging and, thereby, treatment. Although SPECT/CT has some potential for improved specificity for the identification of benign skeletal lesions, the evidence is still lacking.[41] In particular, there have been some promising reports regarding the assessment of infection within the bone and the evaluation of the diabetic foot, an area that continues to be challenging.

There may be a number of other applications where SPECT/CT may be of clinical value on a case-by-case basis. SPECT/CT has also shown promise with respect to lung imaging. Specifically, combining CT with tomographic nuclear ventilation and perfusion (V/Q) imaging may be more accurate than CT angiography or V/Q scans performed separately.[42] As in other applications, the direct combination of anatomy and function has the potential for identifying features essential to the diagnosis of pulmonary embolism. SPECT/CT may also add value in certain cases when using a variety of tumor imaging agents. There have been some reports that SPECT/CT can be of very useful in the clinical evaluation of neuroblastoma or pheochromocytoma using ^{123}I metaiodobenzylguanidine (MIBG). SPECT/CT has also been used to image ^{111}In-labeled radiopharmaceuticals, such as prostiscint of octeotide.

10.4.2 Dose Optimization

SPECT/CT is used for a wide variety of clinical applications, as discussed in the previous section. As in all nuclear medicine procedures, the acquisition parameters must be tailored to address the particular clinical question at hand. On the SPECT side, the proper radiopharmaceutical and the administered activity must be considered. As described above, the administered activity is the primary factor in the determining the radiation dose to the patient in the use of radiopharmaceuticals. Thus, this activity should be optimized to provide the least radiation dose that will yield a study of proper diagnostic quality. Professional medical societies such as the Society of Nuclear Medicine and Molecular Imaging (SNMMI), the European Association of Nuclear Medicine, and the American College of Radiology have published practice guidelines for many of the procedures where SPECT/CT may be applied, and, as part of these guidelines, a range of recommended radiopharmaceutical administered activities. However, each site must determine the proper administered activity for their patients based on their specific patient population, the equipment at their clinic, and the manner in which they acquire and interpret the studies. For example, for a certain procedure, a site might routinely image larger patients using older equipment using a specific collimator. The site may also decide that the first phase of the study must be acquired in dynamic rather than static mode with a particular framing regimen. All of these factors should be considered when determining the proper administered activity, and the activities used at a particular site may therefore differ from those published practice guidelines. Of course, none of this is unique to SPECT/CT, and such considerations should be applied across all nuclear medicine procedures. On the other hand, if SPECT/CT is routinely applied for a particular clinical application or even if it is only applied in specific cases, the clinic may need to modify their acquisition parameters to meet the needs of this technology.

For the CT portion, all of the acquisition parameters that affect radiation dose as discussed previously must be considered, including tube voltage, current, duration, collimation and pitch. Again, the variety of types of studies described above, including imaging of the skeleton, lungs, head and neck and abdomen suggests that a "one size fits all" approach is most likely not appropriate. The acquisition parameters should be tailored to meet the specific needs of each procedure. For example, for a particular application the CT may be used primarily for attenuation correction and a very limited study for anatomical correlation may be adequate. In this case, it may be possible to reduce the tube voltage and current considerably as well as increasing the pitch. Such modifications could lead to a substantial reduction in radiation dose to the patient. On the other hand, a near-diagnostic-quality study may be required in other cases, leading to a higher radiation dose but subsequent greater benefit to the patient. The extent of the CT scan and the region of the body being scanned must also be taken into account when considering the radiation dose. In SPECT/CT, the axial field of view is often limited to a specific region of diagnostic interest. Limitation of the FOV to the proper region can lead to a considerable reduction in radiation dose to the patient.

10.5 Dose Optimization in Hybrid PET

10.5.1 Clinical Applications of PET/CT

A key benefit of PET imaging over SPECT imaging is that PET relies on positron-emitting radioisotopes. Several of these isotopes (^{11}C, ^{13}N, ^{15}O) have analogues in nearly all biologic molecules, and therefore the labeling with these isotopes does not change the natural structure and biological behavior of the molecules. PET radioisotopes have found applications for many conditions and are commonly used in oncology, neurology, and cardiology.

While not the original motivation for PET imaging, the most common application by far is ^{18}F-FDG imaging for detecting and staging cancer and metastatic disease.[43–46] As a brief summary, FDG PET is commonly used in the assessment of patients with recurrent or residual disease, especially colorectal cancer and lymphoma. FDG PET is also used for staging and restaging of patients with advanced melanoma and single pulmonary nodules. Likewise, FDG PET has proven value in preoperative staging of non-small-cell lung cancer.[47] Emerging applications of PET in oncology include tumor grading and evaluation of tumor response to therapy.[48] These later applications are moving PET beyond binary detection tasks and into quantitative tasks that rely on accurate and precise assessment of tracer uptake.[49]

Neurologic PET imaging is commonly used to assess neuropsychiatric diseases, including seizures, brain tumors, movement disorders, and dementia.[50–52] Alzheimer's disease is the most prevalent cause of dementia, and PET can elucidate mechanisms of this disorder by evaluating metabolic and beta-amyloid plaque abnormalities.[53–56] Along with common clinical applications, PET has had a long history of imaging a multitude of physiologic processes in the brain.[57–59]

PET has an established role for imaging cardiovascular physiology and pathology. In this domain, PET is most commonly used to assess myocardial perfusion using ^{82}Rb. PET offers the added value over cardiac SPECT in being able to quantify myocardial blood flow under conditions of rest and pharmacologic stress.[60,61] This quantification has become fairly widespread, with advances in dynamic data processing software offering reproducible estimates of coronary flow reserve.[62,63] There are a plethora of other cardiovascular applications including evaluation of myocardial metabolism,[64,65] cardiac innervation,[66,67] and atherosclerotic plaque evaluation.[68]

10.5.2 Dose Optimization

As in SPECT/CT, an optimal PET/CT imaging protocol would result in images with sufficient diagnostic quality at a minimum of risk. For PET/CT, images often must be of sufficient image quality to perform a variety of tasks (detection, staging, and quantitative monitoring), not just a single binary detection task of a single disease. Images that can perform all of these tasks, without compromising clinically relevant information, have full diagnostic utility. In general, increased scan duration and injected activity leads to increased, favorable count levels, although it is very challenging to determine the "image quality" of an acquisition protocol and therefore challenging to determine if it provides sufficient image quality.[69] On the other hand, there are risks associated with the scan duration and injected activity. Specifically, longer scan durations are associated with issues such as motion artifacts, patient discomfort, and complications from potential sedation. Likewise, increased injected activity is linked to a potentially increased risk of radiation-induced cancer.

Several groups have presented recommendations for PET protocols and injected activities.[23,70–72] Considering it is essentially impossible to predict if sufficient diagnostic quality will be achieved prior to an examination, most centers determine this quality based on experience and local preference. Once the criteria for sufficient quality are accepted, the most important factors to determine appropriate injected activity include: (1) acquisition duration, (2) patient size, and (3) scanner sensitivity.

A common method to assess the quality of PET data and scanners is to calculate the noise equivalent count (NEC) density. Studies have shown that increased NEC density is closely related to improved lesion detection.[73] The NEC is computed from the true (T), random (R), and scattered (S) event counts as

$$NEC = \frac{T^2}{T + S + R}$$

In brief, in the absence of count-rate limitations, the true and scattered count rates will increase linearly with injected activity, and the random coincidence rate will increase with the square of injected activity. In contrast to SPECT imaging, where increased injected activity leads to improved imaging quality, in PET imaging the rapid increase in random counts limits improvements with increased injected activity.

The NEC levels will increase linearly with acquisition duration, and this supports the reality that the longer the acquisition, the better the quality. The only *caveat* to this logic is that long acquisition durations run the risk of increased patient motion. Likewise, long acquisition durations will eventually lead to decreased activity because of the physical decay of the PET tracer. In common PET imaging with ^{18}F, with a 110-minute half-life, and with total acquisition durations on the order of 10–35 minutes, this is not a significant concern.

Patient size should be taken into account when deciding on the injected activity. A dominant factor of PET image quality is attenuation; increasing patient thickness exponentially decreases the number of photons detected. Consequently, NEC density decreases exponentially with increasing patient size. Studies have shown that there is no way to normalize total NEC across all patient sizes because the

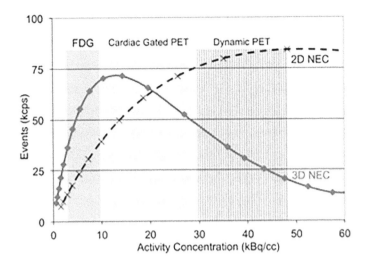

FIGURE 10.3 Comparison of 2D (dashed) and 3D (solid) noise equivalent count (NEC) curves versus activity concentration in phantom for a single scanner. Typical dosing levels during static FDG imaging, cardiac gated PET imaging, and dynamic PET imaging are highlighted to demonstrate that the optimal mode (2D vs 3D) is dependent on the activity concentration and that increased activity does not equate to increased quality.

injected activity would need to be excessively high for larger patients.[74,75] Despite this, patient size should be taken into account in an attempt to improve image quality for larger patients.[72,76]

Scanner sensitivity plays an important role in image quality and therefore optimal dosing protocols. Figure 10.3 presents the NEC curves for 2D and fully 3D acquisition modes for a single scanner. Fully 3D mode, unlike 2D mode, is performed without septa separating detector rows and provides as much as an eightfold increase in sensitivity compared to 2D mode. For common static FDG imaging acquired 60–120 minutes post-injection, 3D mode provides higher NEC levels and is widely accepted as the preferred acquisition mode for most diagnostic FDG tasks.[71,77] Dynamic acquisitions often have very high activity levels in the field of view because they typically image immediately after injection of the radiotracer. For these acquisitions, 2D mode may be advantageous because its septa limit the total count levels resulting in less deadtime compared to 3D mode. However, because of the dominance of static FDG imaging, most commercial scanners only support 3D acquisitions. Along with acquisition mode, different scanners have varying levels of effective sensitivity. Conventional PET system designs offer an axial FOV of 15–20 cm. Some systems offer longer axial FOVs providing substantial gains in NEC levels and the potential for significantly reduced injected activity.[78,79]

10.5.3 PET/MR

Hybrid PET/Magnetic Resonance Imaging (MR) devices have recently been developed and brought to market. Simultaneous systems provide the opportunity to acquire both PET and MR images at the same time,[80] yielding the functional information from PET and soft-tissue detail from MR. The devices offer the opportunity to interrogate new physiologic questions providing a valuable research tool and a potentially valuable clinical tool.[81,82]

For dose optimization, PET/MR systems offer the added value of providing attenuation correction from the MR images without the need for an x-ray-based CT acquisition. While the generation of attenuation correction factors is more challenging from MR images than CT images, there are numerous viable methods for MR-based attenuation correction, and they are generally accepted as sufficient for most PET applications.[83–85] Attenuation correction from MR images will spare the patient from the added radiation dose from the CT.[86] This savings should be put in context of the overall dose from the examination. As presented in Table 10.1, an attenuation correction-only CT acquisition imparts a radiation dose of only 0.5–1 mSv. Considering the FDG-PET acquisition imparts 7–14 mSv, the reduction of 1 mSv from the overall exam is less than a 12% dose savings.

10.5.4 Pediatric Considerations

Nuclear medicine including SPECT and PET has been shown to be of considerable value in the pediatric population.[87] SPECT has been used in children in a wide variety of applications including the fields of oncology, neurology, urology, endocrinology, and orthopedics. Although the availability of SPECT/CT for most pediatric applications is limited, many of the applications being considered for the adult population are adaptable to children. These include the use of SPECT/CT for skeletal imaging, localization of parathyroid adenomas, evaluation of thyroid carcinoma, and tumor imaging with ^{123}I MIBG. It is expected that the application of SPECT/CT to pediatrics will continue to grow and its proper use to be better defined in the years to come.

PET/CT use in children has been well established over the past 10 years. This is particularly true in pediatric oncology, where the combination of anatomical CT imaging combined with the metabolic imaging provided by ^{18}F FDG is invaluable. It is now considered the standard of care, for example, to follow children with lymphoma using FDG PET/CT. The same is true for a variety of other childhood cancers, including rabdomyosarcoma, osteosarcoma, and neuroblastoma. FDG PET/CT has also been found to be very useful in children with epilepsy to assist in the localization of seizure foci.[88] Some centers have also found skeletal PET/CT with ^{18}F sodium fluoride (NaF) to be very useful. For example, NaF PET has been found to be very helpful in the evaluation of infants and very small children who are suspected to be the victims of child abuse.[89] The greater sensitivity and excellent spatial resolution have proven quite beneficial in these very challenging patients.

Children are considered to be more sensitive to the risks of ionizing radiation than adults. According to the risk estimates provided from the report of the U.S. National Academy of Sciences on the Biological Effects of Ionizing Radiation (BEIR VII Phase 2 Report), children may be at two- to threefold higher risk of radiation-induced cancer than adults, and young girls may be 30%–40% more sensitive that young boys.[90] For these reasons, nuclear medicine practitioners should be particularly prudent when imaging these most sensitive of patients.

On the other hand, small children and, in some cases, even adolescents can be among the most challenging patients to image. SPECT/CT and PET/CT studies typically take tens of minutes to complete, and, for the best results, the patient must lie still for the duration of the study. For smaller children, this may require sedation or, in some cases, the use of general anesthesia. Even in older children, some coaching is typically necessary to get the patient to lie still for 20–30 minutes. Therefore, technical improvements in hardware or image processing that have been applied to make SPECT/CT or PET/CT more efficient may lead to either lower administered activity or a faster scan time. A balance must be struck between dose optimization and minimizing patient motion or the need for sedation or anesthesia.

For SPECT, dual-detector gamma camera systems are recommended for use with children as they provide good sensitivity and thus acceptable image quality in a reasonable imaging time. For imaging ^{99m}Tc, there is typically a choice of parallel-hole collimation ranging from ultra-high-resolution to high-sensitivity that can be considered. In general, it has been shown that the collimation with the better spatial resolution (high- or ultra-high-resolution) will provide the best image quality for the same imaging time, even if it provides fewer counts.[91,92] This is partly due to the fact that the high-resolution collimators maintain their spatial resolution at the greater distances typically encountered over the course of a SPECT acquisition. Focused collimation may also be considered, since it can provide substantially higher sensitivity (50%–300%) as compared to parallel-hole collimation with the same spatial resolution. In the past decade, a number of dedicated SPECT units for cardiovascular imaging have been developed. These systems tend to utilize multiple detectors that are specifically focused on the heart and thereby provide high sensitivity. Several of these systems have options to incorporate CT, either solely for attenuation correction or with the capability of diagnostic-quality scanning. Although these systems were specifically designed to improve throughput (i.e., a scan could be completed in 5 rather than 20 minutes), the higher sensitivity could also be used to perform the study in a reasonable amount of time (e.g., 10 minutes) with a lower administered activity. Even though these devices have been specifically designed for cardiovascular SPECT, they could potentially also be used for imaging small children.

With respect to PET, the consideration for selecting a system that will be primarily used for imaging children may be different than one for adult imaging. The introduction of time-of-flight technology has led to a notable improvement in PET image quality, particularly in large patients. However, this advance may be of limited use with small children, where the image-quality improvements are considerably less. On the other hand, the extension of the axial FOV from the standard 15–16 cm to 20 or even 25 cm leads to a considerable increase in sensitivity that can be particularly valuable when imaging children in regards to both the potential for lower administered activity and faster scan times. A smaller ring diameter (e.g., 80 rather than 100 cm) will also lead to greater sensitivity. Although the smaller ring diameter may lead to the loss of spatial resolution on the periphery of larger patients due to the depth-of-interaction effect, this may be of less significance in smaller patients.

Reconstruction algorithms for SPECT and PET have improved considerably over the past 20 years. The largest improvement has been the shift from using filtered back-projection to iterative approaches and, more recently, to the inclusion of imaging characteristics of the device directly into the reconstruction process. For SPECT, this has involved the incorporation of the collimator response as a function of distance specifically for the particular collimator being used into the reconstruction algorithm. This approach is often referred to as resolution recovery. A similar approach has also been implemented for PET that takes into account, for instance, the depth-of-interaction effect that leads to a loss of resolution towards the periphery of the field of view. These enhancements to the reconstruction algorithm can lead to considerable improvements in image quality, which may allow for a reduction in the administered activity while still providing an appropriate level of diagnostic information to the clinician. Several studies have demonstrated this to be the case in pediatric SPECT.[93] The high-resolution anatomical CT image could also be used to constrain the reconstruction of the functional image through the incorporation of information shared between the two modalities, such as the location of edges. This continues to be an area of active research, but the challenge is to ensure that the resulting reconstructed function image is not overly biased by the anatomical data.

With respect to the CT component of hybrid imaging, one must consider how the CT might be used for each specific imaging protocol and, in some cases, with regards to a particular patient. The CT technique a particular study should be selected to address the specific clinical question at hand. For example, the inclusion of diagnostic quality CT in a PET/CT study of child with lymphoma may be appropriate whereas the CT may only be used for attenuation correction for brain imaging since the anatomical modality of choice is MR rather than CT. Therefore, the acquisition parameters for the latter scan can be adjusted to yield a considerably reduced radiation dose to the patient.[23,24] Even in the cases where diagnostic quality CT is considered to be essential, it may make sense to limit its application only to the region of the body where it will be most useful and to use dose-reducing acquisition parameters in the other portions of the body. SPECT/CT may be used in addition to the initial SPECT study in cases where anatomical correlation is deemed to be particularly useful. In these cases, limiting the application of CT only to those regions that are most pertinent will lead to a considerable reduction in the radiation dose to the patient. If CT-based attenuation correction is known to improve the SPECT data quality considerably, but there is no real need for anatomical correlation, then a very low-dose CT protocol can be used.

When imaging children with CT, in the context of hybrid imaging or otherwise, the acquisition parameters must be adjusted based on the size of the patient. For the same CT acquisition parameters, a child can receive two to three times the radiation dose as an adult due to the reduced attenuation of the x-ray beam in the smaller patient. Schemes have been developed for adjusting the acquisition parameters for pediatric CT in general and also for PET/CT[23,94] and SPECT/CT.

PET/MR scanners have been recently introduced into the clinic, as described above, leading to a reduction in radiation dose. Several clinical applications of both PET and SPECT utilize MR rather than CT as the structural modality of choice (e.g., brain imaging and neuroblastoma). Another factor to consider for children who require sedation or anesthesia is that PET/MR allows both modalities to be acquired in a single session. This is particularly true for the PET/MR systems that acquire both studies simultaneously. There is also the potential for utilizing the dynamic MR signal to provide on-the-fly motion correction that can be directly incorporated into the PET reconstruction algorithm. This may be particularly pertinent for pediatric imaging, possibly allowing light sedation rather than general anesthesia for smaller patients. For all these reasons, PET/MR may have a considerable impact on pediatric imaging in the near future. However, the issues regarding MR safety are not to be taken lightly, and the ability to perform PET/CT in a considerably shorter time may be most appropriate for specific hybrid imaging protocols and in certain children.

10.6 Conclusions

Hybrid imaging with PET/CT or SPECT/CT scanners has been shown to substantially enhance the diagnostic information provided for a variety of clinical applications, including oncology, neurology, cardiology, and musculoskeletal imaging. With these imaging devices, the patient is exposed to ionizing radiation from both the administered radiopharmaceutical and the CT component. Therefore, it is essential that the radiation dosimetry of both of these sources be well understood in order to optimize the procedure to utilize the least amount of radiation dose necessary while providing the diagnostic information to the clinician essential for the patient's care. Each procedure should be considered separately taking into account the institutions specific patient population, the available instrumentation, and the clinical task at hand. Pediatric patients require special consideration, as they are considered to be at higher risk from exposure to radiation than adults. However, lower radiation dose and faster imaging times must be balanced when imaging children. The introduction of PET/MR hybrid imaging to the clinic may have a considerable impact with respect to dose optimization, particularly in children.

References

1. Cherry S, Sorensen J, Phelps M. *Physics in Nuclear Medicine*. Philadelphia, PA: W. B. Saunders Company, 2003.
2. Hsieh J. *Computed Tomography*. Bellingham, WA: SPIE, 2009.
3. Lang TF, Hasegawa BH, Liew SC, Brown JK, Blankespoor SC, Reilly SM, Gingold EL, Cann CE. Description of a prototype emission-transmission computed tomography imaging system. *J Nucl Med* 1992; 33(10), 1881–1887.
4. Beyer T, Townsend DW, Brun T, Kinahan PE, Charron M, Roddy R, Jerin J, Young J, Byars L, Nutt R. A combined PET/CT scanner for clinical oncology. *J Nucl Med* 2000; 41(8), 1369–1379.
5. Kinahan PE, Townsend DW, Beyer T, Sashin D. Attenuation correction for a combined 3D PET/CT scanner. *Med Phys* 1998; 25(10), 2046–2053.
6. Chang L-T. A method for attenuation correction in radionuclide computed tomography. *Nucl Sci, IEEE Trans On* 1978; 25(1), 638–643.
7. King MA, Tsui BM, Pan TS, Glick SJ, Soares EJ. Attenuation compensation for cardiac single-photon emission computed tomographic imaging: Part 2. Attenuation compensation algorithms. *J Nucl Cardiol* 1996; 3(1), 55–64.
8. Kinahan PE, Hasegawa BH, Beyer T. X-ray-based attenuation correction for positron emission tomography/computed tomography scanners. *Semin Nucl Med* 2003; 33(3), 166–179.
9. Alessio AM, Kinahan PE, Cheng PM, Vesselle H, Karp JS. PET/CT scanner instrumentation, challenges, and solutions. *Radiol Clin North Am* 2004; 42(6), 1017–1032, vii.

10. von Schulthess GK, Steinert HC, Hany TF. Integrated PET/CT: Current applications and future directions. *Radiology* 2006; 238(2), 405–422.

11. Buck AK, Nekolla S, Ziegler S et al. SPECT/CT. *J Nucl Med* 2008; 49(8), 1305–1319.

12. Bolch WE, Eckerman KF, Sgouros G, Thomas SR. MIRD pamphlet No. 21: A generalized schema for radiopharmaceutical dosimetry—Standardization of nomenclature. *J Nucl Med* 2009; 50(3), 477–484.

13. Cristy M, Eckerman KF. *Specific Absorbed Fractions of Energy at Various Ages.* ORNL/TM-8381. Oak Ridge, TN: Oak Ridge National Laboratories, 1987.

14. Whalen S, Lee C, Williams JL, Bolch WE. Anthropometric approaches and their uncertainties to assigning computational phantoms to individual patients in pediatric dosimetry studies. *Phys Med Biol* 2008; 53(2), 453–471.

15. Xu XG, Eckerman KF. *Handbook of Anatomical Models for Radiation Dosimetry.* Boca Raton, FL: CRC Press, 2009.

16. Maynard MR, Long NS, Moawad NS, Shifrin RY, Geyer AM, Fong G, Bolch WE. The UF Family of hybrid phantoms of the pregnant female for computational radiation dosimetry. *Phys Med Biol* 2014; 59(15), 4325–4343.

17. Stabin MG. Uncertainties in internal dose calculations for radiopharmaceuticals. *J Nucl Med* 2008; 49(5), 853–860.

18. ICRP. *International Commission on Radiological Protection: ICRP Publication 26.* New York: Pergamon Press, 1977.

19. ICRP. *International Commission on Radiological Protection: ICRP Publication 103.* New York: Pergamon Press, 2007.

20. Fisher D, Fahey F. Appropriate use of effective dose in radiation protection and risk Assessment. *Health Phys* 2017; 113(2), 102–109.

21. Brix G, Lechel U, Glatting G, Ziegler SI, Munzing W, Muller SP, Beyer T. Radiation exposure of patients undergoing whole-body dual-modality 18F-FDG PET/CT examinations. *J Nucl Med* 2005; 46(4), 608–613.

22. Gelfand MJ, Lemen LC. PET/CT and SPECT/CT dosimetry in children: The challenge to the pediatric imager. *Semin Nucl Med* 2007; 37(5), 391–398.

23. Alessio AM, Kinahan PE, Manchanda V, Ghioni V, Aldape L, Parisi MT. Weight-based, low-dose pediatric whole-body PET/CT protocols. *J Nucl Med* 2009; 50(10), 1570–1577.

24. Fahey FH, Palmer MR, Strauss KJ, Zimmerman RE, Badawi RD, Treves ST. Dosimetry and adequacy of CT-based attenuation correction for pediatric PET: Phantom study. *Radiology* 2007; 243(1), 96–104.

25. Xia T, Alessio AM, De Man B, Manjeshwar R, Asma E, Kinahan PE. Ultra-low dose CT attenuation correction for PET/CT. *Phys Med Biol* 2012; 57(2), 309–328.

26. ICRP. Radiation dose to patients from radiopharmaceuticals: (Addendum 2 to ICRP Publication 53) ICRP Publication 80 Approved by the Commission in September 1997. *Ann ICRP* 1998; 28(3).

27. ICRP. Radiation dose to patients from radiopharmaceuticals: (Addendum 3 to ICRP Publication 53) ICRP Publication 106. *Ann ICRP* 2008; 38(1–2).

28. ImPACT Group. *CT Dosimetry Tool.* London, UK: ImPACT, St. George's Healthcare NHS Trust, 2007.

29. Fahey FH, Goodkind A, MacDougall RD, Oberg L, Ziniel SI, Cappock R, Callahan MJ, Kwatra N, Treves ST, Voss SD. Operational and dosimetric aspects of pediatric PET/CT. *J Nucl Med* 2017; 58(9), 1360–1366.

30. Sureshbabu W, Mawlawi O. PET/CT imaging artifacts. *J Nucl Med Technol* 2005; 33(3), 156–161; quiz 163–164.

31. Lodge MA, Mhlanga JC, Cho SY, Wahl RL. Effect of patient arm motion in whole-body PET/CT. *J Nucl Med* 2011; 52(12), 1891–1897.

32. Kinahan PE, Hasegawa BH, Beyer T. X-ray-based attenuation correction for positron emission tomography/computed tomography scanners. *Semin Nucl Med* 2003; 33(3), 166–179.

33. Berthelsen AK, Holm S, Loft A, Klausen TL, Andersen F, Hojgaard L. PET/CT with intravenous contrast can be used for PET attenuation correction in cancer patients. *Eur J Nucl Med Mol Imaging* 2005; 32(10), 1167–1175.

34. Mawlawi O, Erasmus JJ, Munden RF, Pan T, Knight AE, Macapinlac HA, Podoloff DA, Chasen M. Quantifying the effect of IV contrast media on integrated PET/CT: Clinical evaluation. *AJR Am J Roentgenol* 2006; 186(2), 308–319.

35. Visvikis D, Costa DC, Croasdale I, Lonn AH, Bomanji J, Gacinovic S, Ell PJ. CT-based attenuation correction in the calculation of semi-quantitative indices of [18F]FDG uptake in PET. *Eur J Nucl Med Mol Imaging* 2003; 30(3), 344–353.

36. Abella M, Alessio AM, Mankoff DA, Macdonald LR, Vaquero JJ, Desco M, Kinahan PE. Accuracy of CT-based attenuation correction in PET/CT bone imaging. *Phys Med Biol* 2012; 57(9), 2477–2490.

37. Garcia EV. SPECT attenuation correction: An essential tool to realize nuclear cardiology's manifest destiny. *J Nucl Cardiol* 2007; 14(1), 16–24.

38. Schaap J, de Groot JA, Nieman K et al. Added value of hybrid myocardial perfusion SPECT and CT coronary angiography in the diagnosis of coronary artery disease. *Eur Heart J Cardiovasc Imaging* 2014; 15(11), 1281–1288.

39. Wong KK, Fig LM, Gross MD, Dwamena BA. Parathyroid adenoma localization with 99mTc-sestamibi SPECT/CT: A meta-analysis. *Nucl Med Commun* 2015; 36(4), 363–375.

40. Chen L, Luo Q, Shen Y, Yu Y, Yuan Z, Lu H, Zhu R. Incremental value of 131I SPECT/CT in the management of patients with differentiated thyroid carcinoma. *J Nucl Med* 2008; 49(12), 1952–1957.

41. Utsunomiya D, Shiraishi S, Imuta M, Tomiguchi S, Kawanaka K, Morishita S, Awai K, Yamashita Y. Added value of SPECT/CT fusion in assessing suspected bone metastasis: Comparison with scintigraphy alone and nonfused scintigraphy and CT. *Radiology* 2006; 238(1), 264–271.

42. Toney LK, Wanner M, Miyaoka RS, Alessio AM, Wood DE, Vesselle H. Improved prediction of lobar perfusion contribution using technetium-99m-labeled macroaggregate of albumin single photon emission computed tomography/computed tomography with attenuation correction. *J Thorac Cardiovasc Surg* 2014; 148(5), 2345–2352.

43. Bomanji JB, Costa DC, Ell PJ. Clinical role of positron emission tomography in oncology. *Lancet Oncol* 2001; 2(3), 157–164.

44. Delbeke D, Martin WH. Positron emission tomography imaging in oncology. *Radiol Clin North Am* 2001; 39(5), 883–917.

45. Hustinx R, Benard F, Alavi A. Whole-body FDG-PET imaging in the management of patients with cancer. *Semin Nucl Med* 2002; 32(1), 35–46.

46. Weber WA, Schwaiger M, Avril N. Quantitative assessment of tumor metabolism using FDG-PET imaging. *Nucl Med Biol* 2000; 27(7), 683–687.

47. Vansteenkiste J, Fischer BM, Dooms C, Mortensen J. Positron-emission tomography in prognostic and therapeutic assessment of lung cancer: Systematic review. *Lancet Oncol* 2004; 5(9), 531–540.

48. Young H, Baum R, Cremerius U, Herholz K, Hoekstra O, Lammertsma AA, Pruim J, Price P. Measurement of clinical and subclinical tumour response using [18F]-fluorodeoxyglucose and positron emission tomography: Review and 1999 EORTC recommendations. European Organization for Research and Treatment of Cancer (EORTC) PET Study Group. *Eur J Cancer* 1999; 35(13), 1773–1782.

49. Westerterp M, Pruim J, Oyen W, Hoekstra O, Paans A, Visser E, van Lanschot J, Sloof G, Boellaard R. Quantification of FDG PET studies using standardised uptake values in multi-centre trials: Effects of image reconstruction, resolution and ROI definition parameters. *Eur J Nucl Med Mol Imaging* 2007; 34(3), 392–404.

50. Assessment: Positron emission tomography. Report of the therapeutics and technology assessment subcommittee of the American Academy of Neurology. *Neurology* 1991; 41(2 (Pt 1), 163–167.

51. Herholz K, Salmon E, Perani D et al. Discrimination between Alzheimer dementia and controls by automated analysis of multicenter FDG PET. *Neuroimage* 2002; 17(1), 302–316.

52. Hoffman JM, Welsh-Bohmer KA, Hanson M, Crain B, Hulette C, Earl N, Coleman RE. FDG PET imaging in patients with pathologically verified dementia. *J Nucl Med* 2000; 41(11), 1920–1928.

53. Buckner RL, Snyder AZ, Shannon BJ et al. Molecular, structural, and functional characterization of Alzheimer's disease: Evidence for a relationship between default activity, amyloid, and memory. *J Neurosci* 2005; 25(34), 7709–7717.

54. Mosconi L, Tsui WH, Herholz K et al. Multicenter standardized 18F-FDG PET diagnosis of mild cognitive impairment, Alzheimer's disease, and other dementias. *J Nucl Med* 2008; 49(3), 390–398.

55. Rowe CC, Ackerman U, Browne W et al. Imaging of amyloid beta in Alzheimer's disease with 18F-BAY94-9172, a novel PET tracer: Proof of mechanism. *Lancet Neurol* 2008; 7(2), 129–135.

56. Verhoeff NP, Wilson AA, Takeshita S, Trop L, Hussey D, Singh K, Kung HF, Kung MP, Houle S. In-vivo imaging of Alzheimer disease beta-amyloid with [11C]SB-13 PET. *Am J Geriatr Psychiatry* 2004; 12(6), 584–595.

57. Wagner HNJ, Burns HD, Dannals RF et al. Imaging dopamine receptors in the human brain by positron tomography. *Science* 1983; 221(4617), 1264–1266.

58. Duara R, Grady C, Haxby J, Ingvar D, Sokoloff L, Margolin RA, Manning RG, Cutler NR, Rapoport SI. Human brain glucose utilization and cognitive function in relation to age. *Ann Neurol* 1984; 16(6), 703–713.

59. Hubner KF, Purvis JT, Mahaley SMJ, Robertson JT, Rogers S, Gibbs WD, King P, Partain CL. Brain tumor imaging by positron emission computed tomography using 11C-labeled amino acids. *J Comput Assist Tomogr* 1982; 6(3), 544–550.

60. Beller GA, Bergmann SR. Myocardial perfusion imaging agents: SPECT and PET. *J Nucl Cardiol* 2004; 11(1), 71–86.

61. Johnson NP, Gould KL. Integrating noninvasive absolute flow, coronary flow reserve, and ischemic thresholds into a comprehensive map of physiological severity. *JACC Cardiovasc Imaging* 2012; 5(4), 430–440.

62. Nesterov SV, Deshayes E, Sciagra R et al. Quantification of myocardial blood flow in absolute terms using (82)Rb PET imaging: The RUBY-10 study. *JACC Cardiovasc Imaging* 2014; 7(11), 1119–1127.

63. Lortie M, Beanlands RS, Yoshinaga K, Klein R, Dasilva JN, DeKemp RA. Quantification of myocardial blood flow with 82Rb dynamic PET imaging. *Eur J Nucl Med Mol Imaging* 2007; 34(11), 1765–1774.

64. Brown M, Marshall DR, Sobel BE, Bergmann SR. Delineation of myocardial oxygen utilization with carbon-11-labeled acetate. *Circulation* 1987; 76(3), 687–696.

65. Lerch RA, Bergmann SR, Ambos HD, Welch MJ, Ter-Pogossian MM, Sobel BE. Effect of flow-independent reduction of metabolism on regional myocardial clearance of 11C-palmitate. *Circulation* 1982; 65(4), 731–738.

66. Caldwell JH, Link JM, Levy WC, Poole JE, Stratton JR. Evidence for pre- to postsynaptic mismatch of the cardiac sympathetic nervous system in ischemic congestive heart failure. *J Nucl Med* 2008; 49(2), 234–241.

67. Bengel FM, Schwaiger M. Assessment of cardiac sympathetic neuronal function using PET imaging. *J Nucl Cardiol* 2004; 11(5), 603–616.

68. Rudd JH, Warburton EA, Fryer TD et al. Imaging atherosclerotic plaque inflammation with [18F]-fluorodeoxyglucose positron emission tomography. *Circulation* 2002; 105(23), 2708–2711.

69. Badawi RD, Dahlbom M. NEC: Some coincidences are more equivalent than others. *J Nucl Med* 2005; 46(11), 1767–1768.

70. Halpern BS, Dahlbom M, Quon A, Schiepers C, Waldherr C, Silverman DH, Ratib O, Czernin J. Impact of patient weight and emission scan duration on PET/CT image quality and lesion detectability. *J Nucl Med* 2004; 45(5), 797–801.

71. Lartizien C, Comtat C, Kinahan PE, Ferreira N, Bendriem B, Trebossen R. Optimization of injected dose based on noise equivalent count rates for 2- and 3-dimensional whole-body PET. *J Nucl Med* 2002; 43(9), 1268–1278.

72. Jacobs F, Thierens H, Piepsz A, Bacher K, Van de Wiele C, Ham H, Dierckx RA. Optimised tracer-dependent dosage cards to obtain weight-independent effective doses. *Eur J Nucl Med Mol Imaging* 2005; 32(5), 581–588.

73. Lartizien C, Kinahan PE, Swensson R, Comtat C, Lin M, Villemagne V, Trebossen R. Evaluating image reconstruction methods for tumor detection in 3-dimensional whole-body PET oncology imaging. *J Nucl Med* 2003; 44(2), 276–290.

74. Kinahan PE, Cheng PM, Alessio AM, Lewellen TK. A quantitative approach to a weight-based scanning protocol for PET oncology imaging. *Nuclear Science Symposium Conference Record, 2005 IEEE Nuclear Science Symposium Conference Record, 2005 IEEE* 2005; 45 p.

75. Accorsi R, Karp JS, Surti S. Improved dose regimen in pediatric PET. *J Nucl Med* 2010; 51(2), 293–300.

76. Halpern BS, Dahlbom M, Auerbach MA, Schiepers C, Fueger BJ, Weber WA, Silverman DH, Ratib O, Czernin J. Optimizing imaging protocols for overweight and obese patients: A lutetium orthosilicate PET/CT study. *J Nucl Med* 2005; 46(4), 603–607.

77. Strother SC, Casey ME, Hoffman EJ. Measuring PET scanner sensitivity: Relating count rates to image signal-to-noise ratios using noise equivalents counts. *IEEE Trans Nucl Sci* 1990; 37(2), 783–788.

78. Poon JK, Dahlbom ML, Moses WW, Balakrishnan K, Wang W, Cherry SR, Badawi RD. Optimal whole-body PET scanner configurations for different volumes of LSO scintillator: A simulation study. *Phys Med Biol* 2012; 57(13), 4077–4094.

79. Jakoby BW, Bercier Y, Watson CC, Bendriem B, Townsend DW. Performance characteristics of a new LSO PET/CT scanner with extended axial field-of-view and PSF reconstruction. *Nucl Sci, IEEE Trans* 2009; 56(3), 633–639.

80. Rischpler C, Nekolla SG, Dregely I, Schwaiger M. Hybrid PET/MR imaging of the heart: potential, initial experiences, and future prospects. *J Nucl Med* 2013; 54(3), 402–415.

81. Torigian DA, Zaidi H, Kwee TC, Saboury B, Udupa JK, Cho ZH, Alavi A. PET/MR imaging: Technical aspects and potential clinical applications. *Radiology* 2013; 267(1), 26–44.

82. Weber WA. PET/MR imaging: A critical appraisal. *J Nucl Med* 2014; 55(Supplement 2), 56S–58S.

83. Aznar MC, Sersar R, Saabye J, Ladefoged CN, Andersen FL, Rasmussen JH, Lofgren J, Beyer T. Whole-body PET/MRI: The effect of bone attenuation during MR-based attenuation correction in oncology imaging. *Eur J Radiol* 2014; 83(7), 1177–1183.

84. Burgos N, Cardoso MJ, Thielemans K et al. Attenuation correction synthesis for hybrid PET-MR scanners: Application to brain studies. *IEEE Trans Med Imaging* 2014; 33(12), 2332–2341.

85. Visvikis D, Monnier F, Bert J, Hatt M, Fayad H. PET/MR attenuation correction: Where have we come from and where are we going? *Eur J Nucl Med Mol Imaging* 2014; 41(6), 1172–1175.

86. Purz S, Sabri O, Viehweger A, Barthel H, Kluge R, Sorge I, Hirsch FW. Potential pediatric applications of PET/MR. *J Nucl Med* 2014; 55(Supplement 2), 32S–39S.

87. Treves ST. *Pediatric Nuclear Medicine and Molecular Imaging.* New York: Springer, 2014.

88. Kumar A, Juhasz C, Asano E, Sood S, Muzik O, Chugani HT. Objective detection of epileptic foci by 18F-FDG PET in children undergoing epilepsy surgery. *J Nucl Med* 2010; 51(12), 1901–1907.

89. Drubach LA, Johnston PR, Newton AW, Perez-Rossello JM, Grant FD, Kleinman PK. Skeletal trauma in child abuse: Detection with 18F-NaF PET. *Radiology* 2010; 255(1), 173–181.

90. National Research Council. Health risks from exposure to low levels of ionizing radiation—BEIR VII Phase 2. Washington, DC: National Academies Press, 2006.

91. Muehllehner G. Effect of resolution improvement on required count density in ECT imaging: A computer simulation. *Phys Med Biol* 1985; 30(2), 163–173.

92. Fahey FH, Harkness BA, Keyes JWJ, Madsen MT, Battisti C, Zito V. Sensitivity, resolution and image quality with a multi-head SPECT camera. *J Nucl Med* 1992; 33(10), 1859–1863.

93. Stansfield EC, Sheehy N, Zurakowski D, Vija AH, Fahey FH, Treves ST. Pediatric 99mTc-MDP bone SPECT with ordered subset expectation maximization iterative reconstruction with isotropic 3D resolution recovery. *Radiology* 2010; 257(3), 793–801.

94. Frush DP, Soden B, Frush KS, Lowry C. Improved pediatric multidetector body CT using a size-based color-coded format. *Am J Roentgenol* 2002; 178(3), 721–726.

11

Optimization and Dose Reduction in Pediatric Medical Imaging

Ted S. Treves, Briana
Sexton-Stallone, and
Frederic H. Fahey

11.1 Introduction

Diagnostic nuclear medicine provides important and unique information that plays a critical role in the care of several pediatric disorders. Although radiation exposure to patients from nuclear medicine procedures is low and within the range of common radiological procedures, some physicians remain concerned about radiation exposure to the patient and its potential risk(s), particularly in children. In some instances, clinicians are hesitant to irradiate a patient and would opt for a less reliable procedure in the interest of limiting radiation exposure. This is especially the case in pediatrics because the developing tissues in children are considered to be more sensitive to ionizing radiation than those of adults.

The potential effects of radiation exposure at the low levels used in nuclear medicine are not fully understood. In fact, there is some controversy over whether or not they are detrimental at all.[1] Regarding risk from diagnostic nuclear medicine procedures, one could say that there is "certainty about risk uncertainty." Nevertheless, in an abundance of caution, it would seem prudent to consider approaches that could maintain and even optimize the diagnostic yield of imaging procedures while achieving reduction in radiation exposures in children undergoing these types of examinations. Use of nuclear medicine diagnostic imaging should not focus primarily on radiation exposure reduction, as this may sacrifice image quality and diagnostic information content. Sometimes, more is better for example, in very noncompliant or obese patients. Therefore, in some cases it may be more appropriate to consider dose optimization rather than dose reduction. In addition, any potential risk needs to be considered in the context of the benefit of the procedure for the patient. If the procedure provides clinical information essential for the child's care, then the benefits of the procedure most likely far outweigh any small potential risk.

Factors affecting radiation exposure in pediatric nuclear medicine include: (a) administered radio-pharmaceutical activity, total recorded counts, and imaging time; (b) choice of camera including detector material and thickness, number of detectors, and choice of collimator; (c) experience with pediatric patients; and (d) choice of image processing and reconstruction methods.

In this chapter we will discuss the optimization of radiation exposure though appropriate use, procedure adaptation to specific clinical task, administered activity guideline development, and improvements in image processing and display and also briefly highlight the notable clinical benefits realized from pediatric nuclear medicine procedures.

11.2 Radiation Exposure Reduction

11.2.1 Appropriate Use

There are a number of diagnostic imaging procedures utilized in the care of children including radiography, computed tomography (CT), nuclear medicine, magnetic resonance imaging (MRI), and ultrasonography. The type and number of procedure(s) and sequencing of various imaging procedures in pediatric patients depends on many factors. The clinical condition of the patient, the diagnostic question to be answered, and the available methodology and instrumentation all need to be considered. Is the procedure considered elective or is it an emergency? It is also important to consider the experience of the imaging teams with different imaging procedures as they are applied to their pediatric practice.

Unfortunately, it is not uncommon that in certain practices a relatively large number of procedures are performed on a given patient with the aim to ensure that "nothing is missed." This approach can prove to be lengthy, expensive, and in some cases poses unnecessary distress and risk to the pediatric patient. Simply because a procedure does not involve radiation exposure, may be inexpensive, or easily available does not mean it should be the initial procedure of choice for the particular clinical situation. For example, in the diagnosis of acute pyelonephritis, an ultrasound is a simple, risk-free, and relatively inexpensive procedure to obtain, but it is only about 40% sensitive. Instead, a dimercaptosuccininc acid (DMSA) scan with a >90% sensitivity can be performed, assuring a rapid and efficient diagnosis. In another example, although MRI does not produce ionizing radiation to the patient, it does not mean it should be the initial procedure of choice for certain diagnostic questions. Other considerations should include the actual risks due to sedation, general anesthesia, and contrast agent reactions in MRI or CT versus the theoretical/potential risks from nuclear medicine procedures. The Society of Nuclear Medicine and Molecular Imaging (SNMMI) state that not performing a procedure that is necessary for a patient's care due to fear of radiation can be detrimental to the patient.[2] The primary focus should be the diagnostic question to be answered and knowing which procedure is most likely to answer the clinical question while providing the highest diagnostic yield in the shortest period of time with the least risk and discomfort to the patient. The SNMMI and its Technologist Section also state that "...the right test with the right dose should be given to the right patient at the right time."

11.2.2 Adaptation of Procedures According to Clinical Task

It is useful to consider tailoring routine imaging protocols focusing on the specific clinical task or diagnostic question to be answered.[3,4] For example, in babies with hyperbilirubinemia, the diagnostic goal is to determine if the tracer injected intravenously migrates into the intestine or not. If hepatobiliary scintigraphy shows tracer migrating into the intestine, the patient would likely have hepatic dysfunction. On the other hand, if tracer does not flow into the intestine, the patient would most likely have biliary atresia. Therefore, in order to achieve this differential diagnosis, it is not necessary to obtain images of exquisite anatomical detail. In this example, it is possible to reduce the administered activity even further than that recommended in the literature or in applicable guidelines.[5,6] In this setting, the synergistic combination of hepatobiliary scintigraphy providing information about biliary flow and ultrasonography providing good anatomic information usually achieves the correct diagnosis.

A similar example is in the case of dynamic renal scintigraphy with ^{99m}Tc-MAG3. When determining split renal function, if urine is eliminated freely into the pelvicalyceal system or if there is an obstruction of urine flow, the administered activity can be lowered significantly utilizing noise reduction software.[7]

11.2.3 Development of Guidelines

The goals for every pediatric nuclear medicine study should be to derive the greatest diagnostic information with the highest quality standards and lowest radiation exposure and in the shortest period of time. Until relatively recently, there was no consensus on pediatric radiopharmaceutical administered activities. As a result, there were wide variations in administered activity among institutions. In a survey of 13 major pediatric hospitals in North America, administered activities in patients older than one year varied by a factor of 3–10 and in babies and infants by a factor of 10–20, the most vulnerable population.[8]

The publication of these results sparked wide interest and helped to motivate the development of the North American Expert Consensus Guidelines for Administered Activities in Children and Adolescents.[9] These guidelines were developed to avoid unnecessary radiation exposure, ensure best practices, and promote consistency of imaging techniques, which is important in multicenter clinical investigations using nuclear medicine procedures. In 2014, the North American Guidelines were harmonized with the European Association of Nuclear Medicine (EANM) Dose Card.[10–12] The most recent version of the North American guidelines added six radiopharmaceutical procedures and was also harmonized with the EANM Dose Card.[5] The implementation and distribution of the guidelines has had a positive effect, as many centers have adopted them, and this reduced the previously large variation of administered activities.[13–15] The development and dissemination of dosing guidelines has provided a low-cost and effective means of reducing radiation exposure while maintaining excellent diagnostic image quality.

11.2.4 Improvements in Image Processing and Display

Imaging processing and display improvements have in many instances led to imaging of higher spatial resolution at significantly lower radiopharmaceutical administered activities and resulting radiation exposures. Further improvements in advanced image processing are on the horizon.

In single-photon emission computed tomography (SPECT), image reconstruction using ordered subsets expectation maximization (OSEM) 3D with resolution recovery provides high-quality images using 50% of the counts required by filtered back projection (FBP). This allows the physician to either reduce the administered activity by 50%, reduce imaging time by 50%, or some combination of the two. This has been successfully applied to ^{99m}Tc-DMSA renal cortical SPECT and to ^{99m}Tc-MDP (^{99m}Tc-methylene diphosphonate) skeletal SPECT. Compared to images reconstructed with FBP, these images showed dramatic improvement in spatial resolution.

Studies have demonstrated that significant radiation exposure reduction could be achieved without losing diagnostic value and in fact improve spatial image resolution. In the case of DMSA SPECT, cortical lesions could be more easily detected using OSEM-3D, and in some cases, lesions could only be detected with it and not with FBP (Figures 11.1 and 11.2).[16,17]

Positron emission tomography (PET) can also be optimized for imaging children and lead to substantial improvements in image quality and spatial resolution. Modern PET systems provide obvious advantages for imaging small children, including enhanced spatial resolution across the field of view, improved field uniformity, faster imaging times and improved imaging processing. Their use allows a reduction in radiopharmaceutical administered activities as well as a reduction in the need for sedation or general anesthesia.

99mTc-MDP SPECT

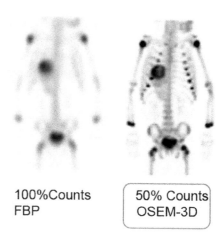

- 17-month-old with a right suprarenal mass

- Neuroblastoma

100%Counts
FBP

50% Counts
OSEM-3D

FIGURE 11.1 99mTc-MDP SPECT Multiple Intensity Projections (MIP) from a 17-month-old with a right supra-renal mass from neuroblastoma. Left: Image containing 100% counts reconstructed using filtered back projection (FBP). Right: Same study with only 50% counts using OSEM 3D with resolution recovery. It is obvious that the image on the right is greatly superior.

99mTc-DMSA SPECT

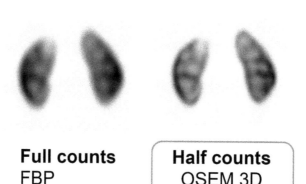

Full counts
FBP

Half counts
OSEM 3D

FIGURE 11.2 99mTc-DMSA SPECT MIP in a young patient suspected of having pyelonephritis. Left: Image reconstructed by FBP using 100% of the acquired counts. Right: Same image with only 50% counts using OSEM 3D with resolution recovery. The image on the right shows improved detail than the image on the left.

For example, it is possible to obtain adequate ^{18}F-2-fluoro-2-deoxy-D-glucose (^{18}F-FDG) brain PET in term and pre-term babies using 0.1 mCi/kg [3.7 MBq/kg], significantly lower than the administered activities routinely used at many centers for this study.[18] Another way to reduce radiation exposure in children who are candidates for a PET/CT scan or a SPECT/CT scan is to consider the use of previously acquired diagnostic CT or MRI for image fusion and not repeat the CT at diagnostic levels of radiation as part of the SPECT or PET examinations. In PET/CT or SPECT/CT acquisitions with CT used only for attenuation correction, CT can be obtained at much lower radiation exposures.[19]

More recent improvements in image quality with reduced administered activities of FDG in children have been reported by Schmidtlein et al.,[20] using sparse representation regularized image reconstruction. This

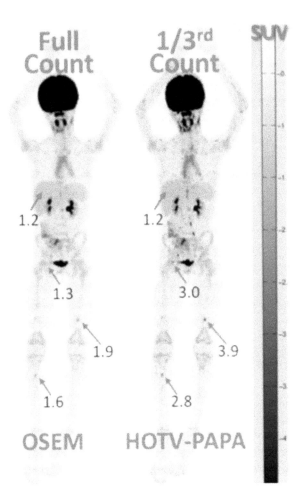

FIGURE 11.3 MIP images from total body ¹⁸F-FDG pediatric PET (266 MBq, 8-year old male, 23 kg) with mela-noma acquired at 66-min post-injection acquired for 3-min./bed (upper body) and 1-min./bed (legs) using a GE D710 PET/CT. The image on the left is a full count OSEM reconstruction using our standard of care parameters. The image on the right used one-third the counts reconstructed using a sparse gradient penalty (HOTV-PAPA). The blue arrows point to metastatic disease (w/maximum SUV), and red arrows indicate mean liver SUV. Increased max SUV in the small lesions arises from improved reconstruction convergence and noise suppression. Note that the pelvic lesion on the left (SUV$_{max}$ 1.3 SOC vs 3.0 HOTV) cannot be detected in the conventional full count MIP image but was readily discernible in the one-third-count sparse gradient images. This lesion was visible in trans-axial slices and confirmed with CT using the conventionally reconstructed images.

form of regularization penalizes small elements in a mathematical transform space in which the noiseless signal would otherwise be compacted into a few (sparse) large elements. One consequence of the random nature of the noise is that it cannot be sparsely represented, and thus it can be easily removed. In Figure 11.3, the maximum intensity projections of a pediatric patient are shown. The image on the left is a full-count image reconstructed using the standard-of-care algorithm (post-filtered OSEM), while the one on the right is reconstructed using one-third the counts with the sparse higher-order gradient penalty (HOTV-PAPA) reconstruction. Since the sparse representation penalty controls noise during the reconstruction process, it can be iterated further allowing small features to more fully realized leading to better resolution recovery. This, in turn, leads to better preservation of edges and enhancement of the Standardized Uptake Value (SUVs) of small lesions that more accurately reflects the underlying tracer distribution.[21]

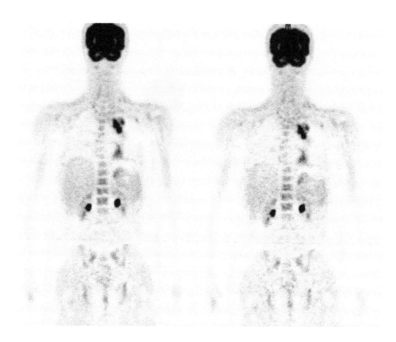

FIGURE 11.4   ¹⁸F-FDG PET images at two activity levels (3.0 and 1.5 MBq/kg) showing a big lesion and a small one, both well preserved in the half-dose image.

Further radiation dose reduction with FDG PET is possible using MRI/PET hybrid systems. Since image acquisition MRI is relatively quite long, there ample time to collect more than sufficient counts for adequate PET. The use of PET/MR scanners can also lead to substantial dose reduction for pediatric molecular imaging. In the first place, they avoid entirely the exposure related to the CT component while still providing diagnostic accuracy comparable to that of PET/CT. The dose reduction varies depending on the CT protocol adopted for PET/CT but can be significant since the CT component typically contributes to 40%–50% or more of the total dose even for optimized PET/CT acquisitions.[22]

The use of PET/MR can also lead to a significant decrease in injected activity, due to the favorable geometry of the PET component in PET/MR integrated scanners. The longer axial FOV and reduced detector diameter provide increased sensitivity and, thereby, higher count rates.[23] Gatidis et al.[24] showed that a reduction from 3.0 to 1.5 MBq/kg has no significant impact on lesion detection, subjective evaluation of image quality, and SUV measurement in children. These data have been confirmed by objective measurements of image quality in pediatric patients (Figure 11.4).[25]

11.3 Benefits of Pediatric Nuclear Medicine[3]

Thus far we have focused on radiation exposures and image quality. It is at least as important to also emphasize the real benefits of nuclear medicine in the care of pediatric patients. Included here are some examples that highlight unique diagnostic information obtained from nuclear medicine studies in children.

11.3.1 Brain

11.3.1.1 Seizure Disorders

Ictal and interictal perfusion brain SPECT using ⁹⁹ᵐTc-ECD (⁹⁹ᵐTc ethyl cysteinate dimer) or ⁹⁹ᵐTc-HMPAO (⁹⁹ᵐTc-hexamethyl propylene amine oxime) are important methods in the detection and localization of ictal cortex in patients with medically refractory seizure disorders. Ictal perfusion brain SPECT provides a powerful tool for the identification of ictal cortex. The radiotracer reflects cerebral perfusion and it is taken up and distribute within the brain according to the first pass. The radiopharmaceutical is administered intravenously while the patient is having a seizure. The tracer is "fixed" within the brain and it does not redistribute with time. Therefore, imaging can take place any time after the injection. The distribution of the tracer during a seizure shows focal increased cerebral perfusion in the brain region responsible for the seizure. A successful ictal SPECT and the subtraction of ictal from interictal perfusion brain SPECT can help localize ictal cortex in most cases of focal seizures. This, along with other imaging tests, can help determine a patient's candidacy for resective surgery.

11.3.1.2 Brain Tumors

In the case of brain tumors, SPECT or PET can provide useful information that can guide therapeutic decisions. Although anatomic imaging can detect the tumor mass, functional nuclear medicine studies can help differentiate scarring from residual or recurrent tumor. The nuclear medicine image is fused to a brain MRI or CT to outline the anatomic features of the tumor mass while the nuclear medicine scan provides information about the metabolic activity of the tumor. This information is useful in deciding what, if any, additional therapy may be needed.

11.3.2 Thyroid

Thyroid imaging with ^{123}I-NaI provides a specific and sensitive method to image functioning thyroid tissue. In newborns with congenital hypothyroidism and young children with hypothyroidism, this method can effectively detect ectopic thyroid tissue, such as lingual or sublingual tissue, helping to determine the most appropriate therapeutic approach.

11.3.3 Lungs

The unique capability of nuclear medicine to image and quantify regional pulmonary perfusion and ventilation is especially useful in the assessment of congenital cardiopulmonary disorders. Along with other cardiac methods, nuclear medicine studies provide useful complementary information about lung function. This is frequently of value in patients before and after therapeutic catheter interventions as well as in pre- and post-operative assessment of regional lung function.

11.3.4 Heart

At present, this is not a frequent indication, but the assessment of regional myocardial perfusion with ^{99m}Tc-MIBI (^{99m}Tc-sestamibi) can be of value in patients with anomalous departure of the coronary artery from the pulmonary artery, heart transplants, and Kawasaki disease, to name a few applications.

11.3.5 Liver

In the newborn with hyperbilirubinemia, imaging with ^{99m}Tc-labeled hepatobiliary agents along with ultrasound can help differentiate biliary atresia from hepatocellular disease, helping to select the best therapeutic approach. In older children, this method can assist in the diagnosis of acalculous cholecistitis.

11.3.6 Spleen

Imaging with ^{99m}Tc-heat-denatured red blood cells can effectively diagnose functional asplenia, polysplenia, heterotaxia, and splenosis.

11.3.7 Gastrointestinal Tract

Dynamic ^{99m}Tc-pertechnetate scintigraphy is a very sensitive method that can effectively diagnose functioning ectopic gastric mucosa in a Meckel's diverticulum. With ^{99m}Tc-sulfur colloid imaging, pulmonary aspiration of gastric content, esophageal transit time, and gastric emptying time can all be reliably identified. In addition, imaging with ^{99m}Tc-heat-denatured red blood cells can help diagnose the site of intestinal bleeding.

11.3.8 Kidneys

^{99m}Tc-DMSA imaging with SPECT can quantify split renal function and diagnose acute and chronic pyelonephritis, renal scarring, duplication, horseshoe kidneys and ectopic kidneys. ^{99m}Tc-MAG3 (^{99m}Tc-mercaptoacetyltriglycine) dynamic imaging can determine split renal function and assess urinary flow through the pelvicalyceal system into the bladder in cases of hydronephrosis or suspected urinary obstruction. These methods are very useful in the evaluation of newborns with prenatal ultrasound diagnosis of hydronephrosis. The addition of diuretic renography in many cases can enhance the diagnosis of obstruction.

11.3.9 Bladder and Ureters

Radionuclide cystography ^{99m}Tc-pertechnetate is a highly sensitive method that can diagnose very small volumes of vesicoureteral reflux at a very low radiation exposure. Depending on the equipment and protocols employed, gonadal radiation exposures are at least 20 times lower than with conventional cystoureterography.

11.3.10 Bone

Assessment of skeletal disorders can be accomplished with ^{99m}Tc-MDP imaging. This method is extremely sensitive although not specific, but it can be valuable in a number of conditions including diagnosis of non-accidental trauma (child abuse) as well as trauma due to fall or physical injury from sport activity. It can be helpful in the assessment of extremity pain when the origin of the pain is not obvious because of referred pain. A whole-body bone image can help diagnose the origin of the pain. With this knowledge, more specific imaging can be applied to the region of interest in order to obtain a more specific diagnosis. In certain oncologic disorders, skeletal imaging can outline the extent of the disease and detect metastatic disease.

11.4 Oncologic Disorders

The role of PET/CT in the assessment of pediatric oncologic disorders has dramatically increased over the past few years. ^{18}F-FDG PET has been found to be effective in the assessment of pediatric lymphoma and brain tumors, among other malignancies. ^{18}F-FET (^{18}F-fluoroethyl-L-tyrosine) and ^{18}F-FLT (^{18}F-Fluorothymidine) have been useful to assess response to therapy in brain tumors. ^{68}Ga-DOTATOC and ^{68}Ga-DOTATATE is very useful in the assessment of neuroblastoma and other neuroendocrine tumors. On-going radiopharmaceutical research and development are rapidly advancing and newer and more specific agents in the pipeline will likely improve the diagnostic sensitivity and specificity in several oncologic disorders.

11.5 Future Considerations, Summary, and Conclusions

During the past few years, significant progress has been made in nuclear medicine resulting in the optimization of diagnostic procedures. Radiopharmaceutical research and development along with improvements in imaging instrumentation and image processing have undoubtedly contributed to this progress. As a result, nuclear medicine has become essential for imaging children as well as adults. The development and dissemination of guidelines on pediatric radiopharmaceutical administered activities has contributed to standardization and improvement in practices, as well as an overall reduction in radiation exposures in children undergoing nuclear medicine examinations.[13]

Despite these improvements, gaps in our knowledge about pediatric patients remain. Radio-pharmaceutical biodistribution and pharmacokinetic data in children are lacking. Most of the estimates of radiation exposures have been derived from the pharmacokinetics of adult patients and animal studies. Due to ethical considerations, it is understandable that normal children cannot be recruited to study radio-pharmaceutical biodistribution and kinetics. In order to understand better actual radiation exposures in children, however, this information is needed. There are current research projects aimed at obtaining this important information so that radiation dose estimates can be improved. Biodistribution and pharmaco-kinetics may depend on patient's age, weight, physiologic maturity, height, and disease state. In addition, it is important to expand our knowledge about the contribution of CT to radiation exposures and how CT methods can be adapted to provide the needed information at the lowest possible radiation exposures. The availability of PET/MRI devices can also contribute to reductions of administered activities. The relatively long acquisition time in MRI provides ample time to acquire ample counts from the PET radiopharmaceu-tical, and geometric factors associated with the PET component can have a positive effect in this regard.

There is no doubt that nuclear medicine provides useful and unique information for clinicians car-ing for children with both simple and complex disease states. Nuclear medicine is a functional imaging method and as such yields information not available through many other imaging techniques. It is safe and lacks toxicity and osmotic or allergic effects. Concerns about potential risks from diagnostic nuclear medicine should be balanced by the clinical benefits of these procedures as well as with the real risks of sedation and reactions to contrast agents from other procedures. Nevertheless, it is prudent to uti-lize procedures with the lowest radiation exposures possible, with approaches that provide the highest diagnostic yield in the shortest period of time and with the greatest patient comfort. While one could state that there is still certainty about risk uncertainty, the application of nuclear medicine diagnostic procedures in pediatric patients should include appropriate patient and caregiver information, not just reduction in radiation exposure.[26] It is important to fully understand and optimize its many clinical benefits and potential risks, allowing for a balance between limiting radiation exposure and obtaining all information necessary for successful diagnosis and treatment.

References

1. Siegel, J.A., C.W. Pennington, and B. Sacks, Subjecting radiologic imaging to the linear no-threshold hypothesis: A non sequitur of non-trivial proportion. *J Nucl Med*, 2017. **58**(1): 1–6.
2. *SNMMI Position Statement of Dose Optimization for Nuclear Medicine and Molecular Imaging.* 2012; Available from: http://snmmi.files.cms-plus.com/docs/SNM_Position_Statement_on_Dose_Optimization_FINAL_June_2012.pdf. Accessed August 8, 2018
3. Treves, S.T., *Pediatric Nuclear Medicine and Molecular Imaging.* 4th ed. 2014, New York: Springer.
4. Treves, S.T. et al., Nuclear medicine in the first year of life. *J Nucl Med*, 2011. **52**(6): 905–925.
5. Treves, S.T. et al., 2016 Update of the North American consensus guidelines for pediatric admin-istered radiopharmaceutical activities. *J Nucl Med*, 2016. **57**(12): 15N–18N.

6. Fahey, F. et al., Beyond current guidelines: Reduction in minimum administered radiopharmaceutical activity with preserved diagnostic image quality in pediatric hepatobiliary scintigraphy. *Eur J Nucl Med Mol Imaging*, 2014. **41**(12): 2346–253.

7. Hsiao, E.M. et al., Reduction in radiation dose in mercaptoacetyltriglycerine renography with enhanced planar processing. *Radiology*, 2011. **261**(3): 907–915.

8. Treves, S.T., R.T. Davis, and F.H. Fahey, Administered radiopharmaceutical doses in children: A survey of 13 pediatric hospitals in North America. *J Nucl Med*, 2008. **49**(6): 1024–1027.

9. Gelfand, M.J. et al., Pediatric radiopharmaceutical administered doses: 2010 North American consensus guidelines. *J Nucl Med*, 2011. **52**(2): 318–322.

10. Lassmann, M., S.T. Treves, and E.S.P.D.H.W. Group, Paediatric radiopharmaceutical administration: Harmonization of the 2007 EANM paediatric dosage card (version 1.5.2008) and the 2010 North American consensus guidelines. *Eur J Nucl Med Mol Imaging*, 2014. **41**(5): 1036–1041.

11. Treves, S.T., M. Lassmann, and E.S.P.D.H.W. Group, International guidelines for pediatric radiopharmaceutical administered activities. *J Nucl Med*, 2014. **55**(6): 869–870.

12. Grant, F.D. et al., Radiation doses for pediatric nuclear medicine studies: Comparing the North American consensus guidelines and the pediatric dosage card of the European Association of Nuclear Medicine. *Pediatr Radiol*, 2015. **45**(5): 706–713.

13. Fahey, F.H. et al., Effects of image gently and the North American guidelines: Administered activities in children at 13 North American pediatric hospitals. *J Nucl Med*, 2015. **56**(6): 962–967.

14. Fahey, F.H. et al., Standardization of administered activities in pediatric nuclear medicine: A report of the first nuclear medicine global initiative project, part 1-statement of the issue and a review of available resources. *J Nucl Med*, 2015. **56**(4): 646–651.

15. Fahey, F.H. et al., Administered activities in pediatric nuclear medicine and the impact of the 2010 North American consensus guidelines on general hospitals in the United States. *J Nucl Med*, 2016. **57**(9): 1478–1485.

16. Sheehy, N. et al., Pediatric 99mTc-DMSA SPECT performed by using iterative reconstruction with isotropic resolution recovery: Improved image quality and reduced radiopharmaceutical activity. *Radiology*, 2009. **251**(2): 511–516.

17. Stansfield, E.C. et al., Pediatric 99mTc-MDP bone SPECT with ordered subset expectation maximization iterative reconstruction with isotropic 3D resolution recovery. *Radiology*, 2010. **257**(3): 793–801.

18. Shi, Y. et al., Brain positron emission tomography in preterm and term newborn infants. *Early Hum Dev*, 2009. **85**(7): 429–432.

19. Fahey, F.H. et al., Dosimetry and adequacy of CT-based attenuation correction for pediatric PET: Phantom study. *Radiology*, 2007. **243**(1): 96–104.

20. Schmidtlein, C.R. et al., Three-fold reduction in pediatric PET dose by advanced image reconstruction without loss of diagnostic and quantitative performance. *J Nucl Med*, 2018. **59**(1): 583–583.

21. Schmidtlein, C.R. et al., Relaxed ordered subset preconditioned alternating projection algorithm for PET reconstruction with automated penalty weight selection. *Med Phys*, 2017. **44**(8): 4083–4097.

22. Alessio, A.M. et al., Weight-based, low-dose pediatric whole-body PET/CT protocols. *J Nucl Med*, 2009. **50**(10): 1570–1577.

23. Delso, G. et al., Performance measurements of the Siemens mMR integrated whole-body PET/MR scanner. *J Nucl Med*, 2011. **52**(12): 1914–1922.

24. Gatidis, S. et al., Defining optimal tracer activities in pediatric oncologic whole-body (18)F-FDG-PET/MRI. *Eur J Nucl Med Mol Imaging*, 2016. **43**(13): 2283–2289.

25. Zuchetta, P. Comments in Discussion about potential administered radiopharmaceutical dose reduction with PET/MRI. Experience from the audience and common indications. 2017 European Association of Nuclear Medicine, Pediatric Interest Group Meeting, Vienna, Austria, October 21–25, 2017.

26. Fahey, F.H., S.T. Treves, and S.J. Adelstein, Minimizing and communicating radiation risk in pediatric nuclear medicine. *J Nucl Med*, 2011. **52**(8): 1240–1251.

12

Optimization and Dose Reduction in Medical Imaging of the Pregnant Patient

Wesley E. Bolch,
Matthew Maynard,
William J. Godwin,
Amy M. Geyer,
and Linda Kroger

12.1 Introduction

Pregnant females may undergo a variety of medical imaging procedures where organs in the developing fetus receive a radiation dose, at various levels, from externally incident x-ray imaging fields (including both in-field attenuated and out-of-field scatter photons) or internally emitted radiations associated with radiopharmaceutical administrations. These exposure scenarios broadly fall within two categories. The first occurs primarily during early stages of gestation where the pregnancy of the patient is ascertained only following the imaging session. In these cases, a dose reconstruction may be warranted to both document the exposure and provide pertinent information for the referring physician in ensuing discussions with the patient regarding fetal health. The second occurs primarily during later stages of gestation whereby imaging is emergent or life-saving in nature. In these cases, the dose estimate may be performed prior to the study to optimize image acquisition parameters to minimize fetal exposure. Retrospective dose reconstructions may also be required to document the exposure as needed for regulatory or clinical purposes. Radiation doses to the fetus during radiographic or diagnostic fluoroscopic imaging of the maternal abdomen are in general exceedingly small, while higher doses may result from nuclear medicine, computed tomography (CT), or interventional fluoroscopy in increasing order. As will be discussed in this chapter, however, the radiation doses received by the fetus, even during the most extreme conditions, are far below scientific consensus threshold levels for acute tissue reactions. Nevertheless, under the linear no-threshold hypothesis, a small yet finite risk of childhood cancer might persist, which should trigger efforts at imaging optimization with clinically reasonable efforts at dose reduction.

12.2 Medical Imaging of the Pregnant Female

12.2.1 Nuclear Medicine

Pregnant women are occasionally administered diagnostic radiopharmaceuticals, either with or without prior knowledge of the pregnancy on the part of her health-care team. Some common studies performed on women with a confirmed pregnancy are pulmonary ventilation–perfusion, thyroid, bone, and renal scans (ACOG 2004). Several cases have been reported in the literature where women in the first trimester of pregnancy were inappropriately administered radiopharmaceuticals, which would have normally been avoided in order to preserve the health of the fetus or embryo. Tran et al. (2010) describes a case where a pregnant woman was treated for Graves' disease using ^{131}I; administration of ^{131}I to pregnant women, especially at therapeutic levels, is particularly problematic. Zanotti-Fregonara et al. (2008, 2010) describe two separate instances of pregnant women being administered ^{18}F-2-fluoro-2-deoxy-D-glucose (^{18}F-FDG) as part of a positron emission tomography/computed tomography (PET/CT) study as a follow-up to Hodgkin lymphoma therapy. Several authors have investigated the radiation dose to the fetus from the administration of a variety of imaging radiopharmaceuticals to pregnant mothers (Russell et al. 1997a, 1997b; Saunders et al. 2002; Keleher et al. 2004; Stabin 2004; Pandit-Taskar et al. 2006; Stabin 2017).

12.2.2 Computed Tomography

A study by Lazarus et al. (2009) has reported a 25% annual increase in the use of CT imaging in pregnancy. Diagnostic CT imaging is particularly necessary in such emergent setting as suspected pulmonary embolism, acute appendicitis, or trauma (Shi and Xu 2004). Acute pulmonary embolism is one of the leading causes of maternal mortality in the United States (Toglia and Weg 1996). Recent studies have shown that multi-detector computed tomography (MDCT) pulmonary angiography is the overwhelmingly favored diagnostic procedure when screening for acute pulmonary embolism in pregnant patients (Doshi et al. 2008; Jha et al. 2010). Consequently, careful consideration must be given to the in utero dose received by the fetus given the higher radiosensitivity of the fetal tissues. Careful attention must be paid to fetal exposure during the first trimester since radiation susceptibility is highest during this gestational stage due to organogenesis. Additionally, maternal breast tissue shows a higher susceptibility to adverse effects of ionizing radiation (Timins 2002). Although there may be alternative non-ionizing diagnostic examinations available (ultrasound, blood tests, magnetic resonance imaging [MRI]), in many cases CT imaging remains the modality of choice even when the patient is known to be pregnant. Acute abdominal pain occurs in 1 in 500 to 1 in 635 pregnancies, while acute appendicitis has an incidence of 1 in 500 to 1 in 2000 during pregnancy (Augustin and Majerovic 2007).

An accurate diagnosis of acute appendicitis is often difficult to reach in pregnant patients given the relocation of the appendix caused by an enlarged uterus (Chen et al. 2008). This difficulty significantly increases when the mother is in the second and third trimester, which leads to an increased rate of perforation surgery at these gestational stages (Weingold 1983; Ueberrueck et al. 2004). Appendicitis occurrence has been shown to have the highest frequency in the third trimester, with the second trimester being a close second (Shetty et al. 2010). Diagnostic tests for appendicitis are not reliable when only blood and ultrasound tests are performed. Blood tests have shown to have a false-negative rate of up to 50% (Maslovitz et al. 2003).

The use of ultrasound imaging is desirable as a non-ionizing alternative to CT; however, diagnosis can be inadequate due to limitations on the positioning of the mother and anatomic movements, which have been shown to be particularly problematic in the third trimester. Consequently, the use of ultrasound techniques to diagnose acute appendicitis becomes unreliable in the later pregnancy stages. As surgical exploration often becomes necessary when only ultrasound diagnoses are used, it becomes imperative to avoid a false-negative appendectomy procedure in order to decrease risk of fetal loss. Perforated appendicitis is the leading cause of fetal loss during surgery (Parangi et al. 2007). Although diagnosis using MRI has been shown to provide comparable results and precision to CT imaging, the necessary equipment is often not available in emergency situations that are typical for acute appendicitis (Shetty 2010; Shetty et al. 2010). The need for prompt diagnosis, often during clinical after-hours, typically demands the use of CT imaging. Even though MRI is desired for its lack of ionizing radiation, surveys have shown that CT imaging is the preferred method of diagnosis and provides the most accurate diagnosis of acute appendicitis during pregnancy (Oto et al. 2005; Lazarus et al. 2007). These unavoidable diagnostic CT scans performed in the abdominal region need to be carefully planned and optimized.

12.2.3 Fluoroscopically Guided Interventions

Exposure of the pregnant female to ionizing radiation during fluoroscopically guided interventions (FGI) is fortunately rare when the pregnancy is known or suspected. However, a variety of life-saving procedures are still performed on pregnant women (NCRP 2010). As with all FGIs, radiation exposure levels may be high enough to result in skin doses above tissue reaction threshold levels (Balter et al. 2010) and to raise serious concerns regarding the health of the developing fetus. Dose reconstructions are typically performed rather crudely using uterine doses as a surrogate organ or extrapolating maternal skin dose via exponential attenuation factors. Emergency FGIs that are performed on pregnant patients include (1) percutaneous nephrostomy for renal obstruction, (2) embolization of organs and blood vessels in the abdomen and pelvis for bleeding due to trauma, (3) embolization of visceral renal or splenic aneurysms, (4) percutaneous drainage of an abdominal or pelvic abscess, and (5) peripartum spontaneous coronary artery dissection (NCRP 2010). The latter intervention is typically performed (although rarely) shortly before or after the delivery, and the condition can be lethal if left untreated. Fluoroscopy guidance is used for the latter when ultrasonic guidance is not clinically possible. Occasionally, urology procedures are performed on pregnant females under fluoroscopic guidance, including both ureteral stent placements and retrograde pyelograms, as well as some gastrointestinal (GI) procedures such as endoscopic retrograde cholangiopancreatography (ERCP) (Tang et al. 2009).

12.3 Tissue Reactions Following Fetal Radiation Exposure

Exposure to ionizing radiation may induce biological effects that may be broadly categorized as either *tissue reactions* or *stochastic effects*. The former involve the destruction of sufficient numbers of cells within a tissue so as to manifest organ-level damage or failure. Tissue reactions are characterized by a dose threshold below which the radiobiological effect does not present. The magnitude of the dose threshold will vary with the specific effect and the organ impacted, and is modified by variations in individual radiosensitivity. In the context of embryonic and fetal exposure during medical imaging of the mother, these threshold doses are critically important to know so that they may be compared to the doses received during medical imaging. If the dose threshold is sufficiently larger than the imaging dose, then decisions on the health of the unborn child need not consider potential risks of tissue reactions. This then leaves one to consider only stochastic cancer risks following in utero exposures. These risks are addressed in the following section.

A variety of radiogenic effects may arise following in utero exposure of the embryo/fetus including mental retardation, neurobehavioral effects, convulsive disorders, congenital malformations, fetal growth retardation, and embryonic death. Excellent and extensive reviews of both animal and human exposure data underlying our understanding of these effects may be found in both ICRP Publication 90 (ICRP 2003) and NCRP Report No. 174 (NCRP 2013). Their impact on the embryo or fetus is strongly dependent upon both the stage of development, the dose rate, and the dose received. During the preimplantation and presomite stages of development (<14 days post-conception), the cells of the embryo are pluripotential (stem cell in nature) and are highly sensitive to the lethal effects of ionizing radiation. However, if these cells survive, the resulting embryo/fetus will develop normally with no increased risk of anatomic malformation at birth. This early period is thus referred to as the *all-or-none period*. During the period 14–40 days post-conception, the embryo is undergoing the early stages of organogenesis, and this is a very vulnerable period for the production of major anatomic malformations. From 40 days post-conception until birth, the vulnerability of the fetus steadily declines with the exception of serious radiation effects to the central nervous system, gonads, and overall fetal growth if the dose is sufficiently high.

Table 12.1, adopted from a similar table in the review of NCRP Report No. 174, nicely summarizes our current state of knowledge of dose thresholds for tissue reactions to the developing child at each stage of embryonic development. All doses are expressed in units of milligray (mGy), the same unit typically used to report imaging doses during in utero imaging exposure. The lowest dose threshold proposed in this table is for the minimum lethal dose to the embryo at one to two weeks post-conception (150–200 mGy). In the conclusions of their report, the authors of NCRP Report No. 174 state that limited human epidemiological data, coupled with extensive mammalian animal data, suggest that *"the no-adverse-effect level from acute exposure for birth defects, growth retardation, pregnancy loss, and other tissue reactions is 200 mGy at the most vulnerable stage of pregnancy."* This threshold represents a substantially higher threshold than the previous tissue reaction dose threshold of 50 mGy recommended in NCRP Report No. 54 (1977). While the 50 mGy threshold is cited in recent literature (e.g., Patel et al. 2007; Litmanovich et al. 2014), the 200 mGy threshold is the current recommended value and should take precedence.

TABLE 12.1 Estimates of the Risks of Radiation Exposure to the Embryo and Fetus as a Function of Effect and Embryonic Age (Data are based on both rodent embryological studies and human epidemiological studies.)

Embryonic Age (Weeks Post-Conception)	Minimum Lethal Dose (mGy)	Approximate LD$_{50}$ (mGy)	Minimum Dose for Permanent Growth Retardation in Adults (mGy)	Minimum Dose for Gross Anatomic Malformations (mGy)	Increased Incidence of Mental Retardation (mGy)	Minimum Dose for Stochastic Effects (mGy)
1–2 wk	150–200	<1,000	No effect	No effect		Unknown
3–5 wk	250–500	1,400–2,000	200–500	>200 (but most >500)		Unknown
6–13 wk	>1,000	>2,000	250–500	Cannot be produced		Unknown
8–25 wk					>500 (Lower 95% CI, 300)	
14 wk–Term	>1,500	Same as for mother	>500	Cannot be produced		Unknown

Source: Table 5.3 from NCRP, *NCRP Report No. 174: Preconception and Prenatal Radiation Exposure: Health Effects and Protective Guidance*, National Council on Radiation Protection and Measurement, Bethesda, MD, 2013.

LD$_{50}$: lethal dose 50%—the dose required to kill 50% of a test population.

12.4 Stochastic Risks Following Fetal Radiation Exposure

As shown in Table 12.1, what is unknown at present is the dose required to yield a stochastic radiobiological risk to the embryo or fetus. Two key stochastic effects are to be considered: (1) hereditary disease following irradiation of the fetal gonads and (2) postnatal cancer, either during childhood or even early adulthood following in utero exposure. The former is not specifically addressed in NCRP Report No. 174, but that report does comment on the risk of hereditary disease following preconception exposure of the parents, noting *"there is no convincing direct evidence of germline mutation manifest as heritable disease in the offspring of humans that is attributable to preconception exposure to ionizing radiation…"*.

As will be noted below in Section 12.5, radiation doses to the embryo and fetus during routine diagnostic and even interventional imaging are not likely to exceed 100 mGy, and thus according to Table 12.1, the only radiobiological effect to consider in medical imaging of the pregnant mother is a possible cancer risk in childhood following in utero fetal exposure. Childhood cancers are in general rare, and represent <1% of total cancer incidence in the U.S. population (http://seer.cancer.gov/csr/1975_2006). Childhood cancers differ from adult cancers of epithelial origin, and are clustered (~70%) as leukemias, lymphomas, and brain tumors. The major source of risk estimates for childhood cancer following in utero exposures are human radiation epidemiological studies.

The possible relationship between in utero medical diagnostic exposures and increased risk of childhood cancer was first described some 60 years ago in the Oxford Survey (Giles et al. 1956; Stewart et al. 1958; Bithell and Stewart 1975). The key finding from this case-control study, along with subsequent reports from studies in Canada, China, Finland, Germany, Italy, Japan, Sweden, The Netherlands, the United Kingdom, and the United States are summarized in Table 5.14 of NCRP Report No. 174 (NCRP 2013). These studies include a minimum of 200 total childhood cancer cases, at least 100 cases of acute lymphoblastic leukemia, or at least 80 cases for each of other childhood cancer types. Relative risks (RR)[1] for leukemia reported in these *case-control* studies in which medical records were reviewed vary from a low of 1.0 (95% Confidence Interval (CI) of 0.5–1.9) (Salonen and Saxen 1975) to a high of 1.5 (95% CI of 1.2–2.0) (Monson and MacMahon 1984). In Table 5.15, NCRP Report No. 174 additionally summarizes key findings of various *cohort-based* radiation epidemiological studies of childhood cancer following maternal diagnostic x-ray imaging. Relative risks for total cancer were shown to range from a low of 0.7 (95% CI of 0.3–1.8) (Ray et al. 2010) to a high of 4.6 (95% CI of 0.9–25.1) (ICRP 2003), while relative risks for leukemia were shown to range from a low of 0.4 (95% CI of 0.1–2.6) (Lewis 1960) to a high of 1.6 (95% CI of 0.6–4.6) (Diamond et al. 1973). In many of these epidemiological studies, however, no specific information on the magnitude of the imaging dose was available in the statistical analysis. Consequently, investigators have debated whether the statistical associations seen between childhood cancer and radiation exposure are causal (e.g., why was the maternal imaging performed in the first place, and was there a predisposition to cancer irrespective of the imaging dose). Also, with no specific dose information, these studies are inconclusive regarding the magnitude of the cancer risk per unit fetal dose.

Preston et al. (2008), in their analysis of the Japanese atomic bomb survivals, compared solid cancer incidence risks among those exposed in utero to those exposed postnatally under the age of six years. The study—referred to as the Life Span Study or LSS—includes perhaps the most extensive cohort of in utero exposed individuals with long-term and continuous follow-up. High-quality dose estimates are also available across a range of gestational ages and fetal doses. The study reported excess relative risk (ERR) values per unit fetal dose.[2] The study found that the ERR, and 95% confidence intervals, at attained age of 50 years

---

[1] Relative risk is defined as the incidence rate of a disease (e.g., cancer) in an exposed group divided by the incidence rate in an unexposed group. An RR of 1.2 then implies a 20% increase in the risk of childhood cancer for the child that was exposed in utero to maternal imaging radiation when compared to the unexposed population's childhood cancer incidence.

[2] An excess relative risk is defined as the ratio of the excess risk of a specified disease (e.g., cancer) to the probability of the same effect in the unexposed population. For example, ERR of 0.3 implies a 30% increase in the lifetime cancer risk for the child that was exposed in utero to radiation when compared to the unexposed population's childhood cancer incidence.

were 0.3 (0.0–2.0) Gy^{-1} and 0.5 (0.0–2.4) Gy^{-1} for males and females exposed in utero, but were 1.3 (0.6–2.2) Gy^{-1} and 2.2 (1.3–3.4) Gy^{-1} for males and females exposed in early childhood. The study confirmed that the risks are higher for females than for males, and are higher for the postnatally exposed individuals than for the in utero exposure cohort members. Furthermore, no variation in ERR by trimester of exposure was found. Excess absolute risks (EARs) increased markedly with attained age among those exposed in early childhood, but showed very little change with time for those exposed in utero. Further follow-up of the LSS in utero exposed cohort was recommended. It is also noted that the dosimetry systems used in this analysis did not include computational models of pregnant female. The dose to the uterus in the non-pregnant adult female was used as a surrogate for the fetal dose target regardless of gestational age (RERF 2005).

Two of the more recent studies of in utero cancer risks following medical imaging of the mother are those of Ray et al. (2010) (a population-based cohort study), and Rajaraman et al. (2011) (a case-control study). In the study by Ray et al., the authors studied 1.8 million maternal-child pairs in Ontario from 1991 to 2008. The cohort included 5,590 mothers exposed to radiodiagnostic x-rays in pregnancy and a control group of 1,829,927 unexposed mothers. Of the exposed group, imaging exams included 73% CT scans and 27% nuclear medicine scans. After a median follow-up of 8.9 years, only 4 childhood cancers arose in the exposed group, while the control group displayed 2,539 cancers. The study thus reported a hazard ratio (HR)[3] of only 0.69 (95% CI of 0.26–1.82), thus strongly suggesting that diagnostic x-ray exposures in pregnancy are not carcinogenic. The authors note, however, that since the upper confidence limit of the risk may be as high as 1.8 times that in the unexposed group, they could not fully exclude the possibility that fetal exposures to diagnostic x-ray imaging are carcinogenic.

In the case-control study by Rajaraman et al. (2011), 2,690 childhood cancer cases and 4,858 age, sex, and region matched controls were selected from the United Kingdom Childhood Cancer Study all born between 1976 and 1996. The study focused on both the cancer risks from diagnostic x-ray exposures and ultrasound examination during pregnancy. The study found a slight increase in risk after in utero exposure to diagnostic x-rays for all cancer types (odds ratio [OR][4] of 1.14, 95% CI of 0.90–1.45) and for leukemia (odds ratio of 1.36, 95% CI of 0.91–2.02), but these were not statistically significant. Exposure to diagnostic x-rays in early infancy (birth to 100 days) was associated with small, non-significant excess risks for all cancers and for leukemia. The data showed no evidence of increased childhood cancer risk in children following in utero ultrasound imaging.

Another exposed cohort of potential importance in quantifying in utero cancer risks are those women who lived in the villages of the Techa River in the Southern Ural Mountains of the former USSR and worked in the Mayak Plutonium Production Facility during peak times of radionuclide release (1948–1955). In a recent study by Schuz et al. (2017), an analysis was conducted of cancer incidence and mortality risk following in utero radiation exposure on a pooled cohort of women exposed to either environmental radionuclide sources along the Techa River, or to radiation sources within the Mayak facility. In both cases, radiation exposure to the fetus was computed from external gamma-ray fields, and from biokinetic and dosimetric models of ingested radionuclides. Both prenatal and postnatal doses were computed and considered. The combined cohort totaled 19,546 subjects with 700,504 person-years at risk over the period of incidence, with slightly larger numbers for the mortality follow-up. Of these subjects, offspring displayed 58 cancer cases and 36 cancer deaths up to age 61. The ERR of cancer incidence was reported to be 1.27 (95% CI of –0.20–4.71) at in utero fetal doses exceeding 80 mGy. This risk increased consistently per 100 mGy of continuous exposure in utero. The study concluded that while a positive association between in utero exposure to ionizing radiation and the risk of hematological malignancies was found, the small number of observed cases, inconsistent incidence and mortality findings, and limited follow-up precluded firm conclusions. Half of the combined cohort are still alive, and thus the study suggested another 10 years of follow-up.

---

[3] Hazard ratios differ from relative risks and odds ratios in that RRs and ORs are cumulative over an entire study, using a defined endpoint, while HRs represent instantaneous risk over the study time period.

[4] The *odds ratio* is a measure of the odds of an event happening in one group compared to the odds of the same event happening in another group. In case-control studies, and in cohort studies in which the outcome occurs in less than 10% of the unexposed population, the OR provides a reasonable approximation of the RR.

In summary, it is difficult to draw conclusions regarding the risk of childhood cancer following in utero exposures to radiation from diagnostic x-ray procedures. Data from case-control studies and meta-analysis of cohort studies support a small increased cancer risk (NCRP 2013). The risk of cancer in offspring exposed to <100 mGy in utero, a value rarely exceeded in diagnostic procedures carried out today, is controversial and has not been fully resolved. Under the linear no-threshold theory of radiogenic cancer risk, however, any radiation dose carries with it a potential risk of cancer. We now consider the magnitude of radiation fetal doses received under different modalities used to image the pregnant mother.

12.5 Radiation Absorbed Dose to the Fetus from Diagnostic/Interventional Imaging

The vast majority of diagnostic radiology procedures deliver doses lower than 20 mGy to the uterus (ACR 2008). Fetal doses remain low (<10 mGy) for those diagnostic procedures where the fetal dose is mainly due to internal scatter radiation, or, in some cases, even if the fetus is partially irradiated by the primary beam (McCollough et al. 2007; Doshi et al. 2008; Lazarus et al. 2009; Bural et al. 2012; NCRP 2013). Maximum fetal doses from radiographs and diagnostic fluoroscopy procedures have been estimated at only 8–10 mGy, even if the conceptus is directly irradiated (Helmrot et al. 2007; NCRP 2013).

Doshi et al. (2008) estimated that maternal chest CT examinations deliver fetal doses of 0.06–0.23 mGy. Anthropomorphic phantom dosimetry studies of pulmonary angiograms using 4-, 16-, and 64-slice MDCT scanners yielded fetal dose estimates of 0.77, 0.54, and 0.33 mGy, respectively (Gilet et al. 2011). Abdominal CT scans of pregnant patients have been estimated to deliver fetal doses of approximately 15–21 mGy, depending on trimester (Gilet et al. 2011). The highest fetal doses due to diagnostic CT imaging are associated with aortic CT angiography, with estimated uterine doses of 34 mGy (McCollough et al. 2007). Among all diagnostic radiology procedures, the highest fetal doses (38 mGy) were associated with [67]Ga Citrate studies investigated by Russel et al. (1997b).

Radiologically guided interventional and radioiodine-based nuclear medicine therapy procedures can result in fetal doses much higher in magnitude than those associated with diagnostic imaging. Fluoroscopy-guided interventional procedures of the pelvic region must be governed by strict imaging protocols due to the potential, however small, for exposures to exceed 100 mGy (ACR 2008; Dauer et al. 2012). Diagnostic and therapeutic nuclear medicine studies involving radioiodine, particularly ^{131}I, must be avoided during pregnancy (Weetman 2007; Bushberg et al. 2012; Nguyen and Goodman 2012). Therapeutic nuclear medicine procedures involving ^{131}I could lead to extremely high fetal whole-body doses approaching 1,000 mGy (Bushberg et al. 2012). Of particular concern is the dose to the fetal thyroid, which can be significant, even for diagnostic administrations to the pregnant mother (Zanzonico and Becker 1991). Risks include in utero hypothyroidism and, as a result, cretinism in the exposed child—a condition of severely stunted physical and mental growth owing to untreated congenital deficiency of thyroid hormone. Stabin (2017) reports that the fetal thyroid dose per maternal administration of ^{131}I ranges from 230 to 270 mGy/MBq over the gestational age range of three to nine months.

With the exception of therapeutic nuclear medicine doses, the estimated ranges of fetal doses for diagnostic procedures and pelvic interventional procedures are summarized in Table 12.2. These doses are visually compared in Figure 12.1 to the NCRP 150 and 200 mGy dose thresholds and corresponding dose-dependent health risks, including *additional lifetime cancer risk* (ALCR). Aside from rare cases of interventional procedures of the pelvis, fetal doses do not approach any estimated tissue effect thresholds. However, because no dose threshold potentially exists for additional lifetime cancer risk, a dose-dependent increase in probability for this effect can be assumed. Adopting data reported by Preston et al. (2008), a fetal exposure of 100 mGy yields an ALCR of 4%, which is comparable to the lifetime cancer-free probability of 95% reported by McCollough et al. (2007).

Although fetal doses resulting from diagnostic and interventional radiography are typically well below dose thresholds for fetal tissue effects and potential conceptus death, a dose-dependent increase in lifetime cancer risk may be conservatively assumed for all non-zero doses. Thus, it is clearly prudent for radiologists

TABLE 12.2   Maximum Estimated Fetal Dose for Various Procedures Administered to the Pregnant Patient

Type of Procedure	Maximum Estimated Fetal Dose (mGy)	References
Computed tomography (CT)	34	McCollough et al. (2007)
Radiography and diagnostic fluoroscopy	8	Helmrot et al. (2007); UNSCEAR (2000)
Interventional fluoroscopy	Potentially > 100	Dauer et al. (2012)
Diagnostic nuclear medicine	38	Russell et al. (1997b)
Therapeutic nuclear medicine	1,000	Bushberg et al. (2012)

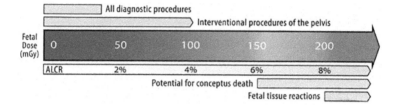

FIGURE 12.1   Comparison of typical ranges of fetal doses from diagnostic and interventional procedures and potential health risks, including additional lifetime cancer risk (ALCR).

and other associated medical personnel to practice reasonable dose reduction and optimization strategies when evaluating pregnant patients with radiologic imaging in order to mitigate the unborn child's additional lifetime cancer risk. As further radiation epidemiology studies continue, these in utero cancer risks will presumably be further refined, including any evidence of non-linearity in the dose response.

12.6 Computational Methods for Dose Assessment

Prospectively or retrospectively estimating radiation doses to the embryo or fetus for all radiologic procedures is an essential component of assessing dose and related health risks. Dose quantification and awareness can assist clinicians with the risk-benefit analysis associated with decisions to avoid or proceed with radiologic imaging of a pregnant patient. The following sections briefly summarize various methods for estimating fetal dose from radiologic procedures.

Wagner et al. (1997) presented comprehensive methods for estimating fetal radiation dose from radiographic, fluoroscopic, and CT procedures, as well as doses from radionuclide studies. Radiographic and fluoroscopic procedures are combined separately from CT, and both categories are further separated into in-field and out-of-field calculation methods. Several presented methods for estimating fetal doses from in-field radiographic and fluoroscopic procedures are based on depth-dose and tissue-air ratios. Two methods are presented for in-field CT doses: one based on free-in-air techniques and the other based on CT dose index (CTDI) measurement techniques presented by Felmlee et al. (1990) (discussed further below). Methods are also presented for estimating nuclear medicine doses from cumulated activity and radionuclide S-values.

Felmlee et al. (1990) developed methods for adjusting CTDI measurements to obtain representative fetal doses. The method involves recording ion chamber measurements over a range of various examination parameters and constructing a library of a dose-to-fetus metric for every slice in a given scan. This metric is termed the "normalized fetal-dose ratio (NFDR)" and is defined as the measured fetal-dose contribution from a single CT scan divided by the CTDI value measured for the same scan parameters. It is important to note that NFDR values from single scans extending well beyond the fetal length are needed for accurate dose estimates. Prospective fetal doses from a given scan can then be estimated by summing the NFDR values over the scan length. Accurate acquisition of reference CTDI measurements is critical. In order for this method to provide reliable fetal dose estimates, the CTDI measurements must be acquired with one's own CT scanner, scanning techniques, ionization chambers, and physical phantoms.

Doses from ionizing radiation can also be estimated by computationally simulating radiation exposures of virtual models of biological systems, including humans (Bolch et al. 2010). These virtual surrogates, or computational phantoms, provide tremendous versatility for calculating doses from a wide range of scenarios, which include radiation protection, radiation therapy, and diagnostic medical imaging. Current computational phantoms are available in three forms or types: (1) stylized phantoms—individual organs and tissues are modeled with shapes composed of mathematically defined surfaces, (2) voxel phantoms—patient-specific anatomy as visualized in medical images (e.g., CT) is contoured and converted into 3D pixels or "voxels," and (3) hybrid/boundary-representation (BREP)—medical images are contoured and converted to deformable non-uniform rational B-spline (NURBS) surfaces and further adjusted to match various anatomical criteria (e.g., patient height and weight).

Computational phantoms representing the anatomies of the pregnant female and fetus are available in all three phantom types, in varying degrees of maternal and fetal anatomical detail. Mathematically stylized pregnant female phantoms were developed by Stabin et al. (1995) (Figure 12.2) and subsequently refined by Chen (2004). In these studies, the fetus is represented by a cylindrical shell. Shi and Xu (2004)

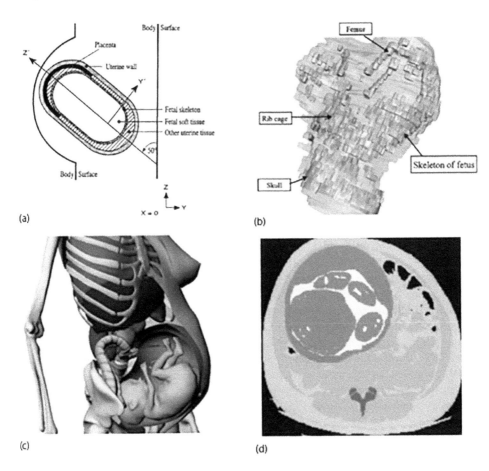

FIGURE 12.2 Graphical survey of previously published fetal and pregnant female computational phantoms: (a) Stylized model of the 36-week fetus. (From Stabin, M. et al., *Mathematical Models and Specific Absorbed Fractions of Photon Energy in the Nonpregnant Adult Female and at the End of Each Trimester of Pregnancy*, ORNL/TM-12907, Oak Ridge National Laboratory, Oak Ridge, TN, 1995.); (b) Voxel model of the 30-week fetus. (From Shi, C., and Xu, X.G., *Med. Phys.*, 31, 2491–2497, 2004.); (c) Hybrid/BREP model of the 24-week fetus. (From Xu, X.G. et al., *Phys. Med. Biol.*, 52, 7023–7044, 2007.); and (d) Voxel model of a 35-week fetus from the 24-patient study. (From Angel, E. et al., *Radiology*, 249, 220–227, 2008.)

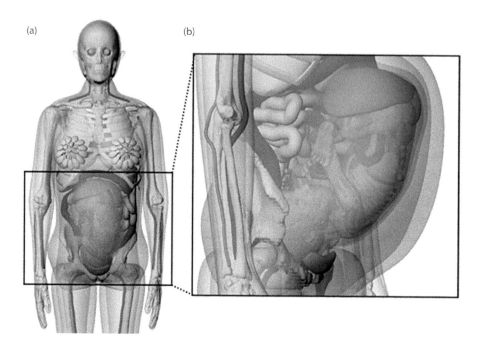

FIGURE 12.3   Hybrid computational phantom of a 38-week pregnant female. (From Maynard, M. et al., *Phys. Med. Biol.*, 59, 4325–4343, 2014.) (a) front view and (b) right-oblique view (magnified).

presented a 30-week fetal voxel phantom, which was refined and incorporated into the development of a series of hybrid/BREP phantoms representing the pregnant female at each trimester (Xu et al. 2007) (Figure 12.2). Angel et al. (2008) reported a series of pregnant female voxel phantoms based on a 24-patient retrospective study, covering fetal ages from 5 to 36 weeks (Figure 12.2). Maynard et al. (2011) developed a series of detailed hybrid fetal phantoms representing eight fetal ages, which were recently incorporated into a series of hybrid computational phantoms of the pregnant female (Maynard et al. 2014) (see Figure 12.3). Becker et al. (2008) presented a voxel model representing the pregnant female at the 24th week of pregnancy.

Computational phantoms can be coupled with Monte Carlo software codes, which use random numbers to simulate the physics of radiation interactions with matter (e.g., human tissues). Such codes allow users to estimate, among many other physical quantities, radiation doses due to exposure from a wide range of sources, including CT, fluoroscopy, and nuclear medicine. Fetal doses due to radiologic imaging of pregnant patients have been estimated using these methods for several modalities. Angel et al. (2008) performed Monte Carlo-based simulations of helical abdominal and pelvic CT scans based on the GE series of multi-slice scanners and observed strong correlations between fetal dose and maternal circumference and fetal depth. Damilakis et al. (2010) developed a method for estimating conceptus dose (<eight weeks) that can be applied to different scanner manufacturers using ratios of weight-CTDI and free-in-air CTDI. Gu et al. (2009) simulated chest and kidney CT scans of pregnant patients for a GE scanner and, later, incorporated tube current modulation and observed fetal dose reductions between 14% and 25% (Gu et al. 2013).

Fetal doses from nuclear medicine procedures can be estimated when appropriate radionuclide biokinetic models are coupled with photon- and electron-specific absorbed fractions (SAFs) (Stabin et al. 1995; Russell et al. 1997a, 1997b; Saunders et al. 2002; Keleher et al. 2004; Stabin 2004; Pandit-Taskar et al. 2006; Shi et al. 2008; Guo et al. 2010; Stabin 2017). Skin doses to non-pregnant patients from fluoroscopic procedures have been investigated by Johnson et al. (2011). Future expansion of these simulations could include internal organ and fetal doses of pregnant patients undergoing fluoroscopic procedures.

12.7 Methods of Dose Optimization and Reduction

Imaging modalities involving ionizing radiation are being increasingly utilized to evaluate pregnant women (Lazarus et al. 2009). While the majority of radiologic imaging doses to the fetus are justified and well below accepted threshold for fetal tissue reactions, the possibility of ALCR is present at all levels of radiation exposure and increases in probability as a function of dose. It is therefore imperative for health-care providers to continue to maintain imaging doses to the fetus as low as reasonably achievable (ALARA). The following section summarizes some of the principle components for accomplishing this goal in a clinical setting, including pre-imaging considerations and modality-specific strategies. Because it is not possible to review in detail all aspects of this complicated issue, the reader is encouraged to explore the wealth of available literature for additional information related to these topics.

12.7.1 Pre-imaging Considerations

Some of the most critical measures for minimizing fetal dose from radiologic imaging should occur well before the patient is considered for a particular imaging study. Recent studies have reported deficiencies in the awareness of many physicians regarding the risks of imaging pregnant patients (Ratnapalan et al. 2004; Groves et al. 2006; Patel et al. 2007), particularly when advising patients of medical termination of pregnancy following inadvertent fetal imaging exposure (Ratnapalan et al. 2004; Brent 2009; NCRP 2013). Many emergency medicine and obstetric physicians scored poorly regarding the safety and appropriateness of radiologic imaging for pregnant women (Johnson et al. 2014). It is essential that physicians and other health-care personnel are properly educated in the appropriate aspects of radiologic imaging of the pregnant patient so that rational, informed decisions regarding patient management can be reached.

Institutional strategies, protocols, and policies governing radiologic imaging, particularly of pregnant patients, should be established and reviewed by pertinent clinical experts well in advance of the imaging study (ACR 2008; Frush et al. 2009; NCRP 2013). Modifications to standard protocols should be based on state-of-the-art literature, local resources, staff experience, and should agree with current guidelines (NCRP 2013). Imaging devices should be regularly tested and maintained (e.g., beam quality and machine output) to ensure optimal performance (Shaw et al. 2011). A key component of minimizing radiation dose to the fetus from radiologic imaging is determination of pregnancy (ACR 2008), as an oversight in this area could lead to higher than necessary fetal doses. With the exception of emergent cases, patients of menstrual age, generally between 12 and 55 years of age, should be questioned about their pregnancy status via a standardized form (ACR 2008; NCRP 2013). Trauma is the leading cause of non-obstetric maternal death, with blunt trauma yielding a significant fetal mortality rate (Baerga-Varela et al. 2000; Grossman 2004). For such emergent cases, radiologic evaluation should occur as rapidly as possible and without hesitation (Patel et al. 2007). The American College of Radiology (ACR 2008) reports several imaging studies that can be safely performed regardless of pregnancy status: mammography, chest radiography during the first and second trimesters, extremity radiography or CT (with the possible exception of the hip), and any diagnostic examination of the head or neck. However, alternative imaging modalities that do not involve ionizing radiation, such as ultrasonography (US) and MRI, should be employed in all cases where they are feasible, available, and do not compromise the clinical care of the mother or unborn child (Patel et al. 2007; ACR 2008; Marx 2010; Shetty et al. 2010; Shaw et al. 2011; NCRP 2013).

12.7.2 Modality-Specific Strategies

In the event that radiologic imaging is the preferred course of action, it is best to establish protocols for imaging pregnant patients in advance to minimize the need for reactive adjustments. It is critical that any efforts to mitigate fetal dose must not interfere with a clinician's ability to acquire the desired diagnostic information (i.e., acquired images have sufficient diagnostic quality) (ACR 2008). As stated previously, modifications to standard protocols should be based on state-of-the-art literature and currently accepted guidelines (NCRP 2013). A qualified medical physicist should provide fetal dose estimates, either prospectively or retrospectively, to facilitate the risk-benefit analysis of potential medical procedures, particularly if multiple CT procedures of the abdomen and pelvis are involved (ACR 2008; NCRP 2013). The following sections summarize some general dose-reduction strategies for each modality.

12.7.2.1 Planar Radiography

As noted, the ACR (2008) reports several radiographic studies that can be safely performed regardless of pregnancy status: mammography, chest radiography during the first and second trimesters, and extremity (with the possible exception of the hip), head or neck radiography. For well-collimated beams, technique parameters for these examinations do not need to be altered from those prescribed for nonpregnant patients (NCRP 2013). For radiography that places the fetus within the primary beam, NCRP Report No. 174 also provides recommendations for appropriately adjusting exposure factors, namely, that they should not be reduced below those prescribed for non-pregnant patients and may need to be increased to account for the increased patient thickness. The NCRP additionally recommends acquiring the minimum number of projections for obtaining the desired diagnostic information. The NCRP also contends that shielding of the abdomen or pelvis will typically provide limited fetal dose reduction but may serve to ease concerns of the patient and family members.

Damilakis et al. (2003) provides several technique recommendations for chest radiography of pregnant patients, many of which could be applied to other radiographic procedures: image the patient in a standing position, tightly collimate the beam, use the most efficient x-ray tube available, increase the tube potential (kVp) to the highest setting that results in acceptable image quality, acquire posterior-anterior (PA) projections rather than anterior-posterior (AP) projections, adopt non-grid techniques and, if clinically permitted, use fast screen-film image receptors. Damilakis et al. (2003) also contends that lead aprons provide limited dose reduction due to internal scatter, yet may provide valuable assurances to patients and family members.

12.7.2.2 Computed Tomography

CT imaging protocols for pregnant patients may include, but are not limited to, reducing the number of images for a given exam, limiting the number of passes over the abdomen in a CT examination, and limiting the scan length of examinations. Multi-phase studies should be avoided (Wagner and Huda 2004; Shetty 2010; Litmanovich et al. 2014). When imaging the abdomen or pelvis, imaging systems with preset automatic noise-control technology can mitigate dose to the fetus by limiting x-ray tube output to the minimum mAs needed for an acceptable signal-to-noise ratio (SNR) (Patel et al. 2007; Shetty 2010; NCRP 2013; Litmanovich et al. 2014). Tube current modulation (TCM), particularly z-axis modulation, should be employed when available (Doshi et al. 2008). Reduction of tube potential, when possible, is advised (NCRP 2013; Litmanovich et al. 2014). It has been shown that a reduction in tube potential from 140 to 100 kV reduced conceptus dose by approximately 60% (Dauer et al. 2012). Pitch values greater than unity can also contribute to decreases in fetal dose (Patel et al. 2007; Litmanovich et al. 2014). Dose-reduction techniques such as these can be implemented in particular in cases where urolithiasis is suspected, as standard CT image quality is not necessary for the detection of high-contrast stones (McCollough et al. 2007; Patel et al. 2007). Similar dose reduction techniques are reported by Dauer (2012) for CT-guided interventional studies.

Wrapping the patient's abdomen or pelvis with lead shielding during non-pelvic CT imaging studies is typically not considered a highly beneficial dose reduction method (NCRP 2013). However, Doshi (2008) recently explored methods for fetal dose reduction in near-term pulmonary CT angiography and reported that mA modulation, lead coat shielding, and a reduction in scan length yielded fetal dose reductions of 10%, 35%, and 56%, respectively. Regardless of potential dose reduction benefits, as in all forms of diagnostic radiology, the presence of shielding may help ease the anxiety of the patient and their family members (NCRP 2013). Users must be especially weary of placing shielding material within the primary beam while simultaneously employing TCM, as this configuration can increase tube output and result in increased fetal doses (NCRP 2013).

12.7.2.3 Diagnostic and Interventional Fluoroscopy

Keeping the fetus out of the direct beam is a highly effective way to minimize fetal dose (Dauer et al. 2012). If the dose to the fetus is due solely to internal scatter radiation, then typically it will remain well below the accepted risk levels (Miller et al. 2010). It is therefore critical to collimate the useful beam as much as possible and minimize primary-beam exposure of the fetus (Dauer et al. 2012). Fluoroscopic procedures and the resulting doses are highly dependent on the experience of personnel administering the examination, leading to several avenues for fetal dose reduction. Of the three components to radiation exposure in fluoroscopic procedures—fluoroscopy, radiographic spot images, and cine angiography—fluoroscopy contributes the most to patient radiation dose (Shaw et al. 2011). Fetal dose reduction strategies for diagnostic and interventional fluoroscopy are similar to those used in CT.

The appropriate characteristics of fluoroscopy equipment (e.g., adequate beam filtration) should be tested by a qualified medical physicist (Miller et al. 2003; Shaw et al. 2011). The use of magnification, exposure-rate setting, fluoroscopic time, and the distance from the patient to the image intensifier should all be minimized (Marx 2010; Shaw et al. 2011; Dauer et al. 2012; NCRP 2013). Tube potential should be maximized and tube current minimized to achieve a reasonable compromise between image quality and radiation dose (Dauer et al. 2012). Removal of the grid should be considered for smaller patients or if the image intensifier cannot be positioned closely to the patient (Dauer et al. 2012). Distance from source to patient should be maximized, and, if available, pulsed fluoroscopic technology should be used (Marx 2010; Shaw et al. 2011; Dauer et al. 2012; NCRP 2013). PA beam projections are preferred when possible (Theocharopoulos et al. 2006; Dauer et al. 2012). Video recording or image-storage archives should be used in the place of spot films (Dauer et al. 2012; NCRP 2013). Exposure from digital subtraction angiography (DSA) should be minimized (Marx 2010; Dauer et al. 2012).

Though fetal doses associated with the majority of interventional cases will remain lower than 100 mGy, more complicated cases may result in doses far above this value (Miller et al. 2003). Due to the increased risk and complexity of these procedures, a greater level of active involvement by appropriate personnel, including a qualified medical physicist, will likely be necessary (NCRP 2013). As with other diagnostic radiology modalities, appropriate placement of lead shielding during non-pelvic fluoroscopic or interventional studies can promote patient well-being but will likely have little impact in terms of dose reduction due to internal scatter (Marx 2010; Dauer et al. 2012).

12.7.2.4 Nuclear Medicine

Dose estimations reported in the current literature conclude that the benefits of nuclear medicine examinations greatly outweigh fetal dose risks when administered correctly (Bural et al. 2012). Aggressive intravenous hydration and frequent voiding, possibly facilitated by Foley catheterization, are frequently adopted to reduce dose to the fetus with radiopharmaceuticals that have rapid clearance via the kidneys (Bural et al. 2012; NCRP 2013). If the patient is able to lie still for extended periods of time, a lower administered activity combined with longer imaging time can help limit fetal dose (NCRP 2013; Astani et al. 2014). Bural et al. (2012) recommended a 50% reduction in administered activity and doubling the imaging time is possible in most cases. For detection of pulmonary embolism, Astani et al. (2014) recommended reducing administered activity by 50%–75% and, under certain circumstances, completely eliminating the ventilation component of the study. Reductions in administered activity should be considered with caution, as patient motion could potentially render the imaging study useless, possibly creating the need for re-imaging (NCRP 2013).

Pregnancy is a contraindication to radioiodine imaging and therapy treatment due to placental transfer and subsequent uptake into the fetal thyroid (Weetman 2007; Nguyen and Goodman 2012). If scintigraphic thyroid imaging of a pregnant patient is necessary, then Tc-99m pertechnetate is preferred (Nguyen and Goodman 2012).

12.8 Summary and Future Research Needs

It has been shown that radiation doses to the embryo and fetus following medical imaging of the pregnant mother—in nuclear medicine, in CT, and in interventional fluoroscopy—are typically well below dose thresholds for tissue reactions. Consequently, the only biological effect of potential concern is the residual risk of cancer. The magnitude and functional shape of the dose-response for cancer induction following in utero fetal exposure is not well known and is a focus of continuing study in radiation epidemiological reviews of exposed cohorts. Under the linear no-threshold hypothesis, these risks should be minimized and the imaging procedures subject to benefit/risk optimization.

While approximate methods are available for rough estimates of whole-body fetal absorbed dose, computational models of the pregnant female, including quite elaborate models of the developing fetal organs, are becoming available and may be applied in Monte Carlo radiation transport simulations to create pre-computed dose libraries for fetal organ dose for all relevant imaging modalities (Ding et al. 2015). These newer-generation models can also be used to alter maternal abdominal girth, thus virtually simulating women of differing body weight prior to pregnancy, or women of differing degrees of weight gain during pregnancy. The capability of matching patient to phantom is particularly important for external exposures of the fetus (CT and fluoroscopy), as increasing maternal tissue thickness can greatly alter fetal doses following photon internal scatter and attenuation of the x-ray fields.

Monte Carlo simulations of fetal dose for CT imaging show that in many cases a single whole-body dose estimate to the fetus, while reliably approximating fetal soft tissue doses, may result in significant underestimates of radiation dose to fetal bone marrow owing to enhanced photoelectric absorption in fetal bone (Gu et al. 2009, 2013). The fetal bone marrow is of particular importance as a target organ for estimating leukemia risk. Existing models of the fetal skeleton, as in the NURBS-based series by Maynard et al. (2011), are constructed using various approximations regarding the density of newly formed fetal bone and the marrow volume fractions of fetal spongiosa. New imaged-based studies are warranted to better define not only the fetal bone microarchitecture but also the bone-specific onset of hematopoiesis. These new data can thus be used to better refine imaging dose to the fetal bone marrow as needed for stochastic risk estimation and as needed for imaging benefit/risk optimization.

References

ACOG 2004 Guidelines for diagnostic imaging during pregnancy (ACOG Committee Opinion No. 299) *Obstet Gynecol* **104** 647–651.

ACR 2008 ACR practice guideline for imaging pregnant or potentially pregnant adolescents and women with ionizing radiation (Reston, VA: American College of Radiology).

Angel E, Wellnitz C V, Goodsitt M M, Yaghmai N, DeMarco J J, Cagnon C H, Sayre J W et al. 2008 Radiation dose to the fetus for pregnant patients undergoing multidetector CT imaging: Monte Carlo simulations estimating fetal dose for a range of gestational age and patient size *Radiology* **249** 220–227.

Astani S A, Davis L C, Harkness B A, Supanich M P and Dalal I 2014 Detection of pulmonary embolism during pregnancy: Comparing radiation doses of CTPA and pulmonary scintigraphy *Nuclear Med Commun* **35** 704–711.

Augustin G and Majerovic M 2007 Non-obstetrical acute abdomen during pregnancy *Eur J Obstet Gynecol Reprod Biol* **131** 4–12.

Baerga-Varela Y, Zietlow S P, Bannon M P, Harmsen W S and Ilstrup D M 2000 Trauma in pregnancy *Mayo Clinic Proc* **75** 1243–1248.

Balter S, Hopewell J W, Miller D L, Wagner L K and Zelefsky M J 2010 Fluoroscopically guided interventional procedures: A review of radiation effects on patients' skin and hair *Radiology* **254** 326–341.

Becker J, Zankl M, Fill U and Hoeschen C 2008 Katja—The 24-week of virtual pregnancy for dosimetric calculations *Pol J Med Phys Eng* **14** 13–19.

Bithell J F and Stewart A M 1975 Prenatal irradiation and childhood malignancy—Review of British data from oxford survey *Br J Cancer* **31** 271–287.

Bolch W, Lee C, Wayson M and Johnson P 2010 Hybrid computational phantoms for medical dose reconstruction *Radiat Environ Biophys* **49** 155–168.

Brent R L 2009 Saving lives and changing family histories: Appropriate counseling of pregnant women and men and women of reproductive age, concerning the risk of diagnostic radiation exposures during and before pregnancy *Am J Obstet Gynecol* **200** 4–24.

Bural G G, Laymon C M and Mountz J M 2012 Nuclear imaging of a pregnant patient: Should we perform nuclear medicine procedures during pregnancy? *Mol Imaging Radionucl Ther* **21** 1–5.

Bushberg J, Seibert J, Leidholdt Jr. E and Boone J 2012 *The Essential Physics of Medical Imaging* (Philadelphia, PA: Lippincott Williams & Wilkins)

Chen J 2004 Mathematical models of the embryo and fetus for use in radiological protection *Health Phys* **86** 285–295.

Chen M M, Coakley F V, Kaimal A and Laros R K, Jr. 2008 Guidelines for computed tomography and magnetic resonance imaging use during pregnancy and lactation *Obstet Gynecol* **112** 333–340.

Damilakis J, Perisinakis K, Prassopoulos P, Dimovasili E, Varveris H and Gourtsoyiannis N 2003 Conceptus radiation dose and risk from chest screen-film radiography *Eur Radiol* **13** 406–412.

Damilakis J, Perisinakis K, Tzedakis A, Papadakis A E and Karantanas A 2010 Radiation dose to the conceptus from multidetector CT during early gestation: A method that allows for variations in maternal body size and conceptus position *Radiology* **257** 483–489.

Dauer L T, Thornton R H, Miller D L, Damilakis J, Dixon R G, Marx M V, Schueler B A et al., Society of Interventional Radiology S, Health C, Cardiovascular and Interventional Radiology Society of Europe Standards of Practice C 2012 Radiation management for interventions using fluoroscopic or computed tomographic guidance during pregnancy: A joint guideline of the society of interventional radiology and the cardiovascular and interventional radiological society of Europe with endorsement by the Canadian interventional radiology association *J Vasc Interv Radiol* **23** 19–32.

Diamond E L, Schmerler H and Lilienfeld A M 1973 The relationship of intra-uterine radiation to subsequent mortality and development of leukemia in children. A prospective study *Am J Epidemiol* **97** 283–313.

Ding A, Gao Y, Liu H, Caracappa P F, Long D J, Bolch W E, Liu B and Xu X G 2015 VirtualDose: A software for reporting organ doses from CT for adult and pediatric patients *Phys Med Biol* **60** 5601–5625.

Doshi S K, Negus I S and Oduko J M 2008 Fetal radiation dose from CT pulmonary angiography in late pregnancy: A phantom study *Br J Radiol* **81** 653–658.

Felmlee J P, Gray J E, Leetzow M L and Price J C 1990 Estimated fetal radiation dose from multislice CT studies *AJR. Am J Roentgenol* **154** 185–190.

Frush D P, Frush K S and Oldham K T 2009 Imaging of acute appendicitis in children: EU versus US... or US versus CT? A North American perspective *Pediatric Radiol* **39** 500–505.

Giles D, Hewitt D, Stewart A and Webb J 1956 Malignant disease in childhood and diagnostic irradiation in utero *Lancet* **271** 447.

Gilet A G, Dunkin J M, Fernandez T J, Button T M and Budorick N E 2011 Fetal radiation dose during gestation estimated on an anthropomorphic phantom for three generations of CT scanners *AJR. Am J Roentgenol* **196** 1133–1137.

Grossman N B 2004 Blunt trauma in pregnancy *Am Fam Physician* **70** 1303–1310.

Groves A M, Yates S J, Win T, Kayani I, Gallagher F A, Syed R, Bomanji J and Ell P J 2006 CT pulmonary angiography versus ventilation-perfusion scintigraphy in pregnancy: Implications from a UK survey of doctors' knowledge of radiation exposure *Radiology* **240** 765–770.

Gu J, Bednarz B, Caracappa P F and Xu X G 2009 The development, validation and application of a multi-detector CT (MDCT) scanner model for assessing organ doses to the pregnant patient and the fetus using Monte Carlo simulations *Phys Med Biol* **54** 2699–2717.

Gu J, Xu X G, Caracappa P F and Liu B 2013 Fetal doses to pregnant patients from CT with tube current modulation calculated using Monte Carlo simulations and realistic phantoms *Radiat Prot Dosimetry* **155** 64–72.

Guo B Q, Xu X G and Shi C Y 2010 Specific absorbed fractions for internal electron emitters derived for a set of anatomically realistic reference pregnant female models *Radiat Prot Dosim* **138** 20–28.

Helmrot E, Pettersson H, Sandborg M and Alten J N 2007 Estimation of dose to the unborn child at diagnostic X-ray examinations based on data registered in RIS/PACS *Eur Radiol* **17** 205–209.

ICRP 2003 ICRP Publication 90: Biological effects after prenatal irradiation (embryo and fetus) *Ann ICRP* **33** 1–200.

Jha S, Ho A, Bhargavan M, Owen J B and Sunshine J H 2010 Imaging evaluation for suspected pulmonary embolism: What do emergency physicians and radiologists say? *AJR Am J Roentgenol* **194** W38–W48.

Johnson A M, Dahn C M and Cadiz L A 2014 Deficiencies in physician knowledge of the risks of imaging in pregnancy *Obstet Gynecol* **123** 58s.

Johnson P B, Borrego D, Balter S, Johnson K, Siragusa D and Bolch W E 2011 Skin dose mapping for fluoroscopically guided interventions *Med Phys* **38** 5490–5499.

Keleher A, Wendt R, 3rd, Delpassand E, Stachowiak A M and Kuerer H M 2004 The safety of lymphatic mapping in pregnant breast cancer patients using Tc-99m sulfur colloid *Breast J* **10** 492–495.

Lazarus E, Debenedectis C, North D, Spencer P K and Mayo-Smith W W 2009 Utilization of imaging in pregnant patients: 10-year review of 5270 examinations in 3285 patients—1997–2006 *Radiology* **251** 517–524.

Lazarus E, Mayo-Smith W W, Mainiero M B and Spencer P K 2007 CT in the evaluation of nontraumatic abdominal pain in pregnant women *Radiology* **244** 784–790.

Lewis T L 1960 Leukaemia in childhood after antenatal exposure to x rays. A survey at Queen Charlotte's Hospital *Br Med J* **2** 1551–1552.

Litmanovich D E, Tack D, Lee K S, Shahrzad M and Bankier A A 2014 Cardiothoracic imaging in the pregnant patient *J Thorac Imaging* **29** 38–49.

Marx M V 2010 Interventional radiology: Management of the pregnant patient *Tech Vasc Interv Radiol* **13** 154–157.

Maslovitz S, Gutman G, Lessing J B, Kupferminc M J and Gamzu R 2003 The significance of clinical signs and blood indices for the diagnosis of appendicitis during pregnancy *Gynecol Obstet Invest* **56** 188–191.

Maynard M R, Geyer J W, Aris J P, Shifrin R Y and Bolch W 2011 The UF family of hybrid phantoms of the developing human fetus for computational radiation dosimetry *Phys Med Biol* **56** 4839–4879.

Maynard M R, Long N S, Moawad N S, Shifrin R Y, Geyer A M, Fong G and Bolch W E 2014 The UF Family of hybrid phantoms of the pregnant female for computational radiation dosimetry *Phys Med Biol* **59** 4325–4343.

McCollough C H, Schueler B A, Atwell T D, Braun N N, Regner D M, Brown D L and LeRoy A J 2007 Radiation exposure and pregnancy: When should we be concerned? *Radiographics* **27** 909–917; discussion 917–908.

Miller D L, Balter S, Cole P E, Lu H T, Schueler B A, Geisinger M, Berenstein A et al. 2003 Radiation doses in interventional radiology procedures: The RAD-IR Study, Part I: Overall measures of dose *J Vasc Interv Radiol* **14** 711–727.

Miller D L, Balter S, Schueler B A, Wagner L K, Strauss K J and Vano E 2010 Clinical radiation management for fluoroscopically guided interventional procedures *Radiology* **257** 321–332.

Monson R and MacMahon B 1984 Prenatal x-ray exposure and cancer in children, In *Radiation Carcinogenesis: Epidemiology and Biological Significance*, J D Boice and J F Fraumeni, Jr. (Eds.) (New York: Raven Press).

NCRP 1977 *NCRP Report No. 54: Medical Radiation Exposure of Pregnant and Potentially Pregnant Women* (Bethesda, MD: National Council on Radiation Protection and Measurement).

NCRP 2010 *NCRP Report No. 168: Radiation Dose Management for Fluoroscopically Guided Interventional Medical Procedures* (Bethesda, MD: National Council on Radiation Protection and Measurement).

NCRP 2013 *NCRP Report No. 174: Preconception and Prenatal Radiation Exposure: Health Effects and Protective Guidance* (Bethesda, MD: National Council on Radiation Protection and Measurement).

Nguyen C P and Goodman L H 2012 Fetal risk in diagnostic radiology *Semin Ultrasound, CT, and MR* **33** 4–10.

Oto A, Ernst R D, Shah R, Koroglu M, Chaljub G, Gei A F, Zacharias N and Saade G 2005 Right-lower-quadrant pain and suspected appendicitis in pregnant women: Evaluation with MR imaging—initial experience *Radiology* **234** 445–451.

Pandit-Taskar N, Dauer L T, Montgomery L, St Germain J, Zanzonico P B and Divgi C R 2006 Organ and fetal absorbed dose estimates from Tc-99m-sulfur colloid lymphoscintigraphy and sentinel node localization in breast cancer patients *J Nucl Med* **47** 1202–1208.

Parangi S, Levine D, Henry A, Isakovich N and Pories S 2007 Surgical gastrointestinal disorders during pregnancy *Am J Surg* **193** 223–232.

Patel S J, Reede D L, Katz D S, Subramaniam R and Amorosa J K 2007 Imaging the pregnant patient for nonobstetric conditions: Algorithms and radiation dose considerations *Radiographics* **27** 1705–1722.

Preston D L, Cullings H, Suyama A, Funamoto S, Nishi N, Soda M, Mabuchi K, Kodama K, Kasagi F and Shore R E 2008 Solid cancer incidence in atomic bomb survivors exposed in utero or as young children *J Nat Cancer Inst* **100** 428–436.

Rajaraman P, Simpson J, Neta G, Berrington de Gonzalez A, Ansell P, Linet M S, Ron E and Roman E 2011 Early life exposure to diagnostic radiation and ultrasound scans and risk of childhood cancer: Case-control study *Br Med J* **342** d472

Ratnapalan S, Bona N, Chandra K and Koren G 2004 Physicians' perceptions of teratogenic risk associated with radiography and CT during early pregnancy *AJR. Am J Roentgenol* **182** 1107–1109.

Ray J G, Schull M J, Urquia M L, You J J, Guttmann A and Vermeulen M J 2010 Major radiodiagnostic imaging in pregnancy and the risk of childhood malignancy: A population-based cohort study in Ontario *PLoS Med* **7** e1000337

RERF 2005 *Reassessment of the Atomic Bomb Radiation Dosimetry for Hiroshima and Nagasaki—Dosimetry System 2002* (Hiroshima and Nagasaki, Japan: Radiation Effects Research Foundation)

Russell J R, Stabin M G and Sparks R B 1997a Placental transfer of radiopharmaceuticals and dosimetry in pregnancy *Health Phys* **73** 747–755.

Russell J R, Stabin M G, Sparks R B and Watson E 1997b Radiation absorbed dose to the embryo/fetus from radiopharmaceuticals *Health Phys* **73** 756–769.

Salonen T and Saxen L 1975 Risk indicators in childhood malignancies *Int J Cancer* **15** 941–946.

Saunders M, Palmer M, Preece A and Millard R 2002 Model-based comparison of maternal and foetal organ doses from Tc-99m pertechnetate, DMSA, DTPA, HDP, MAA and MAG(3) diagnostic intakes during pregnancy *Eur J Nuc Med Molec Imag* **29** 1365–1373.

Schuz J, Deltour I, Krestinina L Y, Tsareva Y V, Tolstykh E I, Sokolnikov M E and Akleyev A V 2017 In utero exposure to radiation and haematological malignancies: Pooled analysis of Southern Urals cohorts *Br J Cancer* **116** 126–133.

Shaw P, Duncan A, Vouyouka A and Ozsvath K 2011 Radiation exposure and pregnancy *J Vasc Surg* **53** 28s–34s

Shetty M K 2010 Abdominal computed tomography during pregnancy: A review of indications and fetal radiation exposure issues *Semin Ultrasound, CT, and MR* **31** 3–7.

Shetty M K, Garrett N M, Carpenter W S, Shah Y P and Roberts C 2010 Abdominal computed tomography during pregnancy for suspected appendicitis: A 5-year experience at a maternity hospital *Semin Ultrasound, CT, and MR* **31** 8–13.

Shi C and Xu X G 2004 Development of a 30-week-pregnant female tomographic model from computed tomography (CT) images for Monte Carlo organ dose calculations *Med Phys* **31** 2491–2497.

Shi C Y, Xu X G and Stabin M G 2008 SAF values for internal photon emitters calculated for the RPI-P pregnant-female models using Monte Carlo methods *Med Phys* **35** 3215–3224.

Stabin M G 2004 Proposed addendum to previously published fetal dose estimate tables for F-18-FDG *J Nucl Med* **45** 634–635.

Stabin M G 2017 Radiation dose and risks to fetus from nuclear medicine procedures *Phys Medica* **43** 190–198.

Stabin M, Watson E, Cristy M, Ryman J, Eckerman K, Davis J, Marshall D and Gehlen M 1995 *Mathematical Models and Specific Absorbed Fractions of Photon Energy in the Nonpregnant Adult Female and at the End of Each Trimester of Pregnancy* (Oak Ridge, TN: Oak Ridge National Laboratory) ORNL/TM-12907.

Stewart A, Webb J and Hewitt D 1958 A survey of childhood malignancies *Br Med J* **1** 1495–1508.

Tang S J, Mayo M J, Rodriguez-Frias E, Armstrong L, Tang L, Sreenarasimhaiah J, Lara L F and Rockey D C 2009 Safety and utility of ERCP during pregnancy *Gastrointest Endosc* **69** 453–461.

Theocharopoulos N, Damilakis J, Perisinakis K, Papadokostakis G, Hadjipavlou A and Gourtsoyiannis N 2006 Fluoroscopically assisted surgical treatments of spinal disorders: Conceptus radiation doses and risks *Spine* **31** 239–244.

Timins J 2002 Pregnancy and medical radiation *J Women's Imaging* **4** 31–32.

Toglia M R and Weg J G 1996 Venous thromboembolism during pregnancy *New Engl J Med* **335** 108–114.

Tran P, Desimone S, Barrett M and Bachrach B 2010 I-131 treatment of graves' disease in an unsuspected first trimester pregnancy; the potential for adverse effects on the fetus and a review of the current guidelines for pregnancy screening *Int J Pediatr Endocrinol* **2010** 858359.

Ueberrueck T, Koch A, Meyer L, Hinkel M and Gastinger I 2004 Ninety-four appendectomies for suspected acute appendicitis during pregnancy *World J Surg* **28** 508–511.

UNSCEAR 2000 United Nations Scientific Committee on the Effects of Atomic Radiation, Sources and Effects of Ionizing Radiation, *UNSCEAR 2000 Report to the General Assembly, with Scientific Annexes, Volume II: Effects*, No. E00.IX.4 (New York, United Nations Publications).

Wagner L K and Huda W 2004 When a pregnant woman with suspected appendicitis is referred for a CT scan, what should a radiologist do to minimize potential radiation risks? *Pediatr Radiol* **34** 589–590.

Wagner L, Lester R and Saldana L 1997 *Exposure of the Pregnant Patient to Diagnostic Radiations: A Guide to Medical Management* (Madison, WI: Medical Physics Publishing).

Weetman A P 2007 Radioiodine treatment for benign thyroid diseases *Clin Endocrinol* **66** 757–764.

Weingold A B 1983 Appendicitis in pregnancy *Clin Obstet Gynecol* **26** 801–809.

Xu X G, Taranenko V, Zhang J and Shi C 2007 A boundary-representation method for designing whole-body radiation dosimetry models: Pregnant females at the ends of three gestational periods—RPI-P3, -P6 and -P9 *Phys Med Biol* **52** 7023–7044.

Zanotti-Fregonara P, Champion C, Trebossen R, Maroy R, Devaux J Y and Hindie E 2008 Estimation of the beta plus dose to the embryo resulting from F-18-FDG administration during early pregnancy *J Nucl Med* **49** 679–682.

Zanotti-Fregonara P, Jan S, Taieb D, Cammilleri S, Trebossen R, Hindie E and Mundler O 2010 Absorbed F-18-FDG dose to the fetus during early pregnancy *J Nucl Med* **51** 803–805.

Zanzonico P B and Becker D V 1991 Radiation hazards in children born to mothers exposed to iodine-131, In *The Thyroid and Pregnancy*, C Beckers and D Reinwein (Eds.) (Stuttgart, Germany: Schattauer Publishers) pp 189–202.

III

Radiation Risks in Medical Imaging

13

Risk of Low-Level Exposure to Radiation-Biological Basis

Tatjana Paunesku
and Gayle E.
Woloschak

13.1 Introduction

Despite the fact that radiation-induced injury is categorized into cancer and non-cancer (or normal-tissue injury), the expression "radiation risk" invariably brings to mind the idea of increased cancer incidence associated with radiation exposure. This is because the most common somatic effect of low-dose exposure is cancer induction. In this document, however, risk of low-level exposure to radiation will always mean both the risk of carcinogenesis and the risk of non-cancer complications. We will begin with the overall information about low-dose effects in induction of cancer and non-cancer diseases and subsequently discuss the possible molecular bases for these findings.

There is also considerable question about what the cut-off should be for considering a dose low or high; this, of course, would depend upon the quality of the radiation (high vs. low Linear Energy Transfer, for example) and a variety of other parameters. The U.S. Department of Energy (DOE) considered low doses to be under 100 cGy [1], but other groups consider the transition at different doses (e.g., between 50 and 500 mGy [2]). For this review, we will specify the doses used in the respective studies.

Development of genomics, proteomics and other fields of biology ending in -omics in 1990's (often referred to by neologism OMICs) initiated the most prolific era in low dose radiation research. This period witnessed a shift into high throughput methods, various OMICs approaches and mathematical modeling (e.g. [3-6]). In the US, low-dose radiation research had a great boost during a period of the U.S. DOE funding between 1995 and 2008 [1], while concerns about biomedical imaging drive much of the present-day studies in the US, Canada, and Australia (e.g., [7–13]). In Japan low-dose radiation research has also been prolific (and with much of the biological material stored and awaiting further studies) (e.g., [14–18]).

A further impetus to the programmatic focus occurred after the Fukushima accident in 2011; from that time, collection and analysis of environmental samples became a very prominent part of radiation studies (e.g., [19–20]). European low-dose radiation research in this period has also been strong (e.g., [21–27]), much of it being done in the last decade in response to a report issued by the High Level Expert Group (http://www.hleg.de/fr.pdf). This report identified several research areas for growth of low-dose radiation studies and supported the long-term (until 2030) plan for the study of low-dose radiation effects proposed by a programmatic platform called Multidisciplinary European Low-Dose Initiative (MELODI). Three research areas considered as top priority during the most recent period were questions on (1) the shape of the dose-response curve for radiation induced cancer, (2) non-cancer disease, and (3) individual human sensitivity. Not surprisingly, research techniques used in all of these studies across the world were often similar and subject to the same inherent limitations. For example, the most robust data were obtained when work was done in animals, especially transgenic animals with genetic susceptibilities to accumulation of mutations or other endpoint(s) under consideration (e.g., [28–30]). This suggests that the development of new complex models is necessary in order to expand low dose radiation biology into a wider research sphere.

13.2 Risk of Carcinogenesis Following Low-Level Exposure to Radiation

The role of stem cells in the development of cancer is considered to be covered by the "cancer stem cell" hypothesis [31], although many radiation biologists argue that the term, "cancer clonogen," used since 1980s (e.g., [32]), has the same meaning. The possibility that low-dose radiation may lead to the development of cancer clonogens or cancer stem cells is one of the important topics in low-dose radiation research, as reviewed in [2]. The effects of different radiation doses on stem cells are distinct. At high doses above 5 Gy, responses of stem cells to radiation are similar to "standard" cellular high-dose responses—activation of apoptotic and anti-apoptotic mechanisms such as increased expression of survivin. The mutation spectra are different, nevertheless, with mitotic recombination as a primary cause for the loss of heterozygosity. At medium doses between 500 mGy and 5 Gy, stem cells undergo G2/M cell cycle arrest and, again, apoptosis and numerous changes in mRNA and micro-RNA expression associated with p53, TGFβ, and Wnt signaling occur. Low doses of radiation, between 5 and 500 mGy, did not lead to apoptosis, although in some cases stem cells demonstrated a diminished capacity for differentiation. Gene expression profiles recorded from these low doses were different from those obtained for moderate-dose exposures. It is important to note that several reports found an increase in stem cell proliferation after low-dose exposures. For example, in diabetic rats 75 mGy exposure led to an increased number of bone marrow and circulating stem cells and improved wound healing. Similarly, a 74-mGy exposure triggered growth of bone marrow hematopoietic progenitor cells in BALB/C mice. With regard to neural stem cells exposed to 300 mGy, levels of proteins Wnt1, Wnt3a, Wnt5a, and β-catenin were found to be increased, leading to proliferation and neuronal differentiation. Overall, this review from Manda et al. from 2014 did not suggest that low-dose radiation is likely to cause transformation of stem cells into cancer stem cells. A recent study suggesting that low doses increase therapeutic neovascularization; Ministro et al. [33] makes the same point—regenerative but not cancer-prone activity of stem cells is triggered by low doses of radiation.

Whole-animal studies investigating the effects of low-dose radiation in cancer induction most often use heterozygotes of transgenic knock-outs of tumor suppressor genes or animals with conditional knock-ins of oncogenes. Mice with knock-ins of reporter genes (e.g., Muta mice, with the knock-in of the lacZ gene with an excisable lambda phage genome) are also popular in low-dose radiation studies but with a different goal—to explore the spectra of mutations caused by radiation. While these mice cannot be considered as cancer models per se, they provide information on mutation spectra and repair mechanisms involved in low-dose responses [18]. These animals sometimes carried additional mutations, such as knock-outs of the non-homologous end joining repair enzyme Ku70 [15]—animals for which exposures even to the high doses of radiation (20 Gy) do not lead to lacZ mutations.

A recent study on skin cancer induction by low-dose radiation (50 mGy) used a transgenic mouse model based on the Patched gene (*Ptch*) [34]. Patched encodes a protein that is a transcription regulator involved in the hedgehog pathway and also functions as a tumor suppressor. In this study initially, work was done in wild-type mice and cells of two phenotypes were followed—sebaceous gland (SG) and bulge epidermal stem cells (SCs). DNA damage induced by 50 mGy led to cell death of SG cells, while SCs survived the exposure and underwent a metabolic switch with activation of Hif1α and overexpression of many cellular pathways (with endocytosis as the only pathway that was decreased). In Ptch1 heterozygous mice, low-dose irradiation led to the development of skin cancer similar to human basal cell carcinoma. In this model system, medium doses of radiation had the same but a more pronounced effect—more skin cancer foci per animal developed. Nevertheless, regardless of dose, 100% of these animals developed skin cancer.

The same transgenic animals, Ptch1+/− mice were used to investigate the type of mutation induced by low-dose radiation that results in the loss of heterozygosity (LOH) and the development of medulloblastomas (with the Ptch−/− genotype). Exposure to 500 mGy, the same medium-dose exposures in previous studies by this group, caused interstitial deletions, beginning downstream of the wild-type *Ptch1* gene and extending towards the centromere. Exposure to 100 mGy, on the other hand, led to the same deletion pattern as LOH mutations in unirradiated mice—loss of a portion of chromosome arm from telomere to beyond the *Ptch1* gene resulting from a faulty recombination [35].

Use of cancer prone p53 heterozygous and knock-out mice for low- and medium- dose radiation research has been extensive over the years, coinciding with the beginning of the period of most active low-dose studies worldwide [36]. In some respects, it may be said that the existence of the Muta mouse and p53+/− mice facilitated the initial growth of the low-dose field. Interestingly, irradiation of p53+/− mice sometimes led to apparently unexpected findings. For example, apparently counterintuitive data came from a report on long-term low-dose comparison of p53+/− and wild-type mice [37]. This study lasted 30, 60, or 90 days and exposed mice for five days a week to 0.33 mGy daily at a low dose rate (0.7 mGy/h), totaling 48, 97, or 146 mGy [37]. Wild-type C57BL/6 mice and their p53+/− counterparts were included in this study. Surprisingly, life shortening and increased cancer incidences were found in wild-type mice exposed for 30 or 60 weeks, but not the animals exposed for 90 weeks. At the same time, no differences in cancer incidence were found between non-irradiated or any of the long-term irradiated heterozygote animals [37]. In addition, a more recent study using p53 heterozygotes found that a single 10mGy X-ray exposure significantly delays onset of cancer in irradiated compared to unirradiated p53+/− animals [9]. These studies are considered as proofs of the concept of an "adaptive response" as described by Mitchel and others [37]; while no exact mechanistic explanation for these findings was demonstrated yet, we make an attempt to propose a hypothesis about these phenomena in Section 13.4.1.

Much more frequent in the literature are papers about medium doses of radiation and carcinogenesis in p53+/− mice [36]; in most cases, these cancers are thymic lymphoma or different types of sarcomas and their p53 status is negative at the time of detection (e.g., [38]). Loss of the remaining copy of p53 after 1–6 Gy radiation exposure was found to be caused by homologous recombination or non-disjunctional chromosome duplication of the p53- allele, an event that occurred at least three weeks after irradiation [38].

While other types of mutations are numerous in these animals because of a diminished capacity to respond to radiation exposure by an increase in p53 expression [39], loss of the second copy of p53 gene is still the most important step in the development of cancer. For that reason, crossbreeding of p53+/− mice with animals of different genetic backgrounds was also done in many different permutations and with different results. While no low-dose radiation exposure data are available for some of these studies, they are mentioned here because they could be of interest to low-dose studies in the future. For example, a cross of p53+/− with ATM ataxia-telangiectasia mutated (ATM) +/− animals led to a genotype with increased susceptibility to breast cancer development [40], especially in double heterozygotes exposed to high-dose (5 Gy) radiation exposure. Crossbreeding was also done with different "regular" (non-transgenic) mouse strains with different cancer susceptibilities. For example, p53+/− mice of the 129/Sv strain were crossed with mice of C57BL/6J strain or a different mouse species *Mus spretus* [41]. Exposure to 4 Gy increased the cancer incidence in all of these animals, albeit with a lower frequency in crosses than in the original strain. Nevertheless, thymic lymphoma clones in all three cases still showed LOH for p53.

Further refinement of this model resulted in the development of p53 delta proline mice [42]; homozygote animals with this p53 mutation develop a broad spectrum of tumor types, with considerably more types of solid tumors. Even more importantly, unlike in p53+/− mice, these tumors develop only after irradiation and not spontaneously.

13.3 Risk of Non-cancer Complications Following Low-Level Exposure to Radiation

This area of research has grown significantly in recent years. Nevertheless, up to now the greatest efforts directed at investigation of non-cancer diseases caused by radiation still investigate the same endpoints that were identified through exposure to high and medium doses of radiation. While this approach to low-dose radiation studies is perhaps the most straightforward, it should be acknowledged that this implicitly supports the concept of the linear non-threshold model of radiation effects. Thus, while it is quite possible that many of the effects of low-dose and high-dose irradiation are qualitatively the same (e.g., injury to the cardiovascular system), it may be that the mechanisms involved in these events are different in the two different dose ranges. In fact, when considering damage from many chemical agents, mechanisms at low concentrations are often quite different from those at the higher toxic concentrations [3]. It is even possible that low-dose radiation may have other, yet unexplored effects (some of which may be less toxic or even hormetic) that do not occur after medium- or high-dose radiation exposures.

Many normal tissue studies employing low doses of ionizing radiation include exploration of wild-type or genetically modified stem cells and the impact of low-dose radiation on their capacity to proliferate and differentiate. A study of murine p53 wild-type and p53−/− mesenchymal stem cells *in vitro* under hypoxic and normoxic conditions evaluated the effect of low-dose radiation on senescence [29]. In this study, mesenchymal cells lost some of their capacity to proliferate under normoxia regardless of genetic background; in addition, exposure of p53+/− cells to radiation gradually increased senescence, and this effect reached a plateau for doses of 200 mGy or greater.

Effects of low-dose radiation on human mesenchymal stromal cells were evaluated from the perspective of cell function [43]. Exposure to low doses (40 and 160 mGy) and moderate doses (640 mGy and 2 Gy) did not affect either proliferation or apoptosis index; however, numbers of senescent cells increased with dose, accompanied with a change in autophagic flux.

A recent review by Baselet and others [44] summarized knowledge about low-dose radiation-induced cardiovascular disease, emphasizing that epidemiological studies find an excess risk of cardiovascular disease only for radiation doses greater than 0.5 Gy [45]. In Europe, this research was part of the PROCARDIO program dedicated to cardiovascular studies within MELODI. Projects from different EU countries were included, all committed to study of cardiovascular disease events associated with low-dose radiation, doses below 0.5 Gy, through epidemiological studies [46] and use of cells in culture [47–48], as well as animal models [28–30]. Not surprisingly, cell culture work was made complicated by the necessity to do experiments over short periods of time, and few statistically significant findings were associated with radiation exposures below 0.5 Gy, either in cultured mouse embryonic stem cells [47] or endothelial cells [48]. The effects of 0.5 Gy exposures included reduction of beating embryoid bodies (formed from stem cells *in vitro*) and reversible cell cycle perturbations, senescence, and production of pro-inflammatory molecules (endothelial cells). Experiments done with mice, on the other hand, were conducted over long periods of time (up to two years), allowing that all lasting effects of low-dose radiation become apparent even if they were minor. Importantly, mice exposed to radiation *in utero* showed significant differences in their non-irradiated counterparts at six months and two years, with proteomics profiles between animals exposed to doses of 100 mGy and 1 Gy showing differences; less surprisingly, the number of protein differences between controls and mice exposed to 1 Gy was two times higher than those for mice exposed to 100 mGy [28]. Of all possible broad cellular functions, protein differences were found to be most substantial and consistent for mitochondrial proteins.

The response of mitochondria to fractionated low-dose radiation was also observed in neural progenitor stem cells induced to differentiate [49]. Differentiated immortalized ReNcell VM human neural progenitor cells were exposed to two 10 mGy X-ray fractions per day, five days a week for a total dose of 460 mGy delivered over a period of 31 days. In these cells, mitochondrial mass increased as well as the quantity of mitochondrial DNA; at the same time, similar changes were not observed in non-differentiated cells exposed to a single dose of radiation, until the dose reached 2 Gy.

In order to ascertain functional differences in cardiovascular disease status between irradiated and sham-irradiated animals, researchers from Italy and Japan used a transgenic mouse model, apolipo-lipoprotein E (ApoE) knock-out mice. Proteins rich in ApoE suppress production of cholesterol and promote cardiovascular resilience to atherosclerotic plaques. An increase in atherosclerosis was found in chronically irradiated ApoE-/- mice exposed to dose rates of 1 or 20 mGy/day for a total dose of 0.3 or 6 Gy, respectively [30]. Similarly, a comparison of ApoE-/- and wild-type mice [50] found premature death in mutant animals: more than 40% of them died at 20 weeks after radiation exposure to 0.2 Gy and showed an increased expression of the fibrogenic factors (TGF-β1 and PAI-1) compared to wild-type animals.

Also of interest for the vascular system is a study of human circulating endothelial colony-forming (progenitor) cells acutely exposed to a dose between 60 and 380 mGy [51]. The mononuclear cells fraction was obtained from cord blood samples from healthy human donors and the progenitor cells collected. Exposure to low-dose radiation did not cause cell death, but it inhibited cell growth.

Until recently, cataracts were considered to be one of the radiation effects for which a threshold (2Gy according to most sources) could be set. While more recent data and reevaluation of older data have changed the view on radiation-induced cataracts [52–53], it should be mentioned that Ptch+/− mice were used for cataract studies as well. In an experiment where two-day-old neonates were exposed to 2, 1, or 0.5 Gy, only the 2 Gy exposed animals developed fibrotic changes in the lens, explained by the induction of epithelial-mesenchymal cell transition [54]. No macroscopically detectable cataracts were found in any of the animals.

Retinal cells of developing chicken embryos irradiated with 500 mGy or more were evaluated for cell cycle arrest at different times post-exposure. While all doses of radiation led to an absence of mitotic events at one hour, by the three-hour timepoint, mitosis was absent only in embryos exposed to 2 Gy [55].

Finally, it should be mentioned that healthy wild-type mice most often show no effects from exposures to low-dose radiation, such as in a recent study on bone health (bone microarchitecture, numbers of osteoblasts, and osteoclasts) in BALB/c mice exposed to doses 0.5 Gy and less [56]. Extended experiments, however, often show significant effects of low-dose radiation. For example, female B6C3F1 mice exposed daily to 20 mGy/day for 400 days started to develop, beginning with day 200 post-irradiation, obesity and fatty degeneration of the liver and different degenerative changes in adrenal glands and ovaries. On the other hand, the most significant life-span shortening in these mice was caused by malignant lymphomas [57].

13.4 Biological Effects of Low-Dose Radiation Exposure

Our ability to study biological effects of any type of cell stress is limited by the tools and assays at our disposal. The study of relatively innocuous cell stresses such as low-dose radiation, especially in a complex biological model such as whole animals, is hampered even more by limits imposed by the state of the art in biotechnology. To make the situation more complicated, most expectations of low-dose radiation research studies are rooted in high- and medium-dose radiation research and toxicology, thus for example, the concept of low-dose radiation hyperradiosensitivity [58] initially met with little interest.

13.4.1 Low Dose and Cell Cycle

Very low doses of radiation were investigated in peripheral blood samples, and one of the most interesting studies of this type found that cells in G2 phase of the cell cycle show chromosomal aberrations in numbers that spike with the increase of dose from 100 to 400 mGy and then fall off and continue to grow with different dynamics [59]. While the primary focus of this study is a discussion on the linear-non-threshold model, this work suggests that there exists a molecular mechanism that reduces chromosomal aberrations that "kicks in" at 400 mGy in cells in G2 stage of the cell cycle. Many have hypothesized that a certain number of DNA lesions are needed to put the cells into G2 arrest, that this arrest increases chances of repair and therefore cells with fewer DNA lesions would have a hypersensitivity relative to cells above the threshold dose leading to increased volume of "fixed" aberrations. A similar study [60] testing the role of oxygenation in hyperradiosensitivity found that oxygen concentrations of 5% or below decrease chromosomal defects caused by exposures to 40–400 mGy. These findings echo those where only in normoxia and in the p53+/− genetic background did low-dose radiation lead to increased senescence of mouse mesenchymal stem cells [29]. Therefore, it is possible that the chromosomal damage and/or senescence at low-dose radiation and normoxia require the activity of the p53 protein.

Also, cell cycle differences may hypothetically provide an explanation of a lowered threshold for the "adaptive response" to low-dose radiation in p53+/− animals. Lower expression of p53 in heterozygote mice should correspond to decreased activation of p21 protein upon radiation, which, in turn, allows progression of the S phase of the cell cycle. While this may lead to the accumulation of DNA damage, its quantity is low enough that it could be handled by homologous recombination balancing the damage introduced by unrestrained S phase progression. In p53 wild-type cells, on the other hand, relatively minor DNA damage can lead to an excessive p21 activation response, resulting in collapsed DNA replication forks. These, in turn, may attract the Xeroderma Pigmentosum A protein leading to a DNA double-strand break as in progeria, for example [61].

13.4.2 Low Dose, Gender, and Genetic Background

A popular transgenic animal genotype used for much research on low-dose radiation are p53 heterozygotes; these mice were used both for cancer and non-cancer low-dose radiation studies. In their recent paper, Mao and others have investigated the spectra of spontaneous mutations—genetic signatures in lymphomas from p53+/− and p53−/− mice [62]. Interestingly, spontaneous cancer spectra were different in these two genetic backgrounds. This suggests that the quantity of p53 in cells makes a difference with respect to the types of mutations to be expected. This further suggests that not only the difference between "maximum copy number" (in p53+/+) and "half copy number" (in p53+/− mice) of p53 matters for mutation induction but the difference between "half copy number" and "no copies" of p53 protein (in p53−/− mice) further modulates the mutation spectra.

The bi-transgenic CCSP-rtTA/Ki-ras mice (FVB/N background with doxycycline inducible expression of the Ki-rasG12C gene in lungs) were developed as a model to study promotion of lung cancer [63], with multiple cancer foci in lungs developing after doxycycline exposure. These animals were exposed to low doses of radiation (80–160 mGy), which further increased lung cancer frequency, with statistically significant gender differences after irradiation and higher numbers of cancer foci recorded in female mice [7].

13.4.3 Low-Dose Induced DNA Damage and Gene Expression

An extensive exploration of gene expression modifications in response to low-dose radiation exposures was published recently by Nosel et al. [64], who irradiated CD4+ cells *in vitro* from healthy donors with doses of 5, 10, 25, 50, 100, and 500 mGy of ^{60}Co gamma rays and compared the data obtained from Agilent microarrays using extensive *in silico* analysis. Several findings were conspicuous: genes upregulated in response to radiation overlapped the most at doses <50 mGy, while genes that were downregulated overlapped the most at doses <100 mGy. The time course of these changes was different as well, for example, gene expression differences for 5 mGy dose compared against control steadily dropped off with time, while samples from 100 to 500 mGy exposures showed the greatest differences against control at 150 and 600 minutes post-exposure. However, the most exciting product of this work is a list of novel transcription factors that may be involved in gene regulation at very low doses of radiation; each of these proteins had binding sites in promoters of 11–170 of genes found to be up- or down-regulated by radiation in this study.

A more artificial cell culture model—coronary artery endothelial cells immortalized with telomerase overexpression—was used to study the effects of a single dose of X-rays. Cells were exposed to 50, 100, 500 mGy, and 2 Gy and investigated at timepoints between 1 and 14 days. At the late timepoints, indicators of senescence were noted in all irradiated samples, while cell cycle changes and production of IL6 and CCL2 were notable only at 500 mGy and above at most timepoints. Evaluation of gene expression pattern in these samples suggested upregulation of genes controlled by nuclear factor kappa B RELA pro-inflammatory pathways and p53 pathways [65].

Another *ex vivo* study using 3D skin cell culture was used to evaluate gene expression triggered by exposure to 100 mGy [12]. Interestingly, expression of more than 1,400 genes was altered in the dermis while more that 400 genes were modulated in the epidermis; moreover, the two cell layers had few overlapping responses at the mRNA level. The authors state, "there is increasing evidence that low and high dose exposures result in different signaling events and may have different response mechanisms than higher doses," echoing all other gene expression studies. A different study using an *in vitro* 3D skin system found that genes upregulated at 3h after 100 mGy were downregulated at 8 hours after 10 Gy exposure [66].

Work that used only keratinocytes [67] explored the role of the transcription factor Zn finger protein GATA3 in response to low-dose radiation. This protein is expressed in differentiated cells or *in vivo*, in suprabasal layers of the epidermis; in this study cells with a lentiviral knock-down of GATA3 and wild-type cells were compared with regard to their responses to 10 mGy and 2 Gy. While no differences between cell viability were notable at 0 and 2 Gy, cells exposed to 10 mGy had decreased survival in a clonogenic assay when GATA3 was reduced. Gene expression differences in these cells were studied by microarray, focusing, in this case, on GATA3-responsive genes (EGR1, DUSP1, GLUL, and GJB6). Interestingly, GATA3 also plays a role in regulation of T helper cells and production of interleukin 5 [68] and 3 [69], as well as probably many more targets, as its knock-out is an embryonic lethal that can be rescued by noradrenaline [70].

While *in vitro* studies listed above did not focus on innate genetic differences between subjects, this was the central point of an *in vivo* study on two mouse strains with different susceptibility for development of breast cancer. Six-week-old C57BL/6 (resistant) and BALB/c (sensitive) mice were exposed to low-dose radiation (four weekly doses of 75 mGy) and their breast tissues collected at 4 hours or one month after the final exposure [71]. mRNA array data demonstrated strain and timepoint differences in gene expression. Initial differences between the two strains prior to irradiation were notable, and they became even more pronounced at 1 month post-irradiation. Interestingly, in a breast-cancer-resistant strain, expression of certain cancer associated genes (e.g., EZH2) was decreased, while the same gene was increased in cancer-prone BALB/c mice. While these data are of interest, it is difficult to decide whether these radiation-induced gene differences depend primarily on pre-existing strain differences (baseline difference in stress response and RNA processing genes) and/or additional genetic differences. It should also be mentioned that excessively high-dose radiation exposures (1.8 Gy per week) were done and gene expression profiles compared with those obtained at low doses. No overlap in gene expression was found between low and high doses in either mouse strain.

Many studies similar to the ones described above were done in recent years, always confirming the same point: genes induced by low-, medium-, and high-dose exposures are largely different, regardless of experimental setup.

13.4.4 Low Dose and Repair

Activation of p53 gene at doses as low as 10 mGy was first observed more than a decade ago [72]; this finding was probably the source of the initial inspiration to use p53 knock-out mice and heterozygotes for low-dose radiation studies.

DNA damage and delayed repair have been noted in low-dose irradiations *in vivo*. For example, lymphocytes isolated from patients who underwent interventional imaging have demonstrated DNA damage associated with radiation exposure. Blood was drawn before and after (−2 and +24 hours) exposure; entrance surface dose values were between 90 and 230 mGy for areas between the shoulder and forehead, and the frequency of micronuclei was significantly increased at both post-imaging time-points compared to controls [73].

In human hepatic cell line exposure to 10 mGy led to histone deacetylation of the micro-RNA miR-193b-3p promoter. Change in expression of this miRNA was first found in spleen and liver of mice exposed to 10mGy whole-body irradiation. Suppression of miR-193b-3p leads to a subsequent increase in expression of the DNA double-strand break repair gene Rad51 [74].

13.4.5 Low Dose and Non-coding RNAs

The involvement of micro-RNAs (miRs) in carcinogenesis (or its prevention) is a frequent topic of research, especially when radiation injury is concerned. One of the well-studied miRs is miR34, which has two binding sites in the 3′ end of p53 mRNA and it is often considered as a "pro-tumor suppressor" miR. For example, low-dose radiation (100 mGy 30 kVp X-rays) was shown to increase expression of miR34a in rat mammary glands as well as increase in p53 and a decrease in E2F3 mRNAs [75]. However, despite these associations a recent study found that p53 completely fulfills its roles in miR34 knock-out animals [76]. This may, in part, be a result of multiple regulatory pathways controlled by p53. For example, in peripheral blood mononuclear cells exposed to doses of 0.9 Gy and greater, p53 upregulates several long non-coding RNAs that counteract an even greater number of miRs [77]. In short, effects of miRs and long non-coding (lnc) RNAs may need to be studied in the context of transgenic organisms before final conclusions can be made.

The micro-RNA expression pattern in cerebellum granule cells (sites where medulloblastomas develop in Ptch+/− mice) was found to be the same in heterozygous and wild-type mice without radiation and at medium-dose radiation (1 Gy) with the exception of mmu-miR-19a-5p. Whether this miR is critical for the development of medulloblastoma at 1 Gy and its involvement in etiology of low-dose radiation-induced medulloblastoma remain to be established. In any case, expression of mmu-miR-19a-5p in medulloblastoma samples from irradiated animals is several-fold higher than that in spontaneous medulloblastoma samples from Ptch+/− mice [25].

Long non-coding RNA expression patterns at 20 and 100 mGy and 2 Gy were explored in primary breast epithelial cells [78]. Interestingly, lncRNA expression patterns in 100 mGy and 2 Gy exposed samples partially overlapped, but no such overlap was notable with lncRNAs isolated form cells exposed to 20 mGy. Not surprisingly, lncRNAs increased in former case included those involved in pathways associated with apoptosis, cell cycle regulation and DNA damage repair.

13.4.6 Low Dose, Exosomes, and Secretion

Low-dose radiation at and below 100 mGy triggers secretion of a pro-survival protein secretory clusterin [79]; this protein interferes with TGFbeta1 signaling and is in turn repressed by p53 signaling. Clusterin acts as an extracellular chaperone; 24 hours after 100 mGy *in vivo* irradiation it is expressed in bone marrow, spleen, thymus, and colon. Klokov and others [79] created a low-dose radiation reporter gene made of the clusterin promoter coupled with luciferase. Expression of this construct, as in the original gene, was induced by insulin-like growth factor-1. Interestingly, a recent study with high-dose radiation of salivary glands connects successful gene repair with pretreatment with insulin-like growth factor-1, which in this situation increased expression of histone deactylase Sirtuin-1 (SirT-1) [80].

Extracellular vesicles and exosomes are generating considerable interest in moderate-and high-dose radiation studies (e.g., [26]). However, their involvement in low-dose radiation responses is still underexplored, probably because of the difficulties inherent in exosome research methodology. *In vitro* experiments such as transwell assay confirmed presence of low-dose radiation-induced lncRNAs in exosomes [81].

13.4.7 Low Dose and Immune System

In recent years, attention has also been devoted to the synergy between exposure to high doses of radiation and activation of the immune system, for example, by revealing immunogenic epitopes on tumor cells. Much less is known about the effects of low-dose radiation on immune cells; not surprisingly, here too, effects of low-dose exposures do not recapitulate the effects of high-dose exposures. For example, an investigation of T cell receptor spectra in animals exposed to low-dose (100 mGy) and medium-dose (1 Gy) radiation shows that the effects of these two doses are very different [82]. While exposure to 1 Gy has no irreversible effect on T cells, 100 mGy exposure results in accelerated aging as shown by loss of receptor diversity.

A model of hormetic effects of low-dose radiation was proposed recently [83], with suppression of that part of the immune system that is responsible for autoimmune diseases and enhancement of activity of natural killer cells and other components of the immune system that may be able to control anti-tumor activity. This model has not yet been tested in experimental systems.

13.4.8 Biological Effects—Summary

In summary, most gene expression studies (monitoring mRNAs or proteins) demonstrate differences in expression profiles between low, medium, and high doses of radiation. The volume of data makes this point incontrovertible, but it should be remembered that initially it was assumed that dose-response pattern characteristic of high radiation doses would be preserved for low doses as well.

The great majority of mechanistic low-dose radiation studies point to different aspects of p53 protein activity (although this could be a research bias of investigators studying p53 to the exclusion of other proteins that could be equally important). This tumor suppressor gene is indeed important in the development of cancer as well as senescence, apoptotic cell death, and teratogenesis. Nevertheless, it is still probable that its predominant recurrence in discussions on low-dose radiation effects comes from the fact that mouse animal models with altered p53 function contributed so much data to this field. It is therefore advisable to generate and explore other transgenic models as well, exploring low-dose radiation in mice heterozygous for other tumor-suppressor genes. Simplifying the genetic landscape through the use of knock-out mutations is a long-accepted tool in mechanistic biology; however, data interpretation remains labor intensive even under such circumstances. For example, another tumor-suppressor gene, BRCA1 associated protein (BAP1), is a ubiquitin carboxy (C)-terminal hydrolase critical for DNA double-strand break repair by homologous recombination [84]; this protein is necessary for G1-S checkpoint transition and its absence increases radiosensitivity [85]. Animals heterozygous for BAP1 spontaneously develop different types of cancer [86]; in most cases, cancer cells demonstrate loss of heterozygosity, suggesting that BAP1 is a tumor-suppressor gene that could "...offer key insights into the contribution of carcinogen exposure to enhanced cancer susceptibility." Interestingly, the most recent paper on BAP1 has shown that BAP1 is also a regulator of apoptosis [87]; thus, it may be that it is (at least in part) the increased survival of cells with chromosomal abnormalities that contributes to the apparent DNA damage load in these cells.

It should be mentioned that the list of genes known to be involved in radiation sensitivity is very long, including even some seemingly obscure polymorphisms, such as, for example, promoter mutations causing decreased activity of MnSOD [88]. While investigation of all of these possible genetic differences in combination with low-dose radiation is not feasible, widening the selection of genetic models would improve our ability to grasp the effects of low-dose radiation.

13.5 Conclusions

This chapter summarizes recent data on the biological effects of low-dose radiation exposures, both in terms of cancer induction and normal tissue complications. We find that the majority of endpoints considered by low-radiation dose research were those that were already well-established in high-dose radiation studies. While this approach is a logical beginning for an investigative effort, it should be recognized that this should not be the only focus of low-dose studies. Other not-yet-considered endpoints could be studied in conjunction with low-dose radiation exposures. Biological mechanisms underlying these endpoints may be more than replicates or variation of mechanisms regulating effects of high-dose exposures.

Another issue that requires consideration is the fact that, unlike most other fields of "stress biology," cells *in vitro* rarely provide significant insights into effects of low-dose ionizing radiation. Only work with animals, and among them transgenic mice, provides deep insight into low-dose radiation effects. While this confirms the expectations that investigation of low-dose radiation effects is difficult, it also illuminates the direction to be taken. Additional animal models are needed for low-dose radiation research (e.g., zebrafish is a versatile model for assessment of epigenetic effects, yet its use to assess effects of radiation was initiated only recently [89–90]). While radiation effects have been studied extensively in *C. elegans* [91], that is not the case for low-level radiation effects. Mice with altered expression of other tumor suppressor genes (e.g., BAP1) or transcription factors (e.g., GATA3) provide additional fertile ground for innovative low-dose radiation studies.

References

1. A.L. Brooks, A commentary on: A history of the United States Department of Energy (DOE) low dose radiation research program: 1998–2008, *Radiation Research*, 183 (2015) 375–381.
2. K. Manda, J.N. Kavanagh, D. Buttler, K.M. Prise, G. Hildebrandt, Low dose effects of ionizing radiation on normal tissue stem cells, *Reviews in Mutation Research*, 761 (2014) 6–14.
3. V. Chauhan, B. Kuo, J.P. McNamee, R.C. Wilkins, C.L. Yauk, Transcriptional benchmark dose modeling: Exploring how advances in chemical risk assessment may be applied to the radiation field, *Environmental and Molecular Mutagenesis*, 57 (2016) 589–604.
4. D. Leszczynski, Radiation proteomics: A brief overview, *Proteomics*, 14 (2014) 481–488.
5. W. Ruhm, M. Eidemuller, J.C. Kaiser, Biologically-based mechanistic models of radiation-related carcinogenesis applied to epidemiological data, *International Journal of Radiation Biology*, 93 (2017) 1093–1117.
6. J.H. Mao, J. Li, T. Jiang, Q. Li, D. Wu, J. Perez-Losada, R. DelRosario, L. Peterson, A. Balmain, W.W. Cai, Genomic instability in radiation-induced mouse lymphoma from p53 heterozygous mice, *Oncogene*, 24 (2005) 7924–7934.
7. M.T. Munley, J.E. Moore, M.C. Walb, S.P. Isom, J.D. Olson, J.G. Zora, N.D. Kock, K.T. Wheeler, M.S. Miller, Cancer-prone mice expressing the Ki-rasG12C gene show increased lung carcinogenesis after CT screening exposures, *Radiation Research*, 176 (2011) 842–848.
8. J.S. Murley, R.C. Miller, R.R. Weichselbaum, D.J. Grdina, TP53 Mutational status and ROS effect the expression of the survivin-associated radio-adaptive response, *Radiation Research*, 188(5) (2017) 579–590.
9. J.A. Lemon, N. Phan, D.R. Boreham, Single CT scan prolongs survival by extending cancer latency in Trp53 heterozygous mice, *Radiation Research*, 188(4.2) (2017) 505–511.
10. J.A. Lemon, N. Phan, D.R. Boreham, Multiple CT Scans Extend Lifespan by Delaying Cancer Progression in Cancer-Prone Mice, *Radiation Research*, 188(4.2) (2017) 495–504.

11. R.C. Miller, J.S. Murley, A.W. Rademaker, G.E. Woloschak, J.J. Li, R.R. Weichselbaum, D.J. Grdina, Very low doses of ionizing radiation and redox associated modifiers affect survivin-associated changes in radiation sensitivity, *Free Radical Biology & Medicine*, 99 (2016) 110–119.

12. C. von Neubeck, H. Shankaran, N.J. Karin, P.M. Kauer, W.B. Chrisler, X. Wang, R.J. Robinson, K.M. Waters, S.C. Tilton, M.B. Sowa, Cell type-dependent gene transcription profile in a three-dimensional human skin tissue model exposed to low doses of ionizing radiation: Implications for medical exposures, *Environmental and Molecular Mutagenesis*, 53 (2012) 247–259.

13. P. Sykes, The ups and downs of low-dose ionising radiobiology research, *Australasian Physical & Engineering Sciences in Medicine*, 39 (2016) 807–811.

14. T. Nakajima, B. Wang, T. Ono, Y. Uehara, S. Nakamura, K. Ichinohe, I. Braga-Tanaka, 3rd, S. Tanaka, K. Tanaka, M. Nenoi, Differences in sustained alterations in protein expression between livers of mice exposed to high-dose-rate and low-dose-rate radiation, *Journal of Radiation Research*, 58(4) (2017) 421–429.

15. Y. Uehara, H. Ikehata, J. Komura, A. Ito, M. Ogata, T. Itoh, R. Hirayama et al., Absence of Ku70 gene obliterates X-ray-induced lacZ mutagenesis of small deletions in mouse tissues, *Radiation Research*, 170 (2008) 216–223.

16. G. Vares, Y. Uehara, T. Ono, T. Nakajima, B. Wang, K. Taki, T. Matsumoto et al., Transcription factor-recognition sequences potentially involved in modulation of gene expression after exposure to low-dose-rate gamma-rays in the mouse liver, *Journal of Radiation Research*, 52 (2011) 249–256.

17. B. Wang, K. Tanaka, B. Ji, M. Ono, Y. Fang, Y. Ninomiya, K. Maruyama et al., Total body 100-mGy X-irradiation does not induce Alzheimer's disease-like pathogenesis or memory impairment in mice, *Journal of Radiation Research*, 55 (2014) 84–96.

18. T. Ono, H. Ikehata, P. Vishnu Priya, Y. Uehara, Molecular nature of mutations induced by irradiation with repeated low doses of X-rays in spleen, liver, brain and testis of lacZ-transgenic mice, *International Journal of Radiation Biology*, 79 (2003) 635–641.

19. S.I. Hayama, M. Tsuchiya, K. Ochiai, S. Nakiri, S. Nakanishi, N. Ishii, T. Kato et al., Small head size and delayed body weight growth in wild Japanese monkey fetuses after the Fukushima Daiichi nuclear disaster, *Scientific Reports*, 7 (2017) 3528.

20. S. Takino, H. Yamashiro, Y. Sugano, Y. Fujishima, A. Nakata, K. Kasai, G. Hayashi et al., Analysis of the effect of chronic and low-dose radiation exposure on spermatogenic cells of male large Japanese field mice (*Apodemus speciosus*) after the Fukushima Daiichi nuclear power plant accident, *Radiation Research*, 187 (2017) 161–168.

21. Z. Barjaktarovic, J. Merl-Pham, O. Azimzadeh, S.J. Kempf, K. Raj, M.J. Atkinson, S. Tapio, Low-dose radiation differentially regulates protein acetylation and histone deacetylase expression in human coronary artery endothelial cells, *International Journal of Radiation Biology*, 93 (2017) 156–164.

22. S.J. Kempf, C. von Toerne, S.M. Hauck, M.J. Atkinson, M.A. Benotmane, S. Tapio, Long-term consequences of in utero irradiated mice indicate proteomic changes in synaptic plasticity related signalling, *Proteome Science*, 13 (2015) 26.

23. V.B. O'Leary, S. Hain, D. Maugg, J. Smida, O. Azimzadeh, S. Tapio, S.V. Ovsepian, M.J. Atkinson, Long non-coding RNA PARTICLE bridges histone and DNA methylation, *Scientific Reports*, 7 (2017) 1790.

24. V. Radulovic, T. Heider, S. Richter, S. Moertl, M.J. Atkinson, N. Anastasov, Differential response of normal and transformed mammary epithelial cells to combined treatment of anti-miR-21 and radiation, *International Journal of Radiation Biology*, 93 (2017) 361–372.

25. B. Tanno, G. Babini, S. Leonardi, P. Giardullo, I. De Stefano, E. Pasquali, A. Ottolenghi, M.J. Atkinson, A. Saran, M. Mancuso, Ex vivo miRNome analysis in Ptch1+/− cerebellum granule cells reveals a subset of miRNAs involved in radiation-induced medulloblastoma, *Oncotarget*, 7 (2016) 68253–68269.

26. R. Yentrapalli, J. Merl-Pham, O. Azimzadeh, L. Mutschelknaus, C. Peters, S.M. Hauck, M.J. Atkinson, S. Tapio, S. Moertl, Quantitative changes in the protein and miRNA cargo of plasma exosome-like vesicles after exposure to ionizing radiation, *International Journal of Radiation Biology*, 93 (2017) 569–580.

27. N. Foray, M. Bourguignon, N. Hamada, Individual response to ionizing radiation, *Mutation Research*, 770 (2016) 369–386.

28. M.V. Bakshi, O. Azimzadeh, J. Merl-Pham, T. Verreet, S.M. Hauck, M.A. Benotmane, M.J. Atkinson, S. Tapio, In-utero low-dose irradiation leads to persistent alterations in the mouse heart proteome, *PloS one*, 11 (2016) e0156952.

29. I. Hofig, Y. Ingawale, M.J. Atkinson, H. Hertlein, P.J. Nelson, M. Rosemann, p53-Dependent senescence in mesenchymal stem cells under chronic normoxia is potentiated by low-dose gamma-irradiation, *Stem Cells International*, 2016 (2016) 6429853.

30. M. Mancuso, E. Pasquali, I. Braga-Tanaka, 3rd, S. Tanaka, A. Pannicelli, P. Giardullo, S. Pazzaglia, S. Tapio, M.J. Atkinson, A. Saran, Acceleration of atherogenesis in ApoE-/- mice exposed to acute or low-dose-rate ionizing radiation, *Oncotarget*, 6 (2015) 31263–31271.

31. M.S. Wicha, S. Liu, G. Dontu, Cancer stem cells: An old idea—a paradigm shift, *Cancer Research*, 66 (2006) 1883–1890; discussion 1895–1886.

32. K.H. Clifton, M.A. Tanner, M.N. Gould, Assessment of radiogenic cancer initiation frequency per clonogenic rat mammary cell in vivo, *Cancer Research*, 46 (1986) 2390–2395.

33. A. Ministro, P. de Oliveira, R.J. Nunes, A.D.S. Rocha, A. Correia, T. Carvalho, J. Rino et al., Low-dose ionizing radiation induces therapeutic neovascularization in a pre-clinical model of hindlimb ischemia, *Cardiovascular Research*, 113 (2017) 783–794.

34. T. Revenco, G. Lapouge, V. Moers, S. Brohee, P.A. Sotiropoulou, Low-dose radiation causes skin cancer in mice and has a differential effect on distinct epidermal stem cells, *Stem Cells (Dayton, Ohio)*, 35 (2017) 1355–1364.

35. C. Tsuruoka, B.J. Blyth, T. Morioka, M. Kaminishi, M. Shinagawa, Y. Shimada, S. Kakinuma, Sensitive detection of radiation-induced medulloblastomas after acute or protracted gamma-ray exposures in Ptch1 heterozygous mice using a radiation-specific molecular signature, *Radiation Research*, 186 (2016) 407–414.

36. C.J. Kemp, T. Wheldon, A. Balmain, p53-deficient mice are extremely susceptible to radiation-induced tumorigenesis, *Nature Genetics*, 8 (1994) 66–69.

37. R.E. Mitchel, P. Burchart, H. Wyatt, A lower dose threshold for the in vivo protective adaptive response to radiation. Tumorigenesis in chronically exposed normal and Trp53 heterozygous C57BL/6 mice, *Radiation Research*, 170 (2008) 765–775.

38. K. Tanaka, K. Watanabe, M. Mori, H. Kamisaku, H. Tsuji, Y. Hirabayashi, T. Inoue, K. Yoshida, S. Aizawa, Cytogenetic and cellular events during radiation-induced thymic lymphomagenesis in the p53 heterozygous (±) B10 mouse, *International Journal of Radiation Biology*, 78 (2002) 165–172.

39. A. di Masi, A. Antoccia, I. Dimauro, A. Argentino-Storino, A. Mosiello, R. Mango, G. Novelli, C. Tanzarella, Gene expression and apoptosis induction in p53-heterozygous irradiated mice, *Mutation Research*, 594 (2006) 49–62.

40. S. Umesako, K. Fujisawa, S. Iiga, N. Mori, M. Takahashi, D.P. Hong, C.W. Song et al., ATM heterozygous deficiency enhances development of mammary carcinomas in p53 heterozygous knockout mice, *Breast Cancer Research: BCR*, 7 (2005) R164–R170.

41. M. Hang, Y. Huang, A.M. Snijders, J.H. Mao, Genetic background influences loss of heterozygosity patterns in radiation-induced mouse thymic lymphoma, *Journal of Nature and Science*, 1 (2015) e96.

42. C.J. Adams, J.S. Yu, J.H. Mao, K.Y. Jen, S.V. Costes, M. Wade, J. Shoemake et al., The Trp53 delta proline (Trp53DeltaP) mouse exhibits increased genome instability and susceptibility to radiation-induced, but not spontaneous, tumor development, *Molecular Carcinogenesis*, 55 (2016) 1387–1396.

43. N. Alessio, S. Del Gaudio, S. Capasso, G. Di Bernardo, S. Cappabianca, M. Cipollaro, G. Peluso, U. Galderisi, Low dose radiation induced senescence of human mesenchymal stromal cells and impaired the autophagy process, *Oncotarget*, 6 (2015) 8155–8166.

44. B. Baselet, C. Rombouts, A.M. Benotmane, S. Baatout, A. Aerts, Cardiovascular diseases related to ionizing radiation: The risk of low-dose exposure (Review), *International Journal of Molecular Medicine*, 38 (2016) 1623–1641.

45. a. Advisory Group on Ionising Radiation Subgroup on Circulatory Disease Risk, i.b. Health Protection Agency (Great Britain), Centre for Radiation Chemical and Environmental Hazards. Circulatory disease risk: Report of the Independent Advisory Group on Ionising Radiation/prepared by the Subgroup on Circulatory Disease Risk of the Advisory Group on Ionising Radiation. London, UK: Health Protection Agency; Centre for Radiation, Chemical and Environmental Hazards, October 2010.

46. M. Kreuzer, A. Auvinen, E. Cardis, J. Hall, J.R. Jourdain, D. Laurier, M.P. Little et al., Low-dose ionising radiation and cardiovascular diseases—Strategies for molecular epidemiological studies in Europe, *Mutation Research. Reviews in Mutation Research*, 764 (2015) 90–100.

47. A. Helm, O. Arrizabalaga, D. Pignalosa, I.S. Schroeder, M. Durante, S. Ritter, Ionizing radiation impacts on cardiac differentiation of mouse embryonic stem cells, *Stem Cells and Development*, 25 (2016) 178–188.

48. B. Baselet, N. Belmans, E. Coninx, D. Lowe, A. Janssen, A. Michaux, K. Tabury et al., Functional gene analysis reveals cell cycle changes and inflammation in endothelial cells irradiated with a single X-ray dose, *Frontiers in Pharmacology*, 8 (2017) 213.

49. T. Shimura, M. Sasatani, H. Kawai, K. Kamiya, J. Kobayashi, K. Komatsu, N. Kunugita, A comparison of radiation-induced mitochondrial damage between neural progenitor stem cells and differentiated cells, *Cell Cycle*, 16 (2017) 565–573.

50. V. Monceau, L. Meziani, C. Strup-Perrot, E. Morel, M. Schmidt, J. Haagen, B. Escoubet, W. Dorr, M.C. Vozenin, Enhanced sensitivity to low dose irradiation of ApoE-/- mice mediated by early pro-inflammatory profile and delayed activation of the TGFbeta1 cascade involved in fibrogenesis, *PLoS One*, 8 (2013) e57052.

51. A.V. Kinev, V. Levering, K. Young, F. Ali-Osman, G.A. Truskey, M.W. Dewhirst, D. Il'yasova, Endothelial colony forming cells (ECFCs) as a model for studying effects of low-dose ionizing radiation: Growth inhibition by a single dose, *Cancer Investigation*, 31 (2013) 359–364.

52. E.A. Ainsbury, S. Barnard, S. Bright, C. Dalke, M. Jarrin, S. Kunze, R. Tanner et al., Ionizing radiation induced cataracts: Recent biological and mechanistic developments and perspectives for future research, *Mutation Research*, 770 (2016) 238–261.

53. L.T. Dauer, E.A. Ainsbury, J. Dynlacht, D. Hoel, B.E. Klein, D. Mayer, C.R. Prescott et al., Status of NCRP scientific committee 1–23 commentary on guidance on radiation dose limits for the lens of the eye, *Health Physics*, 110 (2016) 182–184.

54. I. De Stefano, P. Giardullo, B. Tanno, S. Leonardi, E. Pasquali, G. Babini, A. Saran, M. Mancuso, Nonlinear radiation-induced cataract using the radiosensitive Ptch1(+)/(-) mouse model, *Radiation Research*, 186 (2016) 315–321.

55. M. Mayer, N. Kaiser, P.G. Layer, F. Frohns, Cell cycle regulation and apoptotic responses of the embryonic chick retina by ionizing radiation, *PLoS One*, 11(5) (2016) e0155093.

56. F. Lima, J.M. Swift, E.S. Greene, M.R. Allen, D.A. Cunningham, L.A. Braby, S.A. Bloomfield, Exposure to low-dose X-Ray radiation alters bone progenitor cells and bone microarchitecture, *Radiation Research*, 188(4) (2017) 433–442.

57. I.B. Tanaka, J. Komura, S. Tanaka, Pathology of serially sacrificed female B6C3F1 mice continuously exposed to very low-dose-rate gamma rays, *Radiation Research*, 187 (2017) 346–360.

58. B. Marples, B.G. Wouters, S.J. Collis, A.J. Chalmers, M.C. Joiner, Low-dose hyper-radiosensitivity: A consequence of ineffective cell cycle arrest of radiation-damaged G2-phase cells, *Radiation Research*, 161 (2004) 247–255.

59. I. Seth, M.C. Joiner, J.D. Tucker, Cytogenetic low-dose hyperradiosensitivity is observed in human peripheral blood lymphocytes, *International Journal of Radiation Oncology, Biology, Physics*, 91 (2015) 82–90.

60. G.S. Joshi, M.C. Joiner, J.D. Tucker, Effects of low oxygen levels on G2-specific cytogenetic low-dose hyper-radiosensitivity in irradiated human cells, *Environmental and Molecular Mutagenesis*, 56 (2015) 545–555.

61. P.R. Musich, Y. Zou, DNA-damage accumulation and replicative arrest in Hutchinson-Gilford progeria syndrome, *Biochemical Society Transactions*, 39 (2011) 1764–1769.

62. J.H. Mao, D. Wu, J. Perez-Losada, T. Jiang, Q. Li, R.M. Neve, J.W. Gray, W.W. Cai, A. Balmain, Crosstalk between Aurora-A and p53: Frequent deletion or downregulation of Aurora-A in tumors from p53 null mice, *Cancer Cell*, 11 (2007) 161–173.

63. H.S. Floyd, C.L. Farnsworth, N.D. Kock, M.C. Mizesko, J.L. Little, S.T. Dance, J. Everitt, J. Tichelaar, J.A. Whitsett, M.S. Miller, Conditional expression of the mutant Ki-ras(G12C) allele results in formation of benign lung adenomas: Development of a novel mouse lung tumor model, *Carcinogenesis*, 26 (2005) 2196–2206.

64. I. Nosel, A. Vaurijoux, J.F. Barquinero, G. Gruel, Characterization of gene expression profiles at low and very low doses of ionizing radiation, *DNA Repair*, 12 (2013) 508–517.

65. B. Baselet, N. Belmans, E. Coninx, D. Lowe, A. Janssen, A. Michaux, K. Tabury et al., Functional gene analysis reveals cell cycle changes and inflammation in endothelial cells irradiated with a single X-ray dose, *Frontiers in Pharmacology*, 8 (2017) 213.

66. S.C. Tilton, L.M. Markillie, S. Hays, R.C. Taylor, D.L. Stenoien, Identification of differential gene expression patterns after acute exposure to high and low doses of low-LET ionizing radiation in a reconstituted human skin tissue, *Radiation Research*, 186 (2016) 531–538.

67. F. Bonin, M. Molina, C. Malet, C. Ginestet, O. Berthier-Vergnes, M.T. Martin, J. Lamartine, GATA3 is a master regulator of the transcriptional response to low-dose ionizing radiation in human keratinocytes, *BMC Genomics*, 10 (2009) 417.

68. D.H. Zhang, L. Cohn, P. Ray, K. Bottomly, A. Ray, Transcription factor GATA-3 is differentially expressed in murine Th1 and Th2 cells and controls Th2-specific expression of the interleukin-5 gene, *The Journal of Biological Chemistry*, 272 (1997) 21597–21603.

69. J. Zhu, B. Min, J. Hu-Li, C.J. Watson, A. Grinberg, Q. Wang, N. Killeen, J.F. Urban, Jr., L. Guo, W.E. Paul, Conditional deletion of Gata3 shows its essential function in T(H)1-T(H)2 responses, *Nature Immunology*, 5 (2004) 1157–1165.

70. K.C. Lim, G. Lakshmanan, S.E. Crawford, Y. Gu, F. Grosveld, J.D. Engel, Gata3 loss leads to embryonic lethality due to noradrenaline deficiency of the sympathetic nervous system, *Nature Genetics*, 25 (2000) 209–212.

71. A.M. Snijders, F. Marchetti, S. Bhatnagar, N. Duru, J. Han, Z. Hu, J.H. Mao, J.W. Gray, A.J. Wyrobek, Genetic differences in transcript responses to low-dose ionizing radiation identify tissue functions associated with breast cancer susceptibility, *PLoS One*, 7 (2012) e45394.

72. D.E. MacCallum, P.A. Hall, E.G. Wright, The Trp53 pathway is induced in vivo by low doses of gamma radiation, *Radiation Research*, 156 (2001) 324–327.

73. S.A.S. Basheerudeen, K. Kanagaraj, M.T. Jose, A. Ozhimuthu, S. Paneerselvam, S. Pattan, S. Joseph, V. Raavi, V. Perumal, Entrance surface dose and induced DNA damage in blood lymphocytes of patients exposed to low-dose and low-dose-rate X-irradiation during diagnostic and therapeutic interventional radiology procedures, *Mutation Research-Genetic Toxicology and Environmental Mutagenesis*, 818 (2017) 1–6.

74. E.S. Lee, Y.J. Won, B.C. Kim, D. Park, J.H. Bae, S.J. Park, S.J. Noh et al., Low-dose irradiation promotes Rad51 expression by down-regulating miR-193b-3p in hepatocytes, *Scientific Reports*, 6 (2016) 25732.

75. L. Luzhna, O. Kovalchuk, Low dose irradiation profoundly affects transcriptome and microRNAme in rat mammary gland tissues, *Oncoscience*, 1 (2014) 751–762.

76. C.P. Concepcion, Y.C. Han, P. Mu, C. Bonetti, E. Yao, A. D'Andrea, J.A. Vidigal, W.P. Maughan, P. Ogrodowski, A. Ventura, Intact p53-dependent responses in miR-34-deficient mice, *PLoS Genetics*, 8 (2012) e1002797.

77. L. Beer, L. Nemec, T. Wagner, R. Ristl, L.M. Altenburger, H.J. Ankersmit, M. Mildner, Ionizing radiation regulates long non-coding RNAs in human peripheral blood mononuclear cells, *Journal of Radiation Research*, 58 (2017) 201–209.

78. M. Terradas, M. Martin, J. Repulles, M. Huarte, A. Genesca, Distinct sets of lncRNAs are differentially modulated after exposure to high and low doses of X-rays, *Radiation Research*, 186 (2016) 549–558.

79. D. Klokov, K. Leskov, S. Araki, Y. Zou, E.M. Goetz, X. Luo, D. Willson, D.A. Boothman, Low-dose IR-induced IGF-1-sCLU expression: A p53-repressed expression cascade that interferes with TGFbeta1 signaling to confer a pro-survival bystander effect, *Oncogene*, 32 (2013) 479–490.

80. S. Meyer, A.M. Chibly, R. Burd, K.H. Limesand, Insulin-like growth factor-1-mediated DNA repair in irradiated salivary glands is sirtuin-1 dependent, *Journal of Dental Research*, 96 (2017) 225–232.

81. V.B. O'Leary, S.V. Ovsepian, L.G. Carrascosa, F.A. Buske, V. Radulovic, M. Niyazi, S. Moertl, M. Trau, M.J. Atkinson, N. Anastasov, PARTICLE, a Triplex-forming long ncRNA, regulates locus-specific methylation in response to low-dose irradiation, *Cell Reports*, 11 (2015) 474–485.

82. S.M. Candeias, J. Mika, P. Finnon, T. Verbiest, R. Finnon, N. Brown, S. Bouffler, J. Polanska, C. Badie, Low-dose radiation accelerates aging of the T-cell receptor repertoire in CBA/Ca mice, *Cellular and Molecular Life Sciences: CMLS*, 74(23) (2017) 4339–4351.

83. J.W. Cui, G.Z. Yang, Z.Y. Pan, Y.G. Zhao, X.Y. Liang, W. Li, L. Cai, Hormetic response to low-dose radiation: Focus on the immune system and its clinical implications, *International Journal of Molecular Sciences*, 18 (2017) 280.

84. H. Yu, H. Pak, I. Hammond-Martel, M. Ghram, A. Rodrigue, S. Daou, H. Barbour et al., Tumor suppressor and deubiquitinase BAP1 promotes DNA double-strand break repair, *Proceedings of the National Academy of Sciences of the United States of America*, 111 (2014) 285–290.

85. H. Nishikawa, W. Wu, A. Koike, R. Kojima, H. Gomi, M. Fukuda, T. Ohta, BRCA1-associated protein 1 interferes with BRCA1/BARD1 RING heterodimer activity, *Cancer Research*, 69 (2009) 111–119.

86. Y. Kadariya, M. Cheung, J. Xu, J. Pei, E. Sementino, C.W. Menges, K.Q. Cai, F.J. Rauscher, A.J. Klein-Szanto, J.R. Testa, Bap1 is a bona fide tumor suppressor: Genetic evidence from mouse models carrying heterozygous germline bap1 mutations, *Cancer Research*, 76 (2016) 2836–2844.

87. A. Bononi, C. Giorgi, S. Patergnani, D. Larson, K. Verbruggen, M. Tanji, L. Pellegrini et al., BAP1 regulates IP3R3-mediated Ca^{2+} flux to mitochondria suppressing cell transformation, *Nature*, 546 (2017) 549–553.

88. R.C. Martin, J. Ahn, S.A. Nowell, D.W. Hein, M.A. Doll, B.D. Martini, C.B. Ambrosone, Association between manganese superoxide dismutase promoter gene polymorphism and breast cancer survival, *Breast Cancer Research: BCR*, 8 (2006) R45.

89. E.Y. Kong, S.H. Cheng, K.N. Yu, Zebrafish as an *in vivo* model to assess epigenetic effects of ionizing radiation, *International Journal of Molecular Sciences*, 17 (2016) 2108.

90. E.Y. Kong, S.H. Cheng, K.N. Yu, Biphasic and triphasic dose responses in zebrafish embryos to low-dose 150 kV X-rays with different levels of hardness, *Journal of Radiation Research*, 57 (2016) 363–369.

91. T. Sakashita, T. Takanami, S. Yanase, N. Hamada, M. Suzuki, T. Kimura, Y. Kobayashi, N. Ishii, A. Higashitani, Radiation biology of Caenorhabditis elegans: Germ cell response, aging and behavior, *Journal of Radiation Research*, 51 (2010) 107–121.

Review of Current Radiation Risk Models as Related to Medical Imaging

Colin J. Martin

14.1 The Link between Dose and Effect

14.1.1 Measurement of Radiation

Understanding what is meant by radiation dose and how it relates to the potential harm that might be produced has been an ongoing problem. Quantities used for measurement of radiation dose relate to different aspects of the deposition of energy as radiation passes through tissue. The basic scientific quantity, absorbed dose, is a measure of the energy deposited per unit mass. It can describe the dose at any point within a tissue, but this will vary throughout the tissue in most circumstances. One of the main concerns when using any form of radiation is the risk of stochastic effects, especially cancer. It became apparent in the years following the detonation of the atomic bombs in Hiroshima and Nagasaki that excess numbers of cancers were occurring in the Japanese populations who had been exposed to radiation and survived. But how big is this risk, and does it project a non-zero excess down to dose levels encountered in medical imaging? The risk will depend not only on the tissues and organs irradiated, but also on the distribution of dose within the tissues and organs irradiated. Even when an individual is exposed to a uniform radiation field covering the whole body, attenuation in the tissue results in organs that are closer to the surface receiving higher doses. In addition to this, there are variations in the sensitivity of individual tissues to radiation. Add to this variations in the effectiveness of radiations with different energies and other physical properties in damaging tissue, and the possibility of being able to assess the risk from radiation exposures or convey any understanding of the relative harm that may result appears remote. But if health professionals are to make judgments and the public is to have some understanding of the risks, it is a task that needs to be addressed. The international United Nations Scientific Committee on Effects of Atomic Radiation (UNSCEAR) and the committee on the Biological Effects of Ionizing Radiation (BEIR), under the auspices of the National Research Council of the U.S., collate information on radiation biological effects and epidemiological studies of exposed populations and, with the International Commission on Radiological Protection (ICRP), develop models to predict health risks to humans from radiation exposure. In this chapter the approaches used by different organizations to risk analysis and the cancer risk values they obtain will be compared.

14.1.2 Radiation Induction of Cancer

Experiments in which cell cultures have been exposed to large doses of radiation have shown that damage to cells occurs through the production of breaks in DNA molecules. Cells have efficient mechanisms for repair of damage, and if any damage is not repaired, then the affected cell can be removed by a process of programmed death called apoptosis. On very rare occasions, a subtle change in a cell's DNA may not be recognised and removed by these processes. Alterations in certain genes called proto-oncogenes, which control cellular growth, proliferation and differentiation, may increase the risk of malignant transformation of a cell. Cancer is thought to result from mutagenic damage to the DNA of a single cell, although there is evidence from in vitro studies that other targets within the cell may also be involved. Damage to a cell's DNA does not mean the cell will become malignant, but certain changes might be an initiating step in the process. Further changes produced by radiation, chemical carcinogens, or other environmental influences may promote more genetic change and render the cell genetically unstable. During the progression stage, the tumor develops a vascular system and invades adjacent tissues, and some tumor types may produce metastases that invade distant tissues. Cancer progression is thus a multi-step process, and any step may be enhanced by radiation exposure, but the development of a tumor also involves interaction with the immune system and the tissue environment.

The induction of cancer by radiation is a rare event. Among the Japanese atomic bomb survivors, who received high doses of radiation, the excess lifetime cancer incidence only amounted to about 5% (Ozasa et al. 2012). Receiving a high dose of radiation does not mean that the irradiated individual will definitely develop cancer, but rather that there is a small increase in their cancer risk. Cancer induction by radiation appears to obey laws similar to those that govern the chance of winning a lottery, an exposure (a lottery ticket) means that there is a small chance that the individual may develop cancer (win the lottery), but that risk will increase with radiation dose (number of lottery tickets purchased). The risk of any individual developing cancer or winning a lottery are extremely small, but how can we estimate that risk?

14.1.3 Risks of Cancer Induction

The estimation of cancer risks for populations exposed at low doses/dose rates is a critical component of radiological protection. Our knowledge of the risks from radiation exposure are derived from epidemiological studies of populations exposed to large doses of radiation. The groups that can be included in such studies are necessarily limited. Radiation workers in the nuclear industry, predominantly a male population of working age, receive low doses of radiation each year, and there is little evidence of any deleterious effects. Medical radiotherapy exposures are of highly selected populations, and they again have limited ranges of age at exposure. These studies have low statistical power, either because of the small populations exposed (radiotherapy patients) or the very low doses received (occupationally exposed groups). Meta-analyses combining data from many studies can overcome the limitation in population size to some extent, but these do not overcome many of the issues in population selection and still have a low statistical power. The one group that does not involve most of these issues is the Japanese atomic bomb survivors. They represent an unselected population of all age groups and both genders, who received whole-body exposures of a moderate dose at high dose rate. International organizations recommending standards for radiation safety, such as the ICRP, BEIR, and UNSCEAR, use epidemiological data derived predominantly from this group to estimate cancer risks.

14.2 Epidemiological Studies

14.2.1 The Life Span Study of the Japanese Atomic Bomb Survivors

Study of the health of the Japanese persons who survived the atomic bombs dropped on Hiroshima and Nagasaki in 1945 forms the foundation for modeling of radiation cancer risk. There had been concerns about effects from radiation prior to this, but exposures were too few to enable any objective evaluation. After the conclusion of World War II in 1946, the Atomic Bomb Casualty Commission (ABCC) was established to conduct investigations into the late health effects from radiation exposure among the atomic bomb survivors through the U.S. National Academy of Sciences-National Research Council. It became apparent in the late 1940s that more of the survivors were developing leukemia, a comparatively rare form of cancer, than would be expected, and a cancer registry was established in 1948 (Folley et al. 1952). About five years after the bombings, as a result of responses to supplementary questions raised by the 1950 Japanese national census, a group of 195,000 residents present in the two cities at the time of the bombings was identified. Following a reappraisal of the ABCC research program, a Life Span Study (LSS) of a cohort of 120,000 residents of Hiroshima and Nagasaki was set up under the Radiation Effects Research Foundation, a binational organization operated by the United States and Japan. The study cohort was made up of persons alive at the time of the national census on October 1, 1950 and comprised 86,611 survivors in the two cities within 2.5 km of the hypocenters of the atomic bomb detonations, who formed the exposed group, and an age- and sex-matched sample of 35,458 people who were between 2.5 and 10 km from the hypocenters and received doses below 0.005 Sv, together with 25,427 residents not in Hiroshima and Nagasaki at the time, who formed the control group (Ozasa et al. 2012). All subjects were interviewed extensively or completed a series of questionnaires to determine their locations at times of the bombings, their orientations with respect to the hypocenter of each bomb, the shielding provided by neighboring buildings, the time spent in the area after the bombing that would affect their dose from subsequent radioactive fallout, and their personal characteristics.

In parallel with this, doses were assessed, and these have been reevaluated three times over the years (Cullings et al. 2006), the most recent being in 2002 (Figure 14.1), and the results of calculations compared with data from neutron activation measurements in order to improve estimates of the contributions from neutron exposure. The responses to the questions of the Japanese survivors were used to estimate the radiation dose each individual received, although it must be remembered that dose estimates can be highly uncertain when exposure data come from personal recall without documentation. The average body dose for the atomic-bomb survivor studies was 0.27 Sv, ranging up to about 5 Sv. The risk estimates have changed over the years as more of the study population have died and the dosimetry evaluations have been refined (Preston et al. 2004). Each person was followed to assess their health over succeeding years and to contribute to refining the epidemiological data on cancer incidence and mortality. The LSS provides information on cancer risks for a wide range of sites that can serve as the basis for calculating risk to individual tissues and organs. Despite the exposure of a whole population, there is uncertainty about how representative the group is of the Japanese population as a whole, and whether the Japanese survivors may be pre-selected in terms of a "healthy survivor" effect, which could mean they are less likely to develop cancer; however, this is largely discounted (Ozasa et al. 2012). It is apparent from the LSS data that there is a link between radiation dose and risk, and that the risk is greater among those exposed at a younger age.

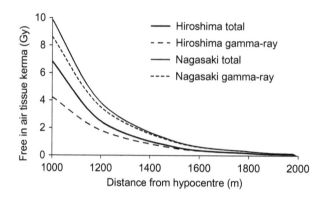

FIGURE 14.1 Variation in free-in-air tissue kerma with distance from the hypocenters of the Hiroshima and Nagasaki atomic bombs. The total is the sum of the γ-ray tissue kerma and that from neutrons multiplied by a radiation biological effectiveness factor of ten, as used in the LSS. (From Cullings et al., *Radiat. Res.*, 166, 219–254.)

14.2.2 Other Epidemiological Studies Used in Derivation of Cancer Risks

Although modeling of cancer risk has been based principally on data from the LSS, results from other radiation epidemiology studies have been compared with LSS estimates for particular tissues. Other studies can provide information on the effects of exposure in a chronic or fractionated manner rather than from an acute exposure and from exposure to high-linear energy transfer (LET) rather than low-LET radiation. They also relate to risks in populations other than Japanese. Examples of groups considered include:

- Patients undergoing medical therapeutic and diagnostic radiation exposures
- Workers exposed to radiation in the course of their job, such as uranium miners
- Persons who received environmental exposures, for example, from fallout or from natural radiation

Studies have been performed on the incidence of second cancers in radiotherapy patients where suitable control populations are available, such as the prostate and cervix where treatment by surgery is an option. Overall excess relative risks per Sv in radiotherapy datasets tend to be less than in comparable age- and sex-matched subsets of the LSS data. Doses to individual organs and tissues are generally much higher in radiotherapy, with doses of 10s of Gy and in some studies >200 Gy, whereas the maximum doses among the atomic-bomb survivors are about 4 Gy. Studies of breast, bladder, and stomach cancer suggest that the incidence increases approximately linearly up to about 5 Gy, and results are generally consistent with the LSS; above 5 Gy, however, the dose-dependent incidence may increase more slowly or level off (Hall and Giaccia 2012). This is likely to occur because the higher doses delivered in radiotherapy kill some cells in which malignant changes have been initiated, termed cell sterilization.

Studies have been carried out on a number of other groups receiving medical exposures. These include radiation therapy of the spine for treatment of ankylosing spondylitis, which created an excess of leukemia and other neoplasms in irradiated tissues (Weiss et al. 1995; Smith 2007). For breast cancer, data from seven cohorts in North America and Western Europe were used by the ICRP in addition to the LSS for estimating risks, since baseline rates of breast cancer in Japan are very low (Little and Boise 1999; Preston et al. 2002). These include women treated with radiotherapy for acute post-partum mastitis and chronic breast diseases and women receiving multiple chest fluoroscopic examinations in the course of therapy for tuberculosis in Massachusetts and Canada (Boice et al. 1991). For thyroid cancer, data from four populations exposed during medical treatment in various countries have been pooled (Ron et al. 1995). These include treatments in Israel of young patients with tinea capitis by irradiation of the scalp, irradiation of infants with enlarged thymus glands, and treatments with X-rays for enlarged tonsils or lymphoid hyperplasia. Follow-up after the Chernobyl accident has provided additional information on the effects of thyroid exposure to radioiodine (UNSCEAR 2011; WHO 2005). The incidence of thyroid cancer in children under 15 years increased rapidly at an earlier stage than in the LSS, among highly exposed groups in neighboring countries. The highest thyroid doses occurred in rural Belarus where 25,000 children aged 0–7 received a mean thyroid dose of 3.1 Gy.

Valuable information on long-term low-dose exposures to internally incorporated radionuclides has been provided by epidemiological studies of workers at the Mayak nuclear complex in the southern Urals of the Russian Federation (Shilnikova et al. 2003; Akleyev et al. 2017) and of the population near the Techa River who have received exposures due to radioactive discharges (Krestinina et al. 2005). Underground hard-rock miners (e.g., uranium, iron, gold, and tin) inhale significant amounts of radon, and a linear relationship between risk of lung cancer and cumulative dose has been demonstrated in many studies (Lubin and Boice 1997). The risk of excess skin cancer induced by ionizing radiation is influenced by exposure to ultraviolet radiation and is dependent on the degree of skin pigmentation. Nominal estimates of risks to skin and bone are used in ICRP models (ICRP 1991). Risk estimates for bone are based on studies of groups having medical exposures, primarily intakes of radium–224 (Spies et al. 1989). Overall, cancer risk estimates from these studies are generally consistent with those obtained from the LSS.

14.2.3 Risks at Low Doses

In order to prove that a radiation effect exists, it is necessary to show that there is a statistically significant increase in cancer among the exposed group. But the numbers of excess cancers of any one type in an exposed population will be relatively small, so the size of population that would be required to provide statistical proof of an increased risk at doses below 100 mSv for many cancers is prohibitive. Thus, estimation of cancer risks from radiation exposure requires very large studies with long-term follow-up, and an absence of significantly elevated risks cannot be taken as evidence that there is no risk. All the solid cancers could be lumped together, but because cancer is widely prevalent in society (50% of the population are diagnosed with cancer and 1 in 4 deaths are the result of cancer), the size of the population required to demonstrate that a dose of 20 mSv is associated with an excess cancer incidence would be about 1 million. However, there is a better chance of proving a statistically significant excess for specific types of cancer that have been found to be linked to radiation. For instance, the natural prevalence of leukemia is relatively low, with about 1 in 300 deaths in the developed world, so the increase in incidence required to prove an excess linked to radiation is smaller. It is more difficult to demonstrate an increase in cancer rate for more common sites such as the lung, colon, and breast, for which natural incidences in the population are between 1 in 25 and 1 in 35. Combined analyses of datasets can help to increase the statistical power. Studies of the nuclear industry workforce, which has good personnel dosimetry records, have been carried out in many countries. Data from studies in 15 countries have been combined to give results for 400,000 workers with a mean cumulative dose of 19.4 mSv and less than 5% of workers receiving cumulative doses over 100 mSv (Cardis et al. 2005). This shows a statistically significant increase in the incidence of leukemia, and results are generally consistent with predictions of the linear no-threshold (LNT) model based on the LSS.

More recently a study of children and young adults in the UK who had undergone computed tomography (CT) scans during the period 1985–2002 has shown that the risk of leukemia is tripled for those receiving cumulative doses over 30 mGy to the bone marrow, and the risk of a brain tumor, another comparatively rare cancer, is tripled by a dose of 50 mGy to the brain (Pearce et al. 2012). However, although subjects developing cancer within five years of the initial exposure were not included, the possibility of reverse causation, whereby the reason for the initial CT scan could be linked to the development of cancer at a later stage, cannot be excluded (Boice 2015).

Information on the effects of low-level radon exposure comes from combined analyses of case-control studies that show raised risks of lung cancer from radon exposure in homes in Europe (Darby et al. 2005), North America (Krewski et al. 2005), and China (Lubin et al. 2004). Results of the European project, which combined data from 13 individual case control studies in member states, show a clear increase in the risk of lung cancer among residents of homes with an enhanced concentration of radon (above 150 Bq m^{-3}). Results indicate that there is synergy between smoking and radon in enhancing risks of lung cancer, and this link is supported by analyses of the LSS data (Pierce et al. 2003). Domestic exposure to radon at a concentration of 130 Bq m^{-3} is considered to equate to an annual dose to the lung of about 10 mSv and an effective dose of 1 mSv (ICRP 2010).

These larger meta-analyses of epidemiological data suggest that the increased risk from radiation exposure of a population may extend down to perhaps organ doses of 20–50 mSv, but the evidence is insufficient to show the form of any dose-effect relationship at these levels. Evidence from radiobiology suggests that the risk might be expected to be lower, but whatever method is used to model the dose-effect relationship needs to include an extrapolation to low doses.

14.3 Modeling Cancer Risks from Epidemiological Results

14.3.1 Absolute and Relative Excess Risk

Data on epidemiological studies have been the subject of numerous publications over the years, and these data have been collated by UNSCEAR and BEIR. The ICRP and BEIR have developed models in an attempt to predict potential risks down to low-dose levels to which workers and patients undergoing medical imaging procedures are exposed. When modeling cancer risk, the first decision to be made is whether the risk should be based on cancer incidence or mortality data. The advantage of the former is that the incidence rate is higher, and of the latter is that, generally, mortality data are more reliable. Recent cancer risk models are based largely on mortality data for leukemia and cancer incidence data for solid tumors (BEIR 2006; ICRP 2007), although previous ICRP recommendations have been derived from mortality data (ICRP 1991). The obvious approach to modeling cancer risk is to use a simple model and assume that the excess absolute risk is proportional to the dose to the tissue (Little et al. 2009). This approach is called the excess absolute risk (EAR) or additive risk model.

$$EAR = \frac{\text{Risk in exposed population} - \text{Risk in unexposed population}}{\text{Organ dose}}$$

However, there is the question of how transferable risks are from a Japanese population to other population groups. Particular populations have diseases that are prevalent and, more specifically, types of cancer that occur more frequently. There are differences between Japanese and Western populations and therefore an excess relative risk (ERR) or multiplicative model has also been developed in an attempt to take these differences into account.

$$ERR = EAR \times \frac{\text{Cancer rate in target population}}{\text{Cancer rate in unexposed study population}}$$

As the study of the health of the Japanese bomb survivors in the LSS continued, it has become apparent from the accumulated data that the risk of cancer induction declines with increasing age at exposure and also, among those exposed at a younger age, it declines with the time since the exposure. These modifying factors are taken into account in the calculations of differences in risk between the sexes (Figure 14.2). Overall, rates of incidence for solid cancers with age at exposure predicted by the two models for a Euro-American population are similar (Figure 14.2a), but there are significant differences for some organs such as the lungs (Figure 14.2b).

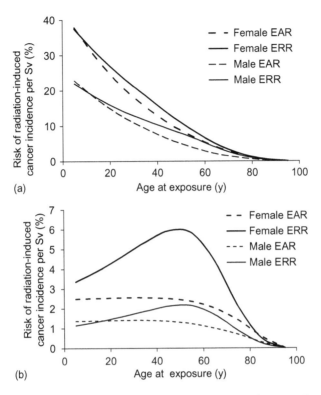

FIGURE 14.2 Plots of the risks of radiation-induced cancer incidence as a function of age at exposure derived from EAR and ERR models for female and male subjects in a Euro-American population for (a) all solid cancers and (b) lung cancer. Risks are based on calculations by R. Haylock and J. Harrison from Public Health England (PHE).

14.3.2 Modeling of Risks to Organs and Tissues by Different Organizations

Three models will be considered in this chapter, (1) those on which the ICRP 103 recommendations are based (ICRP 2007), (2) those in the BEIR VII report on the U.S. population (BEIR 2006), and (3) an online radiation risk assessment tool (RadRAT), which can be used to estimate the lifetime risk of radiation-related cancer from a user-specified exposure history (Berrington et al. 2012). Data will be included from the UK Health Protection Agency (HPA) (now Public Health England [PHE]) (Wall et al. 2011), which follows the ICRP format for evaluation, except that a Euro-American population is used for ERR calculations. The explanation will be based primarily around the ICRP methodology and comparisons made with other models with examples of how risk varies with age at exposure, as age is a primary determinant of the risk. Data showing values for EARs and ERR s have been derived by PHE and will be used in future ICRP risk calculations.

The data from the LSS have been sufficient to allow ERR and EAR models to be developed for esophagus, stomach, colon, liver, lung, breast, ovary, bladder, thyroid, and leukemia (bone marrow). ICRP also use nominal risks for bone and skin cancers (ICRP 1991). The tissues modeled in each of the assessments are listed in Table 14.1. The data for many human tissues and organs indicate that there is an excess risk but are insufficient to allow an assessment of the magnitude for individual organs, so they are consigned to a "remainder" category and ERR and EAR models developed for the whole group. The parameters for most solid cancers in these risk models have been estimated using incidence data from the LSS with follow-up from 1958 through to 1998 (Preston et al. 2007). However, the ICRP have taken account of data from medical exposure studies of breast cancer from North America and Europe (Preston et al. 2002), and of thyroid cancer (Ron et al. 1995), although these data are not used in the BEIR and RadRAT models.

TABLE 14.1 Comparison of Methods Used in Deriving Risks at Low Doses from the Life Span Study

ICRP 103 (ICRP 2007)	BEIR VII (BEIR 2006)	RadRAT (Berrington et al. 2012)
Organ risk models		
12 cancers modeled individually. Bladder, bone surfaces, colon, liver, lungs, oesophagus, ovary, skin, stomach, and leukemia from LSS data, breast and thyroid take data from other studies into account. Group of remainder tissues (13 in each sex) modeled: adrenals, extrathoracic tissue, gall bladder, heart, kidneys, lymphatic nodes, muscle, oral mucosa, pancreas, prostate, small intestine, spleen, thymus, uterus/cervix.	11 cancers modeled individually: bladder, breasts, colon, liver, lungs, ovary, prostate, stomach, thyroid, uterus, and leukemia from the LSS data. Group of remainder tissues and organs modeled for cancer incidence (solid cancers other than the ones modeled individually).	18 cancers modeled individually: bladder, breasts, central nervous system, colon, gallbladder, kidney, liver, lungs, oesophagus, oral cavity and pharynx, ovary, pancreas, prostate, rectum, stomach, thyroid, uterus, leukemia. Group of remainder tissues and organs modeled for cancer incidence (solid cancers other than the ones modeled individually).
Extrapolation models		
Linear relationship with dose for solid cancers that depends on sex, attained age, and age at exposure. Linear–quadratic model used for leukemia applied for all exposures to low doses.	Linear extrapolation applied to solid cancers that depends on sex, attained age, and age at exposure. Linear–quadratic model used for leukemia applied for all exposures to low doses.	Linear extrapolation applied to solid cancers that depends on sex, attained age, and age at exposure. Linear–quadratic model applied to leukemia for acute exposures, but only the linear term for chronic exposures.
Adjustment for period of minimum latency		
Abrupt change from 0 to maximum risk at five years after exposure for solid cancers and two years for leukemia.	Abrupt change from 0 to maximum risk at five years after exposure for solid cancers and two years for leukemia	Changes as S-shaped function of time since exposure with a smooth transition from zero to maximum risk. The midpoint of function is 2.25 years for leukemia, five years for thyroid, and 7.5 years for other solid cancers.
Dose and dose-rate effectiveness factor (DDREF)		
A DDREF equal to 2.0 is applied for all solid cancer sites for exposures to low doses or at low dose rates.	A DDREF equal to 1.5 is applied for all solid cancer sites for acute doses <100 mGy.	A DDREF equal to 1.5 is applied for chronic exposures to low-dose rates. For acute exposure, a DDREF is only applied when a given equivalent dose is < a reference dose, between 0.03 and 0.2 Sv.

Once the modifying effects of sex, age at exposure, and attained age had been incorporated, the models appear to show that there is a linear relationship between cancer incidence and organ dose between 0.1–0.2 and 4 Sv. There is no evidence of a departure from linearity below 0.1 Sv for solid cancer data overall or for individual subtypes (Little 2009). Therefore, a LNT approach, in which data for the Japanese atomic bomb survivors on excess cancer risk versus equivalent dose to the organ at risk are extrapolated linearly to zero dose (ICRP 2007), is applied to evaluate radiation effects. However, as discussed in Section 14.2.3, it is not possible to demonstrate a relationship at lower doses, because the numbers of excess solid cancers that might be expected with an LNT relationship would not be large enough for the size of the study population to achieve the statistical significance required for proof. Leukemia risk estimates are based on a linear-quadratic dose response model using the EAR model allowing for modification by sex, age at exposure, and time since exposure (Figure 14.3), as LSS reports have indicated that this model describes the dose response for leukemia among the LSS more accurately (Preston et al. 1994, 2007; Richardson et al. 2009).

The LSS has provided valuable data on the time course from initiation to the development of a tumor, because it comprised a large number of individuals who were exposed at one time. The time delay is called the latent period, and the LSS has shown that for leukemia the period is about five to eight years, although it can occur in as little as two years after exposure, while for solid tumors the minimum

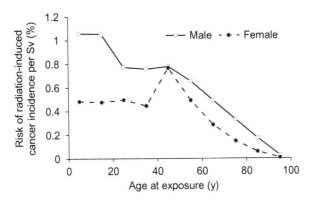

FIGURE 14.3 Plot of the risks of radiation-induced leukemia incidence as a function of age at exposure derived from the EAR model for male and female subjects in a Euro-American population. Risks are based on calculations by R. Haylock and J. Harrison from PHE.

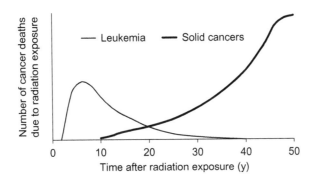

FIGURE 14.4 Schematic plot of the risks of radiation-induced leukemia and solid cancer incidence as a function of time after exposure from atomic bomb radiation.

latent period is five to ten years after radiation exposure (Figure 14.4). While the incidence of leukemia declines to the baseline level after about 30 years, this is not the case for solid cancers, for which the EARs continued to increase with attained age, although the ERR declined with both age and time after the bombing (Ozasa et al. 1012). The approach to risk evaluation in the models allow for a latent period of two years for leukemia and five years for solid cancers, applied as an abrupt change by ICRP and BEIR, and a gradual one for RadRAT (Table 14.1).

14.3.3 Dose and Dose-Rate Effectiveness Factors

Radiobiology experimental investigations tend to show that fractionation or protraction of dose is associated with reduced risk, suggesting that estimates based on high-dose, acute exposure need to be modified for use in evaluation of risk at low doses. Risk estimates for solid cancers are simply divided by a dose and dose-rate effectiveness factor (DDREF), because the epidemiological evidence available is insufficient to provide a definitive answer to the question of how much lower the risk is for low dose rate exposures than for those in the LSS (Table 14.1). Values of DDREF for the induction of gene and chromosomal mutations fall in the range of 2–4 (ICRP 2007). The BEIR Committee (BEIR 2006) combined epidemiological and radiobiological evidence for solid cancer from the LSS and studies in animals using Bayesian statistical analysis. The modal value of DDREF from these analyses was 1.5 with a range of 1.1–2.3 and based on this the BEIR VII Committee use a value of 1.5 in its assessments. ICRP used a DDREF of 2 for an earlier publication (ICRP 1991) and, since this is compatible with analysis of the LSS data, they continued with this approach in their revised recommendations (ICRP 2007). The ICRP uses general terminology regarding the dose level below which the DDREF should be applied. But BEIR VII applies the DDREF to exposures that deliver doses to organs of less than 100 mGy. RadRAT also uses a factor of 1.5 but has a sliding scale of application to avoid discontinuities in calculations (Berrington et al. 2012).

14.3.4 Modeling Transfer between Populations

Patterns of incidence for some cancers vary significantly between populations. If two populations differ with respect to the prevalence of different types of cancer and known risk factors, their responses to radiation exposure might be expected to vary as well. However, it is not obvious how important the prevalence of a particular cancer will be in determining the risk. Estimates of excess risk for a particular organ or tissue can be based on an absolute risk EAR model or a relative risk ERR model (Section 14.3.1), although ERR values can only be derived for populations for which sufficient cancer incidence data are available. For many sites, the difference between Japanese and Western cancer rates is not large, so whether the EAR or ERR transfer model is employed is not that significant. However, the age-specific baseline rates for stomach cancer in Japan are a factor of 12 higher than in the United States; this is apparent in the differences between ERR values for Euro-American and Asian populations used by PHE and ICRP in risk assessments and the differences between the EAR and ERR values for the Euro-American populations (Figure 14.5).

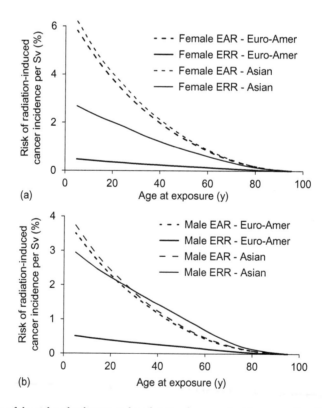

FIGURE 14.5 Plots of the risks of radiation-induced stomach cancer incidence as a function of age at exposure derived from EAR and ERR models for Euro-American and Asian populations for (a) females and (b) males. Risks are based on calculations by R. Haylock and J. Harrison from PHE.

Figure 14.5 plots the risks of radiation-induced stomach cancer incidence as a function of age at exposure derived from EAR and ERR models for Euro-American and Asian populations for (a) females and
(b) males. Risks are based on calculations by R. Haylock and J. Harrison from PHE.

The ICRP and HPA concluded that there was sufficient information to justify choice of specific relative values for weighting of the EAR and ERR model contributions for breast, thyroid, and lung cancer
(ICRP 2007). For breast cancer, pooled analyses of radiation effects by Preston et al. (2002) had provided
strong evidence that the ERR model was not appropriate, and breast cancer risks are therefore based on
an EAR model. However, the study also showed that the LSS data do not provide a good representation
of risks from other exposures; ICRP therefore base breast cancer risks on a combination of the LSS and
other data. For thyroid cancer, the use of EAR models is problematic because variations in the amount of
thyroid screening performed have a marked effect on the rate at which cancers are identified, so lesions
not associated with radiation might be diagnosed and attributed to radiation exposure. Therefore, thyroid cancer risks are based solely on the ERR model developed from the pooled analysis of radiation
associated thyroid cancer risks (Ron et al. 1995) and, for the same reason, a similar approach is employed
for skin cancer. Similar weighting regimes have been adopted by BEIR VII and RadRAT for the breast
and thyroid, but the EAR values have been based solely on the LSS data. This combined with the higher
DDREF value (Section 14.3.3) results in the risk of breast cancer incidence predicted by BEIR VII being
double relative to that based on the HPA/ICRP approach, while the risk for thyroid cancer is lower
(Figure 14.6). Radiation-induced cancer incidence is substantially higher among females for both breast
and thyroid cancer but declines steeply with age.

Figure 14.6 plots the risks of radiation-induced cancer incidence in females as a function of age at
exposure in a Euro-American population using the HPA/ICRP approach (Wall et al. 2011) and a U.S.
population using BEIR VII (BEIR 2006) (Tables 14.1 and 14.2). Both approaches use an EAR model for
the female breast and ERR models for the thyroid.

For lung cancer, the LSS data indicate that the EAR model is more comparable across genders than
the ERR and also that radiation dose and smoking history interact additively as lung cancer risk factors
(Pierce et al. 2003), so they gave the EAR data a weighting of 0.7 and the ERR 0.3. Smoking influences
the form of the ERR versus age at exposure relationship for lung cancer (Figure 14.2b) and the radiation
risk will change as smoking habits of the population alter.

The question remains as to which weighting is appropriate for other organs and tissues for which
there is insufficient data to decide between the two models to make an informed choice. The ICRP
decided to apply equal weightings of 0.5 to all the other organs, apart from the breast, thyroid, lung,
and skin, but in BEIR VII and RadRAT, the ERR data for the other organs is given a greater weighting

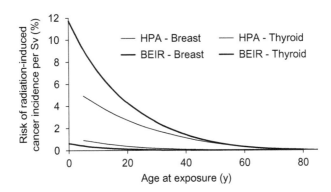

FIGURE 14.6 Plots of the risks of radiation-induced cancer incidence in females as a function of age at exposure
in a Euro-American population using the HPA/ICRP approach (Wall et al. 2011) and a U.S. population using BEIR VII
(BEIR 2006) (Tables 14.1 and 14.2). Both approaches use an EAR model for the female breast and ERR models for
the thyroid.

TABLE 14.2   Comparison of Methods Used to Transfer Risk Model to Other Populations

ICRP 103 (ICRP 2007)	BEIR VII (BEIR 2006)	RadRAT (Berrington et al. 2012)
Application of EAR and ERR in transfer from Japanese to other population		
EAR and ERR projections of the lifetime risk are weighted in logarithmic space. Assigned weights are ERR 0.5 and EAR 0.5 for most cancers, apart from lung (ERR 0.3 and EAR 0.7), thyroid and skin are based on ERR model, and breast on EAR model.	EAR and ERR projections of the lifetime risk are weighted in logarithmic space. Assigned weights are ERR 0.7 and EAR 0.3 for most cancers, apart from lung (ERR 0.3 and EAR 0.7), thyroid is based on ERR model, and breast on EAR model.	EAR and ERR projections of the lifetime risk are weighted in linear space. Assigned weights are ERR 0.7 and EAR 0.3 for most cancers, apart from lung (ERR 0.3 and EAR 0.7), thyroid and gall bladder are based on ERR model, and breast on EAR model.
Population used for cancer prevalence with age in transfer		
Average for four Asian populations, two European and the U.S.	United States population	United States population

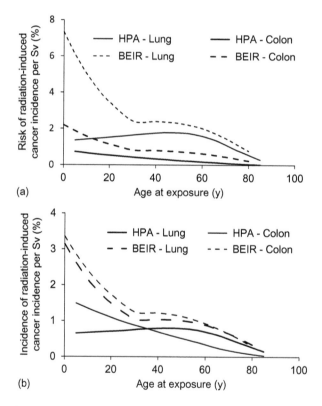

FIGURE 14.7 Plots of the risks of radiation-induced lung and colon cancer incidence as a function of age at exposure for Euro-American populations derived using the HPA/ICRP (Wall et al. 2011) and BEIR VII (BEIR 2006) approaches (Tables 14.1 and 14.2) for (a) females and (b) males.

of 0.7 and the EAR 0.3 (Table 14.2). The use of alternative weightings influences the evaluation of risk for organs such as the stomach (Figure 14.5), for which the magnitudes of the EAR and ERR models are significantly different. Calculations for some organs derived by the BEIR Committee with their model weightings and DDREF values give risks that are 2–4 times higher than those derived using the ICRP/HPA factors for younger age groups (Figure 14.7). These variations in approach by experts from different professional organizations provide an indication of the uncertainties that are present. Either of the alternatives could be valid based on current evidence and so reflect real uncertainties that are present. This should provide a caution to those who try to interpret and apply risk values as if these values are known with certainty.

Figure 14.7 plots the risks of radiation-induced lung and colon cancer incidence as a function of age at exposure for Euro-American populations derived using the HPA/ICRP (Wall et al. 2011) and BEIR VII (BEIR 2006) approaches (Tables 14.1 and 14.2) for (a) females and (b) males.

The populations used by ICRP and BEIR in deriving ERRs differ because their reports are directed at different populations. The ICRP aims to provide global average values and bases its assessments on a composite population comprising four Asian populations (Shanghai, Osaka, Hiroshima, and Nagasaki), two European (Sweden and UK), and the U.S., but ICRP also includes separate risk data for the Euro-American and Asian populations. On the other hand, BEIR and RadRAT base their predictions on the U.S. population (Table 14.2), and the HPA have published risk factors based on the Euro-American component of the ICRP populations.

14.4 Differences and Uncertainties in Estimation of Cancer Risk and Detriment

14.4.1 Radiation Detriment, Application of Risk Estimates, and Effective Dose

The results from the models described provide the foundation on which assessment of the risk of cancer induction is based (UNSCEAR 2000, 2006; BEIR 2006; ICRP 2007; Berrington et al. 2012; Wall et al. 2011). The risk models have been used by ICRP to compute sex-specific estimates of lifetime risk for a range of ages at exposure (0–85 years in five-year intervals) in an Asian and Euro-American composite population. These lifetime risks for exposure ages have then been averaged using weights reflecting the age distribution for full populations and working age populations (18–64 years old). Other organizations have published tables of sex-specific estimates of lifetime risk of cancer incidence for groupings of age at exposure BEIR VII and RadRAT for the U.S. population, and HPA for a Euro-American population. Such data can potentially be used to calculate risks for individuals based on their sex and age. Assessments of cancer incidence by HPA/ICRP and BEIR models differ by more than a factor of 3 for several organs and tissues (Figure 14.7), but overall the total lifetime incidence differs by a factor of less than 2 (Figure 14.8).

Health detriment is a concept used to quantify the harmful effects of radiation exposure in different parts of the body. ICRP defines it as the total harm to health experienced by an exposed group and its descendants as a result of the group's exposure to a radiation source. It is determined from nominal risk coefficients, taking into account the impact of the disease in terms of lethality and years of life lost and includes the probability of severe heritable effects. ICRP has calculated detriment for full populations and working-age populations. The ICRP introduced a protection quantity called the effective dose to provide a tool for dose comparisons in terms of detriment. The total stochastic radiation detriment is equated to the sum of radiation doses to individual organs, weighted according to their aggregate detriment for a population of all ages and averaged for the two sexes. Thus, effective dose applies to a reference person representing a global population average, and so can be used to make comparisons among radiation doses from medical procedures that expose different regions of the body in terms of potential harm. The effective dose provides a single quantity for conveying an indication of radiation dose relating to possible risk to health. It has been widely used in providing a broad understanding of relative risks associated with radiation exposures from different sources and raising awareness of dose levels from diagnostic medical procedures.

Figure 14.8 plots the risks of radiation-induced cancer incidence as a function of age at exposure from a uniform whole-body irradiation for Euro-American populations derived using the HPA/ICRP (Wall et al. 2011) calculation method, and the U.S. population using BEIRVII (BEIR 2006) and RadRAT (Berrington et al. 2012) approaches for (a) females and (b) males.

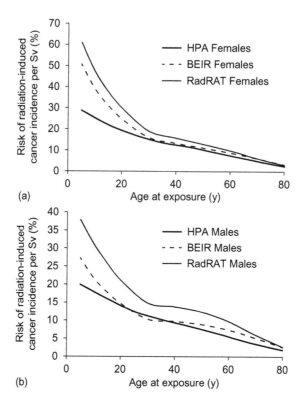

(a)

(b)

FIGURE 14.8 Plots of the risks of radiation-induced cancer incidence as a function of age at exposure from a uniform whole-body irradiation for Euro-American populations derived using the HPA/ICRP (Wall et al. 2011) calculation method, and the U.S. population using BEIR VII (BEIR 2006) and RadRAT (Berrington et al. 2012) approaches for (a) females and (b) males.

14.4.2 Sources of Uncertainty in Estimates of Risk

The sophisticated risk models that have been constructed are based on a theoretical extrapolation of data, and there is a danger that those applying the results may neglect the many underlying uncertainties. First of all, there are uncertainties that arise in developing estimates of risk from the LSS in terms of the accuracy with which dose levels for the exposed individuals could be assessed, based on knowledge of the radiation field and the location of each exposed individual. Then there are the methods used in evaluating the epidemiological data. The issue of methodology is relevant to all epidemiological studies.

As previously discussed, the majority of the evidence available relating to radiation risk is derived from higher dose levels than those encountered in medical imaging, necessitating extrapolation of risks from moderate- or high-dose exposures down to dose levels of interest. A linear extrapolation has been used, and for chronic or low-dose exposures, the risk estimate is divided by a DDREF that reduces the estimate but increases its uncertainty. Evidence from the LSS suggests that linear extrapolation is appropriate when considering an exposed population, but the magnitude of the DDREF is a matter of debate (Table 14.1).

The transfer of risk estimates from one population to another, with adjustments depending on the relative incidences of cancer in the LSS and the population of interest, appears to provide a reasonable approach, but the values of the weightings that should be attached to the ERR and EAR models for organs and tissues for which data are limited is unclear, and different choices have been made. Cancer incidence data are only available for a limited number of populations, so there is uncertainty in how representative any ERR factors are for other population groups for assessment of risk.

Other major sources of uncertainty include possible interaction of radiation exposure with other cancer risk factors, notably smoking history in the case of lung cancer and reproductive history in the case of female breast cancer. The way in which differences in the relative biological effectiveness for radiations of different qualities are accounted for, including medical X-rays in the 10–200 keV range, as well as neutrons and alpha particles, is an approximation. Another source of uncertainty is extrapolation of risks for whole-body exposures to external radiation in the LSS to partial-body exposures typical of radiographic procedures and to non-uniform internal exposures associated with nuclear medicine procedures.

When assessing risk for any individual there are more approximations, as organ doses often can only be assessed through Monte Carlo calculations tracking the movement of photons through anthropomorphic phantoms, which represent idealized anatomical forms in terms of size, shape, and position of each tissue. Conversion coefficients have been derived for X-ray exposures for a reference patient that allow values for organ and tissue doses to be calculated from measurable dose quantities, such as entrance surface air kerma and kerma-area product for radiographic examinations and dose-length product for CT, and adjustments can be made for differences in patient body habitus (Li et al. 2011; Huda and He 2012). However, beam collimation can have a significant effect on the proportions of individual organs and tissues that are exposed, especially where radiosensitive organs lie near the edge of the field. For nuclear medicine procedures, dose coefficients (dose per unit administered activity in Sv/Bq, for example) have been derived for a large number of radiopharmaceuticals. These are based on biokinetic models that calculate the number of radioactive disintegrations occurring within each tissue for a reference patient. The models take into account transfer, retention, and excretion of radioactive material and the type of radiation emitted. The radionuclides are assumed to be distributed uniformly within the respective source organs, and results are combined with data from dosimetric models from which the deposition of energy in relevant target organs is derived. Although every attempt is made to provide as accurate assessments of dose as possible when applying the models, the uncertainties are large, and application to any individual can only be regarded as providing an approximate indication of potential risk.

14.5 Conclusions

The calculation of risk from doses to individual organs and tissues using the models described, based on the age and sex of the individual exposed, provides the best assessment that we can make of the risk of stochastic effects. But there are many sources of uncertainty, and users should be aware of these and not place more weight on the result than the methods of derivation justify. The use of medical imaging and associated population doses have been increasing rapidly over the last 10–20 years, and there is reason to try to reduce numbers of unnecessary exposures. In promotion of this message, claims, some alarmist in tone, have been made quoting large numbers of extra cancers that will result from this increased use of medical imaging (Brenner and Hall 2007; Berrington de González et al. 2009). These numbers are based on the BEIR VII model, but described and presented as fact, with little account taken of the uncertainties in the epidemiological data, the extrapolation to low doses, or the reduced life expectancy of patients because of their illnesses. Such assessment of risks based on exposures of large populations to low doses are inappropriate.

There is sometimes a need to have a knowledge of the radiation exposure that a patient has received, and the factors that have been derived from the models provide the potential to give assessments of risk. Since the factors have been published are linked to sex and age, there has been a trend to make numerical risk calculations for individual exposures; at the low dose level of almost all diagnostic exposures, however, this is inappropriate. The risks linked to age and sex for individual organs may provide a better assessment of potential cancer incidence than is possible using effective dose, but the uncertainties are still large and it is inappropriate to apply them to exposures less than about 50 mSv. In addition to concerns regarding the suitability of the LNT extrapolation, and the other uncertainties discussed above, the genetic variation among individuals is huge, so actual risks to an individual might be a factor of ten higher or lower than a value calculated from the organ dose data. Clinicians who refer patients for medical procedures involving radiation may have little understanding of the potential health detriment from radiation exposure, in part because the risks are so small compared to the benefits. Although there is a need for quantification of radiation exposure, the question remains as to the best way of determining such exposures and expressing the associated risks. The concept of effective dose developed by the ICRP perhaps is the best option currently for expressing relative values of radiation dose. A knowledge of typical values for common medical procedures is used in training medical professionals and informing judgments on relative radiation dose levels (Martin 2007). Risks perhaps can better be described in general terms, such as negligible (<0.1 mSv), minimal (0.1–1 mSv), very low (1–10 mSv), and low (10–100 mSv). Such terms reflect uncertainties in the predictions in a realistic manner.

Acknowledgments

The author wishes to thank Richard Haylock from PHE and John Harrison from ICRP for provision and permission to use data for plots of EAR and ERR in Figures 14.2, 14.3, and 14.5.

References

Akleyev, A., Krestinina, L., Degteva, M. and Tolstykh, E. 2017. Consequences of the radiation accident at the Mayak production association in 1957. *J. Radiol. Prot.* 37: R19.

Berrington de Gonzalez, A., Apostoaei, A.I., Veiga, L.H.S. et al. 2012. RadRAT: A radiation risk assessment tool for lifetime cancer risk projection. *J. Radiol. Prot.* 32: 205–222.

Berrington de González, A., Mahesh, M., Kim, K.P. et al. 2009. Projected cancer risks from computed tomographic scans performed in the United States in 2007. *Arch. Intern. Med.* 169(22): 2071–2077.

Biological Effects of Ionizing Radiation (BEIR). 2006. *Health Risks from Exposure to Low Levels of Ionizing Radiation. U.S. National Academy of Sciences, National Research Council, BEIR VII, Phase 2.* Washington, DC: The National Academies Press.

Boice, J.D. Jr., Preston, D., Davis, F.G. and Monson, R.R. 1991. Frequent chest X-ray fluoroscopy and breast cancer incidence among tuberculosis patients in Massachusetts. *Radiat. Res.* 125: 214–222.

Boice, J.D. Jr. 2015. Radiation epidemiology and recent paediatric computed tomography studies. *Ann. ICRP* 44(1 Suppl): 236–248.

Brenner, D.J. and Hall, E.J. 2007. Computed tomography: An increasing source of radiation exposure. *N. Engl. J. Med.* 357(22): 2277–2284.

Cardis, E., Vrijheid, M., Blettner, M. et al. 2005. Risk of cancer after low doses of ionising radiation: Retrospective cohort study in 15 countries. *Br. Med. J.* 331: 77–80.

Cullings, H.M., Fujita, S., Funamoto, S., Grant, E.J., Kerr, G.D. and Preston, D.L. 2006. Dose estimation for atomic bomb survivor studies: Its evolution and present status. *Radiat. Res.* 166: 219–254.

Darby, S., Hill, D., Auvinen, A. et al. 2005. Radon in homes and risk of lung cancer: Collaborative analysis of individual data from 13 European case–control studies. *Br. Med. J.* 330: 223–227.

Folley, J.H., Borges, W. and Yamawaki, T. 1952. Incidence of leukemia in survivors of the atomic bomb in Hiroshima and Nagasaki, Japan. *Am. J. Med.* 13: 311–321.

Hall, E.J. and Giaccia, A.J. 2012. *Radiobiology for the Radiologist*, 7th ed. Philadelphia, PA: Lippincott Williams & Wilkins.

Huda, W. and He, W. 2012. Estimating cancer risks to adults undergoing body CT examinations. *Rad. Prot. Dosim.* 150: 168–179.

ICRP. 1991. The 1990 Recommendations of the international commission on radiological protection. ICRP Publication 60. *Ann. ICRP* 21(1–3).

ICRP. 2007. The 2007 Recommendations of the international commission on radiological protection. ICRP publication 103. *Ann. ICRP* 37(2–4): 1–332.

ICRP. 2010. Lung cancer risk from radon and progeny and statement on radon. Publication 115. *Ann. ICRP* 40(1).

Krestinina, L.Y., Preston, D.L., Ostroumova, E.V. et al. 2005. Protracted radiation exposure and cancer mortality in the Techa River cohort. *Radiat. Res.* 164: 602–611.

Krewski, D., Lubin, J.H., Zielinski, J.M. et al. 2005. Residential radon and risk of lung cancer. A combined analysis of 7 North American case–control studies. *Epidemiology* 16: 137–145.

Li, X., Samei, E., Segars, W.P., Sturgeon, G.M., Colsher, J.G. and Frush, D.P. 2011. Patient-specific radiation dose and cáncer risk for pediatric chest CT. *Radiology* 259: 862–874.

Little, M.P. and Boice, J.D. Jr. 1999. Comparison of breast cancer incidence in the Massachusetts tuberculosis fluoroscopy cohort and in the Japanese atomic bomb survivors. *Radiat. Res.* 51: 218–224.

Little, M.P., Wakeford, R., Tawn, E.J., Bouffler, S.D. and Berrington-Gonzalez, A. 2009. Risks associated with low doses and low dose rates of ionizing radiation: Why linearity may be (Almost) the best we can do. *Radiology* 251: 6–12.

Little, M.P. 2009. Cancer and non-cancer effects in Japanese atomic bomb survivors. *J Radiol. Prot.* 29: A43–A59.

Lubin, J.H. and Boice, J.D. 1997. Lung cancer risk from residential radon: Meta-analysis of eight epidemiological studies. *J. Natl. Cancer Inst.* 89: 49–57.

Lubin, J.H., Wang, Z.Y., Boice, J.D. Jr. et al. 2004. Risk of lung cancer and residential radon in China: Pooled results of two studies. *Int. J. Cancer* 109: 132–137.

Martin, C.J. 2007. Effective dose: How should it be applied to medical exposure? *Br. J. Radiol.* 80: 639–647.

Ozasa, K., Shimizu, Y., Suyama, A. et al. 2012. Studies of the mortality of atomic bomb survivors, report 14, 1950–2003: An overview of cancer and noncancer diseases. *Radiat. Res.* 177: 229–243.

Pearce, M.S., Salotti, J.A., Little, M.P. et al. 2012. Radiation exposure from CT scans in childhood and subsequent risk of leukaemia and brain tumours: A retrospective cohort study. *Lancet* 380: 499–505.

Pierce, D.A., Sharp, G.B. and Mabuchi, K. 2003. Joint effects of radiation and smoking on lung cancer risk among atomic bomb survivors. *Radiat. Res.* 159: 511–520.

Preston, D.L., Kusumi, S., Tomonaga, M. et al. 1994. Cancer incidence in atomic bomb survivors. Part III. Leukaemia, lymphoma and multiple myeloma, 1950–1987. *Radiat. Res.* 137: S68–S97.

Preston, D.L., Mattsson, A., Holmberg, E. et al. 2002. Radiation effects on breast cancer risk: A pooled analysis of eight cohorts. *Radiat. Res.* 158: 220–235.

Preston, D.L., Pierce, D.A., Shimizu, Y. et al. 2004. Effect of recent changes in atomic bomb survivor dosimetry on cancer mortality risk estimates. *Radiat. Res.* 162: 377–389.

Preston, D.L., Ron, E., Tokuoka, S. et al. 2007. Solid cancer incidence in atomic bomb survivors: 1958–1998. *Radiat. Res.* 168: 1–64.

Richardson, D., Sugiyama, H., Nishi, N. et al. 2009. Ionizing radiation and leukemia mortality among Japanese atomic bomb survivors, 1950–2000. *Radiat. Res.* 172: 368–382.

Ron, E., Lubin, J.H., Shore, R.E. et al. 1995. Thyroid cancer after exposure to external radiation: A pooled analysis of seven studies. *Radiat. Res.* 141: 259–277.

Shilnikova, N.S., Preston, D.L., Ron, E. et al. 2003. Cancer mortality risk among workers at the Mayak nuclear complex. *Radiat. Res.* 159: 787–798.

Smith, P.G. 2007. The 1957 MRC report on leukaemia and aplastic anaemia in patients irradiated for ankylosing spondylitis. *J. Radiol. Prot.* 27: B3–B14.

Spies, H., Mays, C.W. and Chmelevsky, D. 1989. Malignancies in patients injected with radium-224. In: D.M. Taylor, C.W. Mays, G.B. Gerber and R.G. Thomas (Eds.), *Risks from Radium and Thorotrast*, Report 21, pp.7–12. London, UK: BIR.

United Nations Scientific Committee on the Effects of Atomic Radiation (UNSCEAR). 2006. Report to the General Assembly with Scientific Annexes. Effects of Ionizing Radiation. Volume I Report and Annexes A and B. New York: United Nations.

UNSCEAR 2000 Report to the General Assembly. Sources and effects of ionising radiation. Volume II Effects, Annexes F, G, H, I and J, New York: United Nations.

UNSCEAR. 2011. UNSCEAR 2008 Report to the General Assembly. Sources and effects of ionizing radiation. Volume II and Scientific Annexes C, D and E, New York: United Nations.

Wall, B.F., Haylock, R., Jansen, J.T.M., Hillier, M.C., Hart, D. and Shrimpton, P.C. 2011. Radiation risks from medical X-ray examinations as a function of age and sex of patient. HPA Report HPA-CRCE-028. Chilton, UK: HPA.

Weiss, H.A., Darby, S.C., Fearn, T. and Doll, R. 1995. Leukemia mortality after X-ray treatment for ankylosing spondylitis. *Radiat. Res.* 142: 1–11.

World Health Organization (WHO). 2005. Health effects of the Chernobyl accident and special health care programmes. *Report of the UN Chernobyl Forum Expert Group "Health" (EGH)*. Geneva, Switzerland: WHO.

15

Quantitative Benefit-Risk Analysis of Medical Radiation Exposures

Pat B. Zanzonico

15.1 Introduction

The introduction of ionizing radiation to the healing arts revolutionized the diagnosis and treatment of disease and dramatically improved the quality of health care and of human life. In recent years, however, concern over the radiogenic risks associated with medical imaging has grown dramatically, fueled by widespread attention in both the scientific and lay media. Such concern is beneficial in terms of promoting critical evaluation of imaging procedures, with technical optimization, elimination of truly unnecessary procedures, and minimization of imaging doses without compromising the diagnostic information being sought. However, consideration of radiogenic cancer risk, sometimes couched in spectacular terms,[1] can create the misconception, and one that is detrimental to optimum patient care, that radiation is the *only* risk to be considered in medical imaging.

Patients should, of course, become well-informed about the risks and benefits of all aspects of medical care, radiation-related and otherwise, and be active participants in decision-making. When advising patients about these risks, a perfunctory statement is usually offered to the effect that any theoretical risks of the radiation exposures received are far smaller than the direct benefits of the study. Importantly, the word "theoretical" should be used because all risk information is predicated on the "linear no-threshold" (LNT) theory of radiation carcinogenesis, which posits that stochastic, or statistical, risks (primarily of cancer) observed in populations exposed to high doses of radiation can be extrapolated to the much lower doses that are encountered in diagnostic medical procedures, certain occupational exposures, and other circumstances.[2] Such an extrapolation essentially involves drawing a straight line through the excess risk-versus-dose data available at higher doses (typically at least tenfold higher than the highest diagnostic or occupational doses) down to zero dose. It is not known if this extrapolation is actually valid, and it remains unproven and highly controversial. While the LNT model may be a defensible (i.e., appropriately conservative) basis for formulating population-based radiation dose limits (e.g., for occupationally exposed individuals), its application to *individual* patients and their medical management is far less justifiable. For example, the LNT-derived risk of a radiogenic cancer associated with a specific diagnostic procedure for an individual patient conceivably may lead the patient and/or the patient's physician to forego this procedure and, in its place, perform another suboptimal procedure with its own non-radiogenic risks, which may possibly exceed the theoretical radiogenic risk of the procedure originally considered.

There are several reasons why application of population- and high-dose-based radiation risk estimates to individuals is inadvisable. First, the uncertainties in the current estimates of radiogenic cancer risks, largely devised from the A-bomb follow-up data, and of radiation dose,[3] are considerable, and propagating such uncertainties when applying these estimates to individual patients yields a *highly* uncertain estimate of patient-specific cancer risk. Second, radiogenic cancer risks vary widely with dose rate and fractionation, gender, age at exposure, and part or parts of the body exposed, among other factors. Prevailing risk factors, on the other hand, are based on effective dose and thus gender- and age-averaged values for a presumed uniform, whole-body exposure. The effect of such averaging can be mitigated somewhat by the use of age- and gender-indexed risk factors, but this is often ignored in the context of evaluating radiogenic risks of imaging procedures. Third, the latent period for radiogenic cancer induction (i.e., the time interval from exposure to the clinical appearance of the resultant cancer) is generally quite long, varying from years to decades, and inversely related to dose. Conversely, the clinical sequelae of a suspected disease or condition, which precipitated prescription of an imaging procedure, are often far more immediate. Thus, a patient may very well *not* outlive the theoretical radiogenic cancer risks of such procedures, especially if the presenting medical condition is not definitively diagnosed and addressed. This, of course, is quite likely among the vast majority of patients, who are typically older.

Application with certitude of the LNT risk model is unjustified and has led some to derive alarming estimates of excess numbers of cancers as a result of medical imaging, for example, in the thousands annually among the United States population.[4] Such exercises largely ignore the considerable uncertainties (entailing values down to zero excess risk) and the other factors enumerated above in deriving such estimates. To the extent that the LNT-derived risk may overestimate the actual radiogenic risk (if any), it undermines a reasonable benefit-risk calculus and may thus adversely impact the patient's medical management. The LNT model may, in fact, overestimate the actual risks of diagnostic and other low-level exposures, and there is creditable evidence for a threshold for radiogenic cancer induction, that is, a non-zero radiation dose below which there is *no* increased risk of cancer.[5,6] Importantly, the data upon which the LNT model is based begin to show significance only at doses above 100 mSv (10 rem). The Health Physics Society has, in fact, issued a position statement advising against estimation of health risks below an individual dose of 50 mSv in one year or a lifetime dose of 100 mSv above that received from natural sources.[7]

As noted above, the point is often made that the benefits of the uses of radiation in medicine are much greater than any theoretical risks, but *quantitative* estimates of the benefits are not cited alongside any quantitative estimates of risk. This alone—expression of benefit in purely qualitative terms versus expression of risk in quantitative, and therefore seemingly more certain, terms—may well contribute to a skewered sense of the relative benefits and risks of diagnostic imaging among health-care providers as well as patients. The current paper, therefore, quantitatively compares the benefits of diagnostic imaging in several cases, based on actual mortality or morbidity data if ionizing radiation were not employed, with the LNT model-derived (i.e., theoretical) estimates of radiogenic cancer mortality.[8] The case studies below are not intended to provide rigorous estimates of medical benefit but rather order-of-magnitude estimates that illustrate the very large benefit-to-risk ratios typical of diagnostic imaging studies.

15.2 Radiogenic Cancer Risk

The radiogenic risk of cancer can be defined as the excess number of cancers in a population estimated using the LNT model as the number of persons exposed × effective dose (rem or mSv) per person × excess risk (/rem or /mSv). The excess risk (ER) is the number of excess fatal cancers (excess above the naturally occurring rate) predicted by the model in a large (and therefore gender- and age-averaged) population uniformly exposed to a whole-body dose of radiation. The concept of "effective dose" (ED) was developed by the International Commission on Radiation Protection and Measurement (ICRP) to allow the comparison of practices that lead to partial irradiation of people to those resulting in uniform, whole-body irradiation from a risk point of view. In first order, ED will be assumed to be equivalent to uniform, whole-body dose. A widely cited ER value is that recommended by the National Council on Radiation Protection and Measurements (NCRP) Report No 115, 5×10^{-5} per person per mSv (or 5×10^{-4} per person per rem).[9] Thus, if a population of a million people were to each receive a dose of 1 mSv, the expected number of excess fatal cancers observed in this population over the balance of the lifetimes of the individuals in this population would be 5×10^{-5} per person mSv $\times 1 \times 10^6$ per person mSv (collective dose) = 50. This compares to the spontaneous, or background, lifetime incidence of ~200,000 (or 20%) otherwise occurring in such a population. Importantly, even if one concedes the accuracy of the LNT model, it cannot be applied reliably to individuals but only to large populations, that is, populations sufficiently large that differences in radiation sensitivity related to gender, age, and intrinsic biology are effectively averaged out.

15.3 Medical Benefit

One benefit of a diagnostic imaging procedures may be expressed as the lives saved, that is, the number of lives lost by *not* performing the procedure or by performing an alternative, invasive procedure. This benefit may be quantified using clinical outcome data gleaned from the literature regarding the actual or hypothetical mortality or morbidity rates in specific instances where a relevant radiological procedure was not performed. There may be other metrics of benefits, such as improvements in the quality of life, shortening of hospital stays, and reduction of medical care costs. However, in the current analyses, the emphasis was placed on lives saved resulting from the performance of the procedure.

15.4 Case Studies of Benefit Risk Analysis

Use of Fluorine-18-Fluoro-Deoxyglucose Positron Emission Tomography/Computed Tomography in Preoperative Assessment of Suspected Non-Small-Cell Lung Cancer. van Tinteren et al.[10] compared the management of suspected non-small-cell lung cancer (NSCLC) with and without pre-operative fluorine-18-fluoro-deoxyglucose (^{18}F-FDG) positron emission tomography (PET). With the conventional pre-operative evaluation of NSCLC, that is, without FDG PET, 81% of patients underwent thoracotomy, and 41% of those thoracotomies were futile (i.e., not potentially curative) because of the progression of disease disclosed at surgery. In the series of van Tinteren et al, the surgery-related mortality was 6.5%. Adding FDG PET to the pre-operative evaluation reduced the proportion of NSCLC patients undergoing thoracotomy to 65%, with only 21% being futile. Thus, a non-curative operation was avoided in 20% of patients with the addition of FDG PET. If one extrapolates the foregoing data to the U.S. population, with 174,470 new cases of lung cancer per year, the conventional pre-operative evaluation would result in $174,470 \times 0.81 \times 0.41 \times 0.065 = 3,766$ futile surgical deaths each year; the addition of FDG PET to the pre-operative evaluation would reduce the number of *futile* surgical deaths per year to $174,470 \times 0.65 \times 0.21 \times 0.065 = 1,547$—a gross benefit of FDG PET $3,766 - 1,547 = 2,219$ lives saved per year. However, for a 370-MBq administered activity of FDG in the context of PET/computed tomography (CT), the ED is 14 mSv (1.4 rem), and the number of radiogenic cancer-related deaths is therefore $174,470 \times 1.4$ rem $\times 0.0005$/rem = 122, where 0.0005/mrem. Thus, the addition of FDG PET/CT to the pre-operative evaluation of suspected lung cancer would result in a *net* benefit of $2,219 - 122 = 2,097$ lives saved per year in the US.

Use of Multi-slice Computed Tomography versus Conventional Coronary Angiography in Screening for Coronary Artery Disease. Multi-slice computed tomography (MSCT) can reliably identify those patients with clinically significant and treatable coronary artery disease (CAD). The ED from the use of MSCT is ~14 mSv (1.4 rem).[11] Using the LNT model, this would suggest an excess radiogenic cancer death risk of 0.07%. The ED from conventional coronary angiography (CCA) by comparison is ~6 mSv (0.6 rem),[11] suggesting an excess radiogenic cancer death risk of 0.03%. The non-radiogenic, non-contrast mortality risk from CCA is 0.11%, according to Noto et al.[12] The U.S. population of individuals aged 50–55 years (2006) is 18.8 million. The number of excess cancer deaths due to the use of MSCT screening if such screening were done would be 13,160 for a single MSCT examination per individual screened. The number of excess cancer deaths would be 39,480, or 2,627 per year, if each individual were screened every five years from age 55 to age 70.

The potential gross benefit of MSCT CAD screening is estimated at 33,500 per year, assuming the number of lives saved would be 10% of ~335,000 sudden cardiac deaths per year.[13] For an individual having a first myocardial infarction at age 68, more than 90% had greater than 75% stenosis in more than one vessel,[14] which is detectable by MSCT. Conservatively, therefore, the net benefit of MSCT would thus be 33,500 − 2,627 = 30,873 lives saved per year in the US.[15]

Use of Sentinel Node Biopsy versus Standard Axillary Therapy in Operable Breast Cancer. In a multicenter randomized trial comparing quality-of-life outcomes in clinically node-negative invasive operable breast cancer, the number of sentinel node biopsies was 515 and the number of standard axillary therapies 516. The total number of biopsies disclosing node-disseminated breast cancer either by sentinel node biopsy or standard axillary therapy was seven. The incidence of lymphedema and sensory loss at 12 months for sentinel node biopsy was only 37% of that for standard axillary therapy risk.[16]

The use of ^{99m}Tc-labeled sulfur colloid for sentinel node imaging results in an ED of ~0.92 mSv (0.092 rem). The use of the LNT model with this dose predicts a number of excess cancer deaths in the 515 subjects of 0.024, essentially zero. Thus, the net benefit of the use of sentinel node biopsy is a 63% reduction in the incidence of lymphedema and arm-sensory loss without any significant increase in cancer risk in the 515-subject cohort studied.

Use of Scintigraphic Perfusion Imaging to Predict Peri-operative Cardiac Events in Non-cardiac Surgery. The most important cause of peri-operative cardiac mortality and morbidity is myocardial infarction (MI) due to occult CAD. In a Veterans Administration (VA) series, the incidence and mortality of such events associated with vascular surgery (most commonly, carotid endarterectomy) were 13% and 40%–70%, respectively. Based on pre-operative dipyridamole thallium-201 (^{201}Tl) imaging, the incidence of peri-operative cardiac events[17] was 2% for a severity level of 0 and extent of 0 and 100% for a severity level of 3 and extent of 5–6, with 22% of patients have reversible perfusion defects. Thus, perfusion imaging was highly accurate for prediction of peri-operative cardiac events. The number of vascular surgeries (from the VA database) is ~9,500 per year, and the number of peri-operative cardiac deaths (i.e., fatal MIs) is thus estimated as 494 per year. Therefore, the gross benefit of pre-operative perfusion imaging with ^{201}Tl is 109 per year, that is, the number of predictable peri-operative cardiac deaths (fatal MIs) avoided.

The ED from the ^{201}Tl study is ~24 mSv (2.4 rem). The LNT model would predict 11 excess cancer deaths per year, yielding a net benefit of pre-operative myocardial perfusion imaging of 109 − 11 = 98 lives saved per year. If one considers a rest/stress myocardial perfusion study using ^{99m}Tc methoxyisobutylisonitrile (MIBI), the ED is ~12 mSv, leading to only 6 excess cancer deaths per year and possible greater clinical benefit. Performing cardiac PET with rubidium-82 (^{82}Rb)-, rubidium chloride would result in an even lower ED of ~7.5 mSv (0.75 rem)[18], thus resulting in a theoretical risk of only 3 cancer deaths and a net savings of 106 lives per year.

Ventilation-Perfusion Imaging for Diagnosis of Pulmonary Embolism in Pregnancy. Risks of deep vein thrombosis (DVT) and pulmonary emboli (PE) are significant during and after pregnancy. The risk of PE is cited as 1 in 1,500 deliveries, or ~7 × 10^{-4}, and ~70% of patients with a known PE have proximal DVT.[19] Marcus et al.[20] estimated that mortality of untreated and treated PE as 12.8% and 0.7%, respectively. They estimated the fetal dose for ^{99m}Tc-macroaggregated albumin (MAA) and for ^{99m}Tc-diethylene triamine pentaacetic acid (DTPA) aerosol using small intestine as a surrogate for the fetus; Russell et al.[21] subsequently updated their calculations using the Oak Ridge National Laboratory pregnant female phantom series.[22] Their dose estimates (mSv/MBq) are:

Radiopharmaceutical	Early Pregnancy	3 Months	6 Months	9 Months
^{99m}Tc DTPA	1.2×10^2	8.7×10^3	4.1×10^3	4.7×10^3
^{99m}Tc MAA	2.8×10^{-3}	4.0×10^{-3}	5.0×10^{-3}	4.0×10^{-3}

Assuming administration of 185 MBq of DTPA aerosol and 37 MBq of MAA, fetal doses are 2.3 mSv in early pregnancy, 1.8 mSv at three months' gestation, 0.94 mSv at six months and 1.0 mSv at nine months. The ED to the mother is ~1.7 mSv.[23]

The dose to the fetus is far below the well-accepted threshold for deterministic effects.[17,18] Brent notes that "There is little doubt that irradiation of the embryo represents a carcinogenic risk, but the magnitude of the risk has been extensively debated. The oncogenic risk from exposing the embryo to ionizing radiation in the diagnostic range is dwarfed by the spontaneous occurrence of cancer."[17] Given ~6 million pregnancies per year, resulting in about 4 million live births (American Pregnancy Association), there may be ~2,700 cases of pregnancy-associated PE annually, with a related mortality differential of 323 cases per year. The predicted number of cancer deaths in the maternal population based in the LNT model with these pharmaceuticals is 0.24, or less than 1, resulting in ~320 lives saved assuming curative treatment of diagnosed PEs. Matthews,[19] among others, considered the use of CT pulmonary angiography in the diagnosis of PE, which perhaps may deliver a threefold lower dose to the fetus and an ED to the mother of perhaps 20–50 mSv. Assuming the upper-limit estimate of 50 mSv, the LNT model yields 7 maternal cancer deaths, with ~316 lives saved.

Head Computed Tomography in Diagnosis of Clinically Important Traumatic Brain Injury after Low-Risk Trauma in Children. In the U.S., there are 600,000 emergency-room visits, 60,000 hospital admissions, and 7,400 deaths annually associated with pediatric head trauma.[24,25] CT is the reference standard for emergently diagnosing traumatic brain injuries, and ~50% of children assessed in North American emergency rooms for head trauma undergo CT.[26] Not surprisingly, such a high frequency of CT scanning in a pediatric population has raised concerns regarding long-term effects, namely, an increased risk of cancer. In a study of 42,412 children (including 19,718 younger than two years and 31,694 two years and older) presenting to 25 North American emergency rooms within 24 hours of head trauma, Kuppermann et al.[27] found that the clinical criteria for excluding clinically important traumatic brain injury (ciTBI) (normal mental status, no scalp hematoma except frontal, no loss of consciousness or loss of consciousness for less than five seconds, non-severe injury mechanism, no palpable skull fracture, and acting normally according to the parents) in children younger than two years had both a negative predictive value (NPV) and sensitivity for ciTBI of 100%. *Any* non-zero increase in risk of CT-induced cancer, even if theoretical, would not be justified, and therefore CT scanning for diagnosis of ciTBI is contraindicated on the basis of radiogenic cancer risk in children younger than two years. In children two years and older being evaluated for ciTBI, the forgoing clinical criteria have an NPV of 99.95% and a sensitivity of 98.80%. This NPV means that 0.5 out of 10,000 cases of potentially fatal ciTBI would be missed on the basis of the clinical criteria. Brenner et al.[28] estimated the excess lifetime risk of brain cancer mortality for head CT among children two years and older as 2.5 cases per 10,000 CT scans. Thus, at least theoretically, there is *no* net benefit—and actually a net detriment—of CT scanning for ciTBI in children two years and older, since there would be 2.5 − 0.5 = 2 more lives *lost* per 10,000 children with than without CT. This, of course, is a very small, theoretical net detriment, and one that is very likely statistically insignificant. It does, however, illustrate the potential utility of quantitative benefit-risk analyses in clinical decision-making.

15.5 Discussion and Conclusions

The analyses presented are representative "order-of-magnitude" examples from the literature of the benefits and radiogenic risks of medical imaging procedures. Although it is difficult to identify clinical studies with the necessary outcome data (i.e., outcomes with and without the relevant the diagnostic study being performed), the approach presented could be extended to other studies. The current analyses suffer from several limitations. For example, as described, technical aspects of the procedures considered (e.g., the radiopharmaceuticals used, the administered activities, and the CT doses) may be somewhat dated. Further, clinical outcomes and the non-radiogenic mortalities and morbidities in the clinical scenarios considered are specific to the respective studies and are debatable. Despite these, and perhaps other, limitations, the conclusion is clear: *overall, the use of radiation in medicine saves hundreds to thousands of lives every year, while the theoretical risks predicted by the LNT model are typically orders of magnitude smaller.*

Importantly, however, there may be clinical scenarios (as illustrated by the analysis of head CT in ciTBI in children) in which the quantitative analysis of benefit versus radiogenic risk may indicate that a diagnostic imaging procedure is not justified. Thus, our focus on the use of radiation in medicine should not be on radiogenic risk exclusively (as is now typically the case) but on reasonable evaluation of benefit as well as risk. Risk-focused analyses (to the exclusion of comparable consideration of benefit) have led to unwarranted concern in many instances about undergoing needed medical examinations that involve ionizing radiation (http://hps.org/publicinformation/ate/cat5.html). Both the American Association of Physicists in Medicine and the Society of Nuclear Medicine and Molecular Imaging have stated that if a patient foregoes an appropriate medical imaging procedure due to concerns of the radiation, it could be detrimental to the patient's health. Patients should be advised of all risks and benefits of any medical procedure, to be sure. Physicians and physicists can speak in general about risks being "low" or "negligible" and are often understandably reticent to provide numerical estimates of risk. Even if requested, it is inadvisable to provide such numerical estimates for an individual patient from an estimated dose of ionizing radiation, although some practitioners may choose to do so. As shown here, however, any numerical estimate of risk from a medical procedure using ionizing radiation, if accompanied by a quantitative estimate of the benefits of these examinations to the medical population, puts the issue in perspective for the patient. We have outlined numerical values for six kinds of medical studies and encourage extension of this analysis to additional examinations and dissemination of the results of such analyses to the greater medical and radiation protection communities to improve communication of the risks and benefits of medical imaging to patients, their families, and society as a whole.

References

1. Redberg RF, Smith-Bindman R. We are giving ourselves cancer. *The New York Times* (Op-Ed), January 20, 2014.
2. NAS/NRC (National Academy of Sciences/National Research Council). *Health Risks from Exposure to Low Levels of Ionizing Radiation, BEIR VII, Phase 2.* National Academy Press, Washington DC, 2006.
3. Stabin MG. Radiopharmaceuticals for nuclear cardiology: Radiation dosimetry, uncertainties, and risk. *J Nucl Med* 49:1555–1563; 2008.
4. Brenner DJ, Hall EJ. Computed tomography–An increasing source of radiation exposure. *N Engl J Med* 357:2277–2284; 2007.

5. Aurengo A, Averbeck D, Bonnin A et al. Dose-effect relationship and estimation of the carci-nogenic effects of low dose ionizing radiation. Academie des Sciences—Academie national de Medicine; March 30, 2005. http://www.radscihealth.org/rsh/Papers/French AcadsFinal07_04_05.pdf. Accessed June 23, 2008.

6. Raabe OG. Toward improved ionizing radiation safety standards. *Health Phys* 101(1):84–93; 2011.

7. Health Physics Society. *Radiation Risk in Perspective. Position Statement PS010-2 of the Health Physics Society.* Health Physics Society, McLean, VA, Adopted January 1996, revised July 2010.

8. Balter S, Zanzonico P, Reiss GR, Moses JW. Radiation is not the only risk. *Am J Roentgenol* 196(4):762–767; 2011.

9. National Council on Radiation Protection and Measurements. NCRP Report No. 115, Risk esti-mates for radiation protection. NCRP, Bethesda, MD, 1993.

10. van Tinteren H, Hoekstra O, Smit EF, van den Bergh J, Schreurs A, Stallaert R, van Velthoven P, Comans E, Diepenhorst F, Verboom P. Effectiveness of positron emission tomography in the pre-operative assessment of patients with suspected non-small-cell lung cancer: The PLUS multicentre randomised trial. *Lancet* 359(9315):1388–1392; 2002.

11. Coles DR, Smail MA, Negus IS, Wilde P, Oberhoff M, Karsch KR, Baumbach A. Comparison of radiation doses from multislice computed tomography coronary angiography and conventional diagnostic angiography. *J Am Coll Cardiol* 47:1840–1845; 2006.

12. Noto TJ Jr, Johnson LW, Krone R, Weaver WF, Clark DA, Kramer JR, Vetrovec GW. Cardiac cath-eterization 1990: A report of the registry of the Society for Cardiac Angiography and Interventions (SCA&I). *Cathet Cardiovasc Diag* 24(2):75–83; 1991.

13. American Heart Association. *Heart Disease and Stroke Statistics-2005 Update.* American Heart Association, Dallas, TX, 2005.

14. Liberthson RR, Nagel EL, Hirschman JC, Nussenfeld SR, Blackbourne BD, Davis JH. Pathophysiologic observations in prehospital ventricular fibrillation and sudden cardiac death. *Circulation* 49:790; 1974.

15. Zanzonico P, Rothenberg HW, Strauss HW. Radiation exposure of computed tomography and direct intracoronary angiography: Risk has its reward. *J Am Coll Cardiol* 47:1846–1849; 2006.

16. Mansel RE, Fallowfield L, Kissin M et al. Randomized multicenter trial of sentinel node biopsy versus standard axillary treatment in operable breast cancer: The ALMANAC trial. *JNCI* 98(9):599–609; 2006.

17. Gojer B, Williams KA. The role of scintigraphic perfusion imaging for predicting ischemic cardiac events in noncardiac surgery. *Am J Card Imaging* 9(3):213–225; 1995.

18. Stabin MG, Blackwell R, Brent RL, Donnelly E, King VA, Lovins K, Stovall M. Fetal radiation dose calculations. ANSI N13.54-2008. American National Standards Institute, Washington, DC, 2008.

19. Matthews S. Imaging pulmonary embolism in pregnancy: What is the most appropriate imaging protocol? *Brit J Radiol* 79:441–444; 2006.

20. Marcus CS, Mason GR, Kuperus JH, Mena I. Pulmonary imaging in pregnancy: Maternal risk and fetal dosimetry. *Clin Nuc Med* 10:1–4; 1985.

21. Russell JR, Stabin MG, Sparks RB, Watson EE. Radiation absorbed dose to the embryo/fetus from radiopharmaceuticals. *Health Phys* 73(5):756–769; 1997.

22. Stabin M, Watson E, Cristy M, Ryman J, Eckerman K, Davis J, Marshall D, Gehlen K. Mathematical models and specific absorbed fractions of photon energy in the nonpregnant adult female and at the end of each trimester of pregnancy. ORNL Report ORNL/TM 12907, 1995.

23. International Commission on Radiological Protection. Radiation dose to patients from radio-pharmaceuticals. ICRP Publication 80. Elsevier Science, Tarrytown, NY, 1998.

24. Langlois JA, Rutland-Brown W, Thomas KE. Traumatic brain injury in the United States. In: CDC, ed. Atlanta, GA. National Center for Injury Prevention and Control, 2006.

25. National Center for Injury Prevention and Control. Traumatic brain injury in the United States: Assessing outcomes in children. CDC, 2006. http://www.cdc.gov/ncipc/tbi/tbi_report/index.htm.

26. Marr A, Coronado V. *Central Nervous System Injury Surveillance Data Submission Standards–2002*. CDC, Atlanta, GA, 2004.

27. Kuppermann N, Holmes JF, Dayan PS et al. Identification of children at very low risk of clinically-important brain injuries after National Center for Injury Prevention and head trauma: A prospective cohort study. *Lancet* 374(9696):1160–1170; 2009.

28. Brenner D, Elliston C, Hall E, Berdon W. Estimated risks of radiation-induced fatal cancer from pediatric CT. *Am J Roentgenol* 176(2):289–296; 2001.

IV

Medical Imaging: Safety Approach

16

Medical Imaging: Approach of the United Nations Scientific Committee on the Effects of Atomic Radiation

Wolfgang Weiss and
Ferid Shannoun

16.1 Introduction

The United Nations Scientific Committee on the Effects of Atomic Radiation (UNSCEAR) was established by the United Nations (UN) General Assembly in 1955 to collect and evaluate information on levels and effects of ionizing radiation from natural and artificial sources. UNSCEAR has systematically reviewed and evaluated the global and regional levels of public and occupational radiological exposure and also the exposure of patients, known as medical exposure. It has also evaluated the evidence for radiation-induced health effects from studies of Japanese atomic bombing survivors and other exposed groups and has reviewed advances in the mechanisms of radiation-induced health effects.

Information on the use of radiation in medicine and the associated exposure is obtained by reviewing the scientific literature and through population-based surveys. These surveys are used to identify trends in radiological exposure and thus serve as an early indicator of potential safety issues that might require attention. They can also be used to identify gaps in treatment capabilities and possible dose variations for the same procedure. The surveys provide the background information to underpin advocacy initiatives and to develop and implement evidence-based radiation protection policies. This chapter describes UNSCEAR's surveys, including the main findings and the challenges when conducting population-based surveys and future perspective.

16.1.1 Historical Review

As early as its initial report in 1958, UNSCEAR recognized that medical exposure was a major component of the total exposure due to artificial sources of radiation worldwide, a fact that remains true today (UNSCEAR 1958). Since 1962, UNSCEAR has been providing information on medical exposure and presents regular, comprehensive evaluations of the levels of ionizing radiation exposure of the population. The UNSCEAR 1962 Report listed annual frequency data in form of number of procedures per 1,000 individuals from 20 countries (UNSCEAR 1962). In 1982, UNSCEAR published data on annual frequency of specific procedures from 16 countries (UNSCEAR 1982). This report was the first to use a survey, developed in cooperation with the World Health Organization (WHO), to obtain additional information on diagnostic radiology equipment. This was also the first UNSCEAR survey to include information on computed tomography (CT). The UNSCEAR (1988) Report greatly expanded the material on medical exposure and an attempt was made to estimate the global exposure from medical diagnostic procedures in addition to country-specific data. Since the UNSCEAR (2000) Report, the collective effective dose and the per capita effective dose have been used to express population dose estimates. The latest evaluation covering global and regional levels and trends in medical exposure was completed in 2008 and some of its findings are presented below.

16.2 The United Nations Scientific Committee on the Effects of Atomic Radiation Medical Exposure Assessment

Medical exposure is defined by the International Commission on Radiological Protection (ICRP 1991) as exposures of

- Patients as part of their medical diagnosis or treatment
- Individuals as part of health screening programs
- Healthy individuals or patients voluntarily participating in medical, biomedical, diagnostic, or therapeutic research programs

The scope of UNSCEAR's evaluations is to estimate the global frequencies of the use of radiological procedures and the levels of radiation exposure, with a breakdown by radiological procedure or examination, age and sex of patients, countries, and their corresponding health-care level (HCL). The evaluations are based on analyses of the responses to its Global Survey of Medical Radiation Usage and Exposure provided by UN Member States and a periodic review of the scientific literature of the three categories of medical practices using ionizing radiation: diagnostic radiology (including imaging-guided interventional procedures), nuclear medicine, and radiation therapy.

The concepts of the effective dose and the collective effective dose are used for the evaluations of medical exposure with exception for radiation therapy treatments, as the dose levels are too high to be expressed in terms of effective dose (UNSCEAR 2000). Further, the collective effective dose estimation from medical radiological exposure is used only for comparative purposes of similar populations (i.e., patients).

The collective effective dose, expressed in man Sieverts (man Sv), is obtained by multiplying the mean effective dose E_e (Sv) for a radiological procedure by the number of procedures N_e. The numerical value of N_e may be deduced from the annual frequency (number of procedures per 1,000 members of the population) and the estimated population size. The collective effective dose from all procedures S for the entire population is a summation of the effective dose from all procedures:

$$S = \sum E_e N_e$$

16.2.1 Health-Care Level Model

UNSCEAR's evaluations are based mainly on information provided by industrialized countries. The surveys reveal a lack of data on the frequency and type of radiological procedures in more than half of the world's countries and only fragmentary data from another quarter. Therefore, a method was used to extrapolate the existing data and to estimate the availability and frequency of medical use of radiation in countries where relevant data were unavailable. Based on a good correlation between the physician-to-population ratio and the annual frequency of diagnostic radiological procedures, an analytical model was developed to enable the estimation of medical radiation exposure on a worldwide basis by grouping countries with similar resources under a particular HCL (see Table 16.1) (Mettler et al. 1987; UNSCEAR 1988).

UNSCEAR's HCL model enables an estimation of the number and type of procedures for a given country where specific data are unavailable by applying the average annual frequencies of radiological procedures from countries of the same HCL.

16.2.2 Survey Findings

The most recent UNSCEAR (2008) Report evaluated the global use of medical exposure from 1997 to 2007, determined the exposure from various modalities and procedures, and assessed the emerging trends. According to this, approximately 3.6 billion diagnostic X-ray procedures (including approximately 0.5 billion dental procedures) were performed annually worldwide (UNSCEAR 2010). Table 16.2 summarizes the estimated annual frequency of diagnostic X-ray procedures and the corresponding annual collective effective dose and per capita effective dose for this period globally and for each HCL.

TABLE 16.1   Classification of Health-Care Level by the Number of Physicians per Population and the Corresponding Population Sizes

HCL	Physicians per Million Population	Population in Million	Percentage of World Population
I	>1,000	1,540	24
II	333–1,000	3,153	49
III	100–332	1,009	16
IV	<100	744	11

Source: United Nations Scientific Committee on the Effects of Atomic Radiation, Sources and effects of ionizing radiation, Volume I: Sources, UNSCEAR 2008 Report to the General Assembly with Scientific Annexes A and B, United Nations, New York, 2010.

TABLE 16.2   Estimations of Annual Frequencies of Diagnostic X-ray Procedures and Collective Effective Doses Globally and for Each HCL

HCL	Annual Frequency per 1,000 People	Annual Collective Effective Dose (man Sv)	Annual per Capita Effective Dose (mSv)
I	1,332	2,900,000	1.91
II	332	1,000,000	0.32
III	20	33,000	0.03
IV	20	24,000	0.03
Global	488	4,000,000	0.62

Source: United Nations Scientific Committee on the Effects of Atomic Radiation, Sources and effects of ionizing radiation, Volume I: Sources, UNSCEAR 2008 Report to the General Assembly with Scientific Annexes A and B, United Nations, New York, 2010.

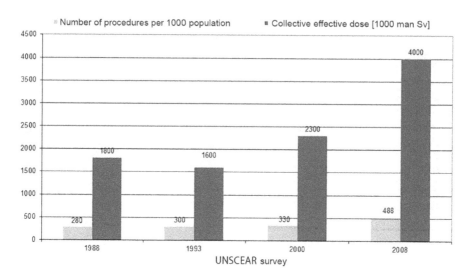

FIGURE 16.1 Trends in diagnostic X-rays over 20 years.

Figure 16.1 summarizes the global evolution of medical radiological use over two decades for the last four UNSCEAR surveys. The total number of diagnostic X-ray procedures has increased continually over 20 years and the total collective effective dose from medical diagnostic procedures has nearly doubled, from 2.3 million man Sv to 4 million man Sv between the periods 1991–1996 and 1997–2007 (UNSCEAR 2000, 2010). The global average annual per capita effective dose has also increased from 0.35 mSv in 1988 to 0.62 mSv in 2008 (UNSCEAR 1988, 2010).

The estimated annual collective effective dose from diagnostic nuclear medicine procedures has increased from 150,000 to 202,000 man Sv between the periods 1991–1996 and 1997–2007 and by a factor 2.7 compared to 74,000 man Sv in the evaluation conducted for the period 1980–1984 as illustrated in Figure 16.2 (UNSCEAR 1988, 2000, 2010). The use of nuclear medicine procedures around the world is quite variable, with 90% of procedures performed in HCL I countries.

Figure 16.3 provides estimates of the annual collective dose for medical exposures by HCL for the evaluation period 1997–2007. Only 1.5% of all worldwide diagnostic procedures were estimated to be performed in HCL III and IV countries, which together account for 27% of the global population. This imbalance

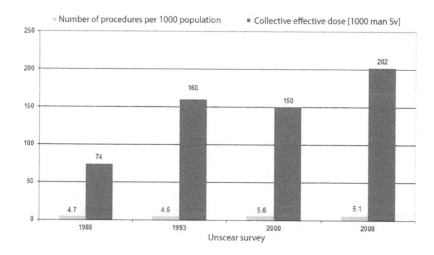

FIGURE 16.2 Trends in nuclear medicine over 20 years.

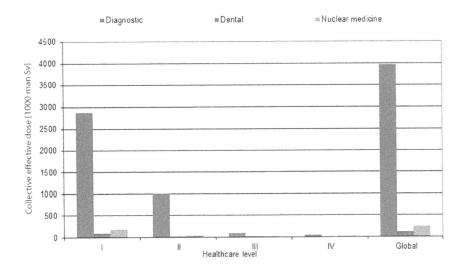

FIGURE 16.3 Global estimate of medical exposure by HCL (1997–2007).

in healthcare practice is also reflected in the availability of radiological equipment and practitioners. The extremely rapid growth in the annual frequency of CT procedures in HCL I countries has also resulted in a marked increase in the collective effective dose from diagnostic medical exposure (UNSCEAR 2010).

Over five million radiation therapy treatments were conducted annually between 1997 and 2007, up from an estimated four million in 1988 (UNSCEAR 1988, 2010). About 4.7 million treatments involved teletherapy and only 0.4 million brachytherapy. The annual frequencies of radiation therapy treatments during the period 1997–2007 were dominated by procedures performed in HCL I countries, accounting for about 70% and 40% of global teletherapy and brachytherapy, respectively (UNSCEAR 2010).

Medical exposure remains by far the largest artificial source of exposure, and it continues to grow considerably. The distribution of medical exposures is uneven among countries and regions. Sixty-six percent of diagnostic radiology procedures, 90% of nuclear medicine procedures, and 70% of radiation therapy treatments are performed in HCL I countries, where only a quarter of the world's population live. In HCL I and II countries, medical uses of radiation have increased from year to year as radiological technology have become more widely available.

16.2.3 Limitations of the United Nations Scientific Committee on the Effects of Atomic Radiation Surveys

While data are readily available from HCL I and II countries, data from HCL III to IV countries are very scarce.

UNSCEAR surveys revealed a range of issues related to participation, survey process, data quality, and analysis. In the past, factors such as the noteworthy differences in equipment, health-care workforce, and systems have resulted in difficulties in obtaining data from many HCL III and IV countries, exacerbated by limited local resources and infrastructure to conduct national surveys. The structure, consistency, complexity, and, to a less extent, the language used in UNSCEAR's surveys were also barriers to data collection.

16.2.4 International Surveys and Activities

Relatively few countries around the world conduct national surveys of medical radiation use and exposure on a regular basis. Examples are Germany, Switzerland, and the United Kingdom, which have institutionalized arrangements to periodically conduct such surveys (Aroua et al. 2002; Brix et al. 2005; Hart and Wall 2004; Nekolla et al. 2017). In addition, focused studies are available in many countries, which sample from the overall population or study certain sets of patients and their exposure in some detail.

Further, member countries of the European Union are required by the European Council Directive 97/43/EURATOM[1] to ensure that doses from medical procedures are determined for the population (Council of the European Union 1997). Thus, in 2005, the European Commission (EC) launched a project to review past surveys of population exposures for medical radiological procedures in Europe in order to clarify the different methodologies used, identify the sources of uncertainty, and develop a harmonized and simplified process for future surveys (EC 2008). This led to the publication of a document entitled, "European Guidance on Estimating Population Doses from Medical X-ray Procedures," which provides guidance on the classification of X-ray procedures, the estimation of the frequency of procedures, and the use of the concept of effective dose. In 2011, the EC launched a follow-up project to study the doses to the European population from medical exposures, which resulted in an average estimate of 1 mSv per capita for the European population (EC 2015). A study conducted in the United States in 2006 estimated an annual per capita effective dose from medical exposure of 3 mSv, which is as high as the annual exposure from background radiation (NCRP 2009).

Other relevant international initiatives include the development of the Nuclear Medicine Database and the Directory of Radiotherapy Centres database by the International Atomic Energy Agency (IAEA) and the collection and publication of data on medical devices by the Nuclear Energy Agency within the Organisation for Economic Co-operation and Development countries.

16.2.4.1 The United Nations Scientific Committee on the Effects of Atomic Radiation Strategy to Improve Data Collection

UNSCEAR developed an improvement strategy to address the existing deficiencies in data quality and collection and to increase participation in future surveys. The major elements of this strategy are an improvement in the questionnaires' structure, consistency, and clarity by applying standardized taxonomy and terminology, a simplified questionnaire on frequency of procedures, and relevant dosimetric information. Further, the introduction of an electronic data collection platform is the main element for improvement as it facilitates the collection, evaluation, and archiving of the submitted data (Figure 16.4). It allows the upload of the collected data in a secure and simple manner. Another element for improvement is the review of the current HCL model, based on the physician/population ratio, with

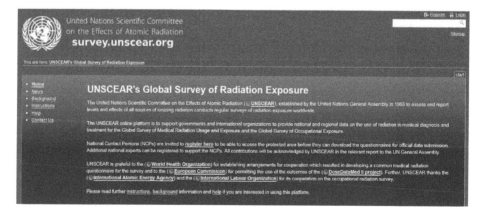

FIGURE 16.4 UNSCEAR's online data collection platform.

---

[1] The European Basic Safety Standards Directive (2013/59/EURATOM)—to be implemented by EC Member States by February 2018—kept this requirement in Article 64 "Estimates of population doses" requesting Member States to ensure that the distribution of individual dose estimates from medical exposure is determined.

a view to adopting the World Bank classification, used also by WHO, which better reflects the impact of a country's economic situation on health-care expenditure. This approach would enable correlation and analysis of medical exposure data and other relevant public health data (Shannoun 2012).

Finally, UNSCEAR will focus on those procedures with substantial contribution to the collective effective dose similar to the methodology used by the European DOSE DATAMED project (EC 2008) and will further obtain relevant data, in a cooperative and sustainable manner, from other reliable sources, such as international and regional organizations and professional associations.

16.3 Health Effects of Medical Radiation Exposure

One of the challenges relating to the interpretation, analysis, and use of radiation exposure data is the uncertainty when attributing cancer risk to ionizing radiation exposure. In epidemiological studies of populations exposed to radiation, there are statistical fluctuations and uncertainties due to selection and information bias; exposure and dose evaluation; and model assumptions used when evaluating data. In addition, transferring the risk estimate based on data from an epidemiological study to a population of interest may not be entirely valid because of differences in location, setting, data collection period, age and gender profile, dose or dose rate, type of radiation, and acute versus protracted exposures. This is especially true for patients, as their age distribution may differ from that of the general population. The uncertainty when estimating cancer risk from exposure to ionizing radiation is often not taken into account. For solid cancers after an exposure of 100 mSv, the uncertainty in risk estimates could differ by a factor of two–three, and the uncertainty of excess risk for a specific cancer type is considerably higher than for all solid cancers. Thus, uncertainties in estimating cancer risk from ionizing radiation exposure need to be considered when attributing health effects to radiation exposure (UNSCEAR 2015).

It is important to distinguish between a manifest "health effect" and "health risk" (likelihood for a future health effect to occur), when describing the health impacts of radiological exposure for an individual or population. When evaluating the health effects and risks from medical radiation exposure for a given population, one of the challenges is statistical fluctuation. For low-linear energy transfer (LET) radiation such as X-rays and gamma rays, the lower the dose is, the higher the uncertainty. The uncertainty increases when extrapolating the probability of effects at moderate doses to that at low and very low doses. Therefore, it is not surprising that a statistically significant increase in radiation-induced cancer can only be observed at exposures of 100 mSv or above (UNSCEAR 2015).

A manifest health effect in an individual can be unequivocally attributed to radiation exposure only if other possible causes for an observable tissue reaction such as skin burns (deterministic effect) are excluded. Malignancies (stochastic effects) cannot be unequivocally attributed to radiation exposure because radiation is not the only possible cause, and there are at present no known biomarkers that are specific to radiation exposure. A manifest increased incidence of stochastic effects in a population can only be attributed to radiation exposure through epidemiological analysis, provided the increased incidence is sufficient to overcome the inherent statistical uncertainties (UNSCEAR 2015).

In general, a manifest increased incidence of health effects in a population cannot reliably be attributed to radiation exposures at levels that are typical of the global average background levels of radiation. The reasons include (1) the uncertainties associated with risk evaluation at low doses, (2) the absence of radiation-specific biomarkers for health effects, and (3) the insufficient statistical power of epidemiological studies (UNSCEAR 2015).

When estimating radiation-induced health effects in a population exposed to incremental doses at levels equivalent to or below natural background, it is not generally recommended to do this simply by extrapolation effects at very low doses to a large number of individuals. However, UNSCEAR recognizes there is a need for estimations to assist health authorities in allocating resources appropriately and in comparing the incidence of health effect from different hazardous sources. This is only valid provided when applied consistently, the uncertainties in the estimations are fully taken into account, and the health effects are hypothetical (UNSCEAR 2015).

16.4 Conclusion

If used appropriately radiation-based medical imaging can be beneficial to patients by improving the detection, diagnosis, and treatment of disease. If applied in an inappropriate way, it can be ineffective, increase the risks of radiation-induced disease, adversely affect health, or even threaten life. UNSCEAR has no direct mandate in radiation protection; its work is strictly focused on the collection, collation, analysis, and publication of medical radiation exposure data to deduce levels, impact, and trends. However, UNSCEAR's surveys and evaluations do provide a scientific foundation to improve the basic understanding of the levels of radiation to which individuals are exposed and of radiation-induced health effects. The scientific findings and data provided by UNSCEAR have been used by other UN agencies to formulate international radiation protection frameworks, such as IAEA's International Action Plan for the Radiological Protection of Patients (IAEA 2002), and WHO's Global Initiative on Radiation Safety in Health Care Settings (WHO 2008) and, more recently, the Bonn Call for Action, which emphasizes the need for increased availability of improved global information on medical exposure and occupational exposure in medicine (IAEA and WHO 2012).

UNSCEAR's evaluations of global medical exposure are based mainly on responses from industrialized countries. From the lessons learned from past surveys, UNSCEAR has developed innovative solutions to improve future collection and analysis of information on diagnostic radiology, nuclear medicine, and radiation therapy. For example, cooperative arrangements with WHO and the IAEA have been established to improve the collection of data on the frequency of medical radiological procedures, particularly in developing countries where this information is scarce.

Increasing medical exposure worldwide is likely associated with increased health benefits to the population as patients receive a direct benefit from their exposures. Explicit comparison of doses resulting from medical exposures with other sources is therefore scientifically incorrect. The age distribution of patients receiving medical radiological exposure is normally older than that of the general population. In contrast, the introduction of new technologies has, in some instances, resulted in an increased use of medical radiation. While the magnitude of medical exposures can be assessed, it is very difficult to estimate the associated health risks. Despite the global increase in demand for diagnostic radiology procedures, their use should be underpinned by evidence-based referral guidelines and optimization of radiation protection (IAEA 2014; ICRP 2007).

UNSCEAR data are widely used to develop and update evidence-based radiation protection and radiation safety recommendations and guidance. To ensure data quality and relevance, the support of national authorities and their participation in surveys are essential. The advocacy by and contribution from other stakeholders, including national and international professional organizations and radiological societies in data collection, is highly encouraged.

UNSCEAR's mandate has not changed over the past decades but its role as the principal focal point for international information exchange on ionizing radiation and its levels and effects have evolved as a result of the revolution in information technology. While communication advances have facilitated data dissemination and information sharing, UNSCEAR's coordinating role and experience in global radiation exposure evaluation have remained invaluable. There is a need to continue to review, digest, and integrate the plethora of relevant data and to establish scientific consensus so that policy-makers and other stakeholders can act accordingly. Despite the current level of knowledge of medical radiation exposure and radiation-induced health effects, more targeted research is needed. There are uncertainties in estimating the cancer risk due to ionizing radiation and in attributing health effects to and inferring risk from medical radiation exposure. An improved understanding in these areas, based on research, evidence, and robust scientific debate, will lead to better application of survey data, more informed decision-making, and targeted action on radiation protection and safety.

References

Aroua A, Burnand B, Decka I et al. (2002) Nation-wide survey on radiation doses in diagnostic and interventional radiology in Switzerland in 1998. *Health Phys* 83(1): 46–55.

Brix G, Nekolla E, Griebel J (2005) Radiation exposure of patients from diagnostic and interventional X-ray procedures. Facts, assessment and trends. *Radiologe* 45(4): 340–349.

Council of the European Union (1997) Council Directive 97/43/Euratom of 30 June 1997 on health protection of individuals against the dangers of ionizing radiation in relation to medical exposure, and repealing Directive 84/466/EURATOM. OJEC 1997/L 180/22.

European Commission (2008) Radiation Protection 154. *European Guidance on Estimating Population Doses from Medical X-ray Procedures*. EU Publications Office, Luxembourg.

European Commission (2015) Radiation Protection 180. *Medical Radiation Exposure of the European Population*. EU Publications Office, Luxembourg.

Hart D, Wall BF (2004) UK population dose from medical X-ray examinations. *Eur J Radiol* 50(3): 285–291.

International Atomic Energy Agency (2002) Measures to strengthen international cooperation in nuclear, radiation, transport and waste safety: International Action Plan for the Radiological Protection of Patients. GOV/2002/36-GC(46)/12. International Atomic Energy Agency, Vienna, Austria.

International Atomic Energy Agency and World Health Organization (2012) *Bonn Call for Action, A Joint Position Statement from the International Conference on Radiation Protection in Medicine—Setting the Scene for the Next Decade*. WHO, Geneva, Switzerland. Accessed from http://www.who.int/ionizing_radiation/medical_exposure/Bonn_call_action.pdf.

International Atomic Energy Agency (2014) Radiation Protection and Safety of Radiation Sources: International Basic Safety Standards. IAEA Safety Standards Series GSR Part 3, IAEA, Vienna, Austria.

International Commission on Radiological Protection (1991) Recommendations of the International Commission on Radiological Protection. ICRP Publication 60. *Annals of the ICRP* 21 (1–3). Pergamon Press, Oxford, UK.

International Commission on Radiological Protection (2007) ICRP Publication 103. The 2007 Recommendations of the International Commission on Radiological Protection. *Annals of the ICRP* 37 (2–4). Elsevier, Oxford, UK.

Mettler FA Jr, Davis M, Kelsey CA et al. (1987) Analytical modeling of worldwide medical radiation use. *Health Phys* 52(2): 133–141.

National Council on Radiation Protection and Measurements (2009) Report No.160 Ionizing radiation exposure of the population of the United States. NCRP, Bethesda, MD.

Nekolla EA, Schegerer AA, Griebel J, Brix G (2017) Häufigkeit und Dosis diagnostischer und interventioneller Röntgenanwendungen. *Radiologe* 57: 555–562. doi:10.1007/s00117-017-0242-y.

Shannoun F (2012) A comparison of classification methodologies of health-care levels used by UNSCEAR and WHO. In: *Dose Datamed II Project Workshop on European Population Doses from Medical Exposure*. Greek Atomic Energy Commission, European Commission, Athens, Greece.

United Nations Scientific Committee on the Effects of Atomic Radiation (1958) Official Records of the General Assembly, Thirteenth Session, Supplement No. 17 (A/3838). UNSCEAR 1958 Report. United Nations, New York.

United Nations Scientific Committee on the Effects of Atomic Radiation (1962) Official Records of the General Assembly, Seventeenth Session, Supplement No. 16 (A/5216). UNSCEAR 1962 Report. United Nations, New York.

United Nations Scientific Committee on the Effects of Atomic Radiation (1982) Ionizing radiation: Sources and biological effects. UNSCEAR 1982 Report to the General Assembly, with annexes. United Nations, New York.

United Nations Scientific Committee on the Effects of Atomic Radiation (1988) Sources, effects and risks of ionizing radiation. UNSCEAR 1988 Report to the General Assembly, with annexes. United Nations, New York.

United Nations Scientific Committee on the Effects of Atomic Radiation (2000) Sources and effects of ionizing radiation. Volume I: Sources. UNSCEAR 2000 Report to the General Assembly, with scientific annexes. United Nations, New York.

United Nations Scientific Committee on the Effects of Atomic Radiation (2010) Sources and effects of ionizing radiation. Volume I: Sources. UNSCEAR 2008 Report to the General Assembly with Scientific Annexes A and B. United Nations, New York.

United Nations Scientific Committee on the Effects of Atomic Radiation (2015) UNSCEAR 2012 Report to the General Assembly with Scientific Annexes A and B. United Nations, New York.

World Health Organization (2008) WHO global initiative on radiation safety in health-care settings technical meeting report. Geneva, Switzerland. Accessed from http://www.who.int/ionizing_radiation/about/med_exposure/en/index1.html.

V

Patient Communications and Shared Decision-Making

17

Informed Decision-Making
in Medical Imaging

17.1 Introduction

It is generally accepted that medical imaging with ionizing radiation has the potential to cause cancer, though this remains uncertain and controversial (ICRP 2007). However, the actual risk to the patient, if any, remains unknown and is heavily dependent on patient characteristics, such as age and gender. As the risks associated with medical imaging radiation exposure have percolated through the popular press, physicians at times are now discussing imaging radiation exposure directly with patients. To document this discussion for medical–legal reasons, an informed consent process has been proposed. However, a more patient-centered approach, known as informed decision-making, may be more appropriate in this setting. This chapter will discuss informed consent and informed decision-making as it applies to medical imaging radiation exposure.

17.1.1 What Is Informed Consent?

The relationship between a patient and physician is sacred. Today, when patients present to their physician, there is a reasonable expectation that the physician will be forthcoming with all clinical information to allow their patients to actively participate in the treatment decision-making process. However, this was not always the case. Hippocrates, the father of ancient medicine, advised a paternalistic approach to treating patients, and physicians made the treatment decisions for patients (Miles 2009). Discussing risks with patients was atypical, because physicians were considered untrustworthy by patients if they disclosed possible complications of a treatment.

One of the earliest examples of informed consent was in the fourteenth century, when patients were asked to sign a "hold harmless document" to protect the physician from legal exposure related to complications from therapies (Leclercq et al. 2010). In 1767, one of the first medical malpractice cases involved a patient who claimed he was not informed about a novel medical treatment for a fractured leg (Annas 2012). Similar case law evolved over the next two centuries until Justice Benjamin Cardozo famously declared in 1914s *Schoendorff vs. Society of New York Hospital*, "Every human being of adult years and sound mind has a right to determine what shall be done with his own body, and a surgeon who performs an operation without the patient's consent commits an assault for which he is liable in damages" (Leclercq et al. 2010). It was not until the 1957 case *Salgo vs. Leland Stanford, Jr. University Board of Trustees* when the term "informed consent" was coined. In that case, a patient became paralyzed after aortography and claimed he was not informed of the procedure risk (Leclercq et al. 2010). Since then, the concept of informed consent has continued to evolve because numerous legal cases have been encountered.

The American Medical Association stipulates the key components of the informed consent process that physicians should discuss with their patients: (1) the diagnosis, if known; (2) the nature and purpose of the proposed treatment or procedure; (3) the risks and benefits of a proposed treatment or procedure; (4) alternatives (regardless of their cost or the extent to which the treatment options are covered by health insurance); (5) the risks and benefits of the alternative treatment or procedure; and (6) the risks and benefits of not receiving or undergoing a treatment or procedure (Brink et al. 2012).

The radiology community has further defined informed consent for image-guided procedures through the American College of Radiology - Society of Interventional Radiology (ACR-SIR) (2011) guidelines, which are composed of the following elements (ACR 2014):

1. The purpose and nature of the procedure or treatment
2. The method by which the procedure or treatment will be performed
3. The risks, complications, and expected benefits or effects of such procedure or treatment
4. The risk of not accepting the procedure or treatment
5. Any reasonable alternatives to the procedure or treatment and their most likely risks and benefits
6. The right to refuse the procedure or treatment

17.1.2 Why Is Informed Consent Flawed with Respect to Radiation Dose?

The crux of the argument against using informed consent with respect to medical imaging radiation dose is that the actual cancer risk of radiation exposure is unknown and therefore cannot exactly be communicated to patients. In addition, risks of diagnostic radiology are very low and likely far less than that of alternative, more invasive procedures, such as an abdominal computed tomography (CT) vs. diagnostic laparoscopy. Therefore, it is not possible to address one of the main requirements of informed consent as established by the American Medical Association (AMA), the American College of Radiology (ACR), and SIR.

The empirical risks of low levels of radiation exposure were highlighted in the late 1990s by the Radiation Effects Research Foundation, which performed a longitudinal study of 35,000 survivors of the atomic bombs of Hiroshima and Nagasaki who were exposed to radiation doses of less than 150 mSv (Pierce and Preston 2000). Exposures of 5–150 mSv were found to be associated with a statistically significant risk of cancer, which prompted the U.S. Food and Drug Administration (2014) to suggest that a 10 mSv exposure from a CT scan could increase the risk of a fatal cancer by 1 in 2000.

The estimates of radiation-induced cancer risks among atomic bomb survivors are based on extrapolations using the linear no-threshold (LNT) hypothesis, a concept well covered in this textbook, and suffers from several weaknesses when applied to medical imaging. First, the type and dose of the atomic bomb radiation differs from those of medical imaging radiation, making it difficult to translate the risks of the former to the latter. Second, the risk of cancer of 1 in 2000 is based on exposure of the general population; however, most CT radiation exposure occurs in the elderly, who are at lower risk of developing cancer from ionizing radiation. Third, there is evidence that the LNT hypothesis is invalid because cell DNA repair occurs with doses up to 50 mSv (Tubiana et al. 2009). Given these uncertainties, it would be impossible to precisely communicate risk within the requirements of the informed consent process.

Besides the lack of precise risk quantification, informed consent has evolved into a legal document that requires a patient's signature, rather than its intended purpose of making patients aware of the risks and benefits of medical treatments and testing. Patients generally demonstrate a poor understanding and retention of information conveyed in the usual informed consent process (Institute of Medicine 2004). For example, Byrne et al. (1988) demonstrated that after informed consent was obtained, 27% of surgical patients were unaware of which organ was being operated on. Patients also have a poor understanding of the risks of medical procedures that are conveyed during the informed consent process. Holmboe et al. (2000) showed that 46% of patients could not recall one complication associated with percutaneous coronary revascularization. In addition, to further reinforce the notion that informed consent has deviated from its intended purpose, another study showed that 62% of patients felt that the pre-operative informed consent process was for legal protection of doctors (Saw et al. 1994).

Given the weaknesses in the informed consent process, a more patient-centered model is emerging that focuses on providing patient's *more* rather than less information to help them make a decision. This model is called "informed decision-making" and is consistent with the new concept of patient-centered care, as emphasized in the Affordable Care Act of 2010 and by the Institute of Medicine.

17.1.3 What Is Informed Decision-Making?

Informed decision-making is a "meaningful dialogue between physician and patient instead of unidirectional, dutiful disclosure of alternatives, risks, and benefits by the physician" (Braddock et al. 1999). Informed decision-making is more formal than traditional methods of clinical decision-making, utilizing guided-discussion tools such as flowcharts, videotapes, and interactive presentations. Ideally, risks are compared to real-life scenarios (such as the risk of dying in a car accident). When a patient presents to their physician to discuss the necessity of an imaging test, the physician will spend a part of the time dedicated for patient counseling to review these tools with the patient.

The role of the radiologist in directly participating in the informed decision-making discussion with patients is unclear, and more research is needed in this area. Until then, radiologists can support their referring physicians by providing the most relevant information for patients and referring physicians. In November 2009, the ACR, the Radiological Society of North America (RSNA), the American Association of Physicists in Medicine (AAPM), and the American Society of Radiologic Technologists (ASRT) addressed the lack of an organized educational initiative regarding the risks of medical imaging with ionizing radiation by launching Image Wisely, a social marketing campaign to improve awareness and provide educational materials about the benefits and risks of adult medical imaging (Brink and Amis 2010). The purpose of the Image Wisely website is to make available the best educational materials related to medical imaging radiation exposure in addition to assisting radiologists with dose reduction strategies for their practice. Image Wisely provides links to a radiology patient education website *Radiologyinfo. org*, a public information site also co-sponsored by the ACR and RSNA. Among other public information related to imaging tests and procedures, *Radiology Info* contains patient education material and videos related to radiation exposure and the associated risks.

There is recent evidence that patients wish to be informed about radiation risks prior to making decisions regarding medical imaging. Robey et al. (2014) surveyed both emergency medicine physicians and patients regarding their attitudes towards radiation risk discussions. This study demonstrated that 66% of surveyed patients wanted to discuss the risk of radiation during all encounters and indicated that a written handout would be the preferred method. Similarly, 75% of emergency medicine physicians indicated that they would like to discuss radiation dose almost all of the time with patients. Physicians in this study were concerned about using a written handout to explain radiation risks because they felt that handouts could be inaccurate and not likely to be read by patients.

Radiologists may also find themselves in situations where they are performing the informed decision-making process directly with patients. For example, in higher-risk situations, such as those involving pediatric and pregnant patients, referring physicians may defer to the expertise of radiologists in discussions regarding imaging these patients. Interventional radiologists, who routinely use ionizing radiation to perform minimally invasive procedures, obtain informed consent for their procedures directly from patients. As part of that discussion, procedural radiation risks can also be discussed, assisted by decision aids. In this case, the consent form can incorporate the informed decision-making process for radiation and procedural risks.

17.2 Special Circumstances

17.2.1 Pediatrics

Pediatric patients are the most vulnerable to radiation exposure, given their rapidly dividing cells and their long life expectancy. In addition, the use of CT imaging in pediatric patients increased threefold in patients aged 5–14 between 1997 and 2005 (Miglioretti et al. 2013). Medical decisions for most pediatric patients are typically made by their parents, so it would be difficult to engage the patient directly. However, parents may be able to be engaged about the risks of radiation exposure. Boutis et al. (2013) demonstrated that nearly 47% of parents were aware of the possible increased risk of lifetime malignancy in their children. In addition, 91% of parents wished to discuss the risks of radiation-induced cancers in their children. Thus, informed decision-making in this population is paramount.

In response to the increased use of CT in pediatric patients and the need to communicate risks to parents, the Alliance for Radiation Safety in Pediatric Imaging was founded by the Society for Pediatric Radiology (SPR), ACR, ASRT, and AAPM in July 2007. In addition to providing strategies to decrease the dose of pediatric exposures, the campaign created a website that provides both referring physicians and parents with information about the risks of medical imaging radiation exposure. These tools can be used to help guide informed decision-making discussions with parents.

17.2.2 Pregnant Patients

Radiation risks to the fetus in pregnant patients pose the highest risk due to the rapidly dividing cells during development and are well covered in this textbook. Weighing the risks and benefits to both the mother and fetus is essential, and it is thus important to perform an informed decision-making discussion with the patient.

The ACR and SPR have released practice guidelines for imaging pregnant, or potentially pregnant, patients with ionizing radiation, which contain comprehensive background material to help physicians guide patient discussions (ACR 2014). These practice guidelines recommend that informed consent be obtained, given the potential risks to pregnant patients. Although the informed-consent discussion does suffer from the limitations described in this chapter, in this case, the documentation provided by obtaining informed consent may avoid malpractice litigation.

17.3 Conclusions

Patient-centered care is a new and important concept. Informed consent suffers from a number of weaknesses, and thus informed decision-making can allow physicians to better communicate with patients about the risks of medical imaging radiation exposure. Future directions should explore the possibility of radiologists discussing these risks directly with patients as a means of improving patient interactions and the perception of radiologists.

References

ACR–SIR Practice Guideline on Informed Consent for Image-Guided Procedures, 2011. https://www.sirweb. org/globalassets/aasociety-of-interventional-radiology-home-page/practice-resources/standards_ pdfs/final_acr_doc7109ed01.pdf (Accessed August 6, 2018).

ACR Practice Guideline for Imaging Pregnant or Potentially Pregnancy Adolescents and Women with Ionizing Radiation. http://www.acr.org/~/media/9e2ed55531fc4b4fa53ef3b6d3b25df8.pdf (Accessed April 11, 2014).

Annas G.J., Doctors, patients, and lawyers–Two centuries of health law, *N. Engl. J. Med.* 367 (2012): 445–450.

Boutis K., Cogollo W., Fischer J., Freedman S.B., Ben D.G., Thomas K.E., Parental knowledge of potential cancer risks from exposure to computed tomography, *Pediatrics* 132 (2013): 305–311.

Braddock C.H., Edwards K.A., Hasenberg N.M., Laidley T.L., Levinson W., Informed decision making in outpatient practice: Time to get back to basics, *JAMA* 282 (1999): 2313–2320.

Brink J.A., Amis E.S. Jr., Image wisely: A campaign to increase awareness about adult radiation protection, *Radiology* 257 (2010): 601–602.

Brink J.A., Goske M.J., Patti J.A., Informed decision making trumps informed consent for medical imaging with ionizing radiation, *Radiology* 262 (2012): 11–14.

Byrne D.J., Napier A., Cuschieri A., How informed is signed consent? *Br. Med. J.* 296 (1988): 839.

Holmboe E.S., Fiellin D.A., Cusanelli E., Remetz M., Krumholz H.M., Perceptions of benefit and risk of patients undergoing first-time elective percutaneous coronary revascularization, *J. Gen. Intern. Med.* 15 (2000): 632–637.

ICRP, 2007. Radiological protection in medicine. ICRP Publication 105. *Ann. ICRP* 37 (6).

Institute of Medicine. *Health Literacy: A Prescription to End Confusion.* Washington, DC: National Academies Press, 2004.

Leclercq W.K., Keulers B.J., Scheltinga M.R., Spauwen P.H., van der Wilt G.J., A review of surgical informed consent: Past, present, and future. A quest to help patients make better decisions, *World J. Surg.* 34 (2010): 1406–1415.

Miglioretti D.L., Johnson E., Williams A. et al. The use of computed tomography in pediatrics and the associated radiation exposure and estimated cancer risk, *JAMA Pediatr.* 167 (2013): 700–707.

Miles S.H., Hippocrates and informed consent, *Lancet* 374 (2009): 1322–1323.

Pierce D.A., Preston D.L., Radiation-related cancer risks at low doses among atomic bomb survivors, *Radiat. Res.* 154 (2000): 178–186.

Robey T.E., Edwards K., Murphy M.K., Barriers to computed tomography radiation risk communication in the emergency department: A qualitative analysis of patient and physician perspectives, *Acad. Emerg. Med.* 21 (2014): 122–129.

Saw K.C., Wood A.M., Murphy K., Parry J.R., Hartfall W.G., Informed consent: An evaluation of patients' understanding and opinion (with respect to the operation of transurethral resection of prostate), *J. Roy. Soc. Med.* 87 (1994): 143–144.

Tubiana M., Feinendegen L.E., Yang C., Kaminski J.M., The linear no-threshold relationship is inconsistent with radiation biologic and experimental data, *Radiology* 251 (2009): 13–22.

U.S. Food and Drug Administration. Radiation-emitting products: What are the radiation risks from CT? http://www.fda.gov/Radiation-EmittingProducts/RadiationEmittingProductsandProcedures/MedicalImaging/Medical X-Rays/ucm115329.htm (Accessed April 11, 2014.)

18

Communicating Risks and Benefits of Medical Imaging: The Patient Perspective

Geoffrey S. Gold
and Jennifer L. Hay

18.1 Introduction

Over the past two decades, technological advances in medical imaging have greatly improved physicians' abilities to diagnose, characterize, and monitor disease (Baerlocher and Detsky 2010). Such advances, including improved resolution and overall image quality as well as enhanced availability and access to diagnostic radiology tests, have been associated with dramatically improved treatment and diagnosis of cancer and neurological diseases, shorter hospital stays, a reduction of exploratory surgeries, and other clinical benefits (Hricak et al. 2010). Successes in medical imaging have led to increased use of these tests; some estimates indicate that use of medical imaging in the United States has increased sixfold in the past 25 years (Mettler et al. 2009; National Council on Radiation Protection and Measurements 2009). At the same time, however, ionizing radiation exposure associated with medical imaging—medical imaging radiation (MIR)—now accounts for one of the two major contributors to overall radiation exposure in the general population, alongside natural background radiation (National Council on Radiation Protection and Measurements 2009).

Given this dramatic increase in MIR exposure, the medical community has increasingly addressed the importance of careful, judicious medical decision-making regarding the clinical necessity of imaging tests that deliver MIR, with consideration of potential ways to reduce this exposure when possible (Hricak et al. 2010; Nievelstein and Frush 2012). Efforts to encourage radiologists to use scientific data and new technology to keep doses "as low as reasonably achievable," while not compromising quality of care, codified as the ALARA principle, have operationalized these considerations (International Commission on Radiation Protection 1991; Ludwig and Turner 2002; National Council on Radiation Protection and Measurements 2009; Baerlocher and Detsky 2010; Dainiak 2013). For example, the ALARA principle promotes physician consideration of test choice, use of better technologies, and the elimination of unnecessary tests to reduce MIR (Kaste 2009; Hricak et al. 2010). The ALARA principle also includes specific consideration of patient characteristics such as age at exposure, gender, body location of the test, and whether that body site has already been heavily imaged, factors that may contribute to higher or unnecessary risk (Hricak et al. 2010). However, not all clinicians consider ALARA an appropriate principle to guide dialogues. The counterargument equates MIR risks with other mundane sources of risk that do not require focused attention and long-term planning.

Another effort to sensitize health-care providers to the radiation risks faced by children, in particular, is the Image Gently Campaign (http://www.imagegently.org). Image Gently is an organized effort by the radiology, imaging, and broader medical community to increase patient knowledge and encourage collaboration regarding MIR. Views vary widely on the campaign's ability to inform patients without cultivating excessive concerns leading to a debate about its future.

Yet rapidly developing scientific literature as well as the mainstream press has begun to address public perspectives about MIR. One important aspect of this debate involves whether, and how, patients themselves should be involved in evaluating the risks and benefits of tests that are recommended to them, and whether full informed consent for medical imaging is a worthwhile or even feasible goal (Karsli et al. 2009; Baerlocher and Detsky 2010; Nievelstein and Frush 2012). In the context of patient-centered approaches to medical care, the patient perspective on medical imaging has become increasingly central to these discussions (Baerlocher and Detsky 2010; Dauer et al. 2011; Nievelstein and Frush 2012). Providers and researchers have begun to adopt a number of approaches to communicating the risks and benefits of MIR. Such approaches include comparison of amount of radiation in an imaging test to that of life exposure, such as during airplane flight (Larson et al. 2007; Jafari and Daus 2013), or use of visuals and text to communicate risks in the form of expected cancer diagnosis rate based on a specific test (Schonberg et al. 2014). The assumption underlying these efforts is that patients should have some understanding of the MIR risks that they face in order to be engaged and empowered participants in their own care (Baerlocher and Detsky 2010; Dauer et al. 2011; Nievelstein and Frush 2012). Complicating such efforts are the large uncertainties regarding the magnitude of risks associated with MIR and the possibility that there are, in fact, no actual risks of MIR.

While well-intentioned, existing approaches to communicating risks and benefits of medical imaging have been developed largely in the absence of literature substantiating patient perspectives, including patient comprehension or satisfaction with the information provided, which may vary based on a range of patient beliefs, attitudes, and personal preferences (Ludwig and Turner 2002; Picano 2004; Viswanath et al. 2006; Larson et al. 2007; Goske and Bulas 2009; Kiviniemi and Hay 2012; Nievelstein and Frush 2012; Busey et al. 2013; Khaliq et al. 2013; McNierney-Moore et al. 2015; Repplinger et al. 2016). Lack of attention to the patient perspective, paired with increasing attention to MIR in the news media, may lead those referred for medical imaging to formulate beliefs about their own personal exposure to MIR that may be grounded in emotional responses to personal risk from MIR rather than the specific statistical likelihood of harm (Slovic 1987; Peters et al. 2004); this has the potential to influence ultimate decisions about imaging (Slovic and Peters 2006). A content analysis of 2010 print and online news showed a relatively neutral presentation of medical imaging (Maloney et al. 2013); overall, most articles reference radiation in a neutral context (40.1%), with roughly one-third of references that are negative (31.4%) and a smaller proportion of positive references (21.8%). Sports and news sections were more likely to present neutral representations of radiation and medical imaging, but health sections presented mostly negatively coded verbiage. Further, for articles that were specifically about risks and benefits of medical imaging radiation, negative sentences were twice as common as positive sentences. Unfortunately, an overestimation of risk could potentially leave patients with unwarranted anxieties or fears or influence them to refuse medical imaging tests critical to their health care.

18.2 Public Knowledge, Attitudes, Beliefs, and Preferences

Public perspectives regarding communication of the risks and benefits of medical imaging likely encompass a broad array of beliefs, attitudes, and preferences. Individuals vary not only on how they understand the form, function, and risks of medical imaging but also on preferred communication needs and patient-centered care practices. Understanding patient perspectives allows those who seek to communicate risks and benefits a greater likelihood of ensuring that the patient experiences a sense of engagement in the process by which medical imaging has been recommended to them, and thus feels sufficiently informed and satisfied with the information received about the benefits and risks of medical imaging.

18.2.1 Knowledge

Evidence suggests that patients are not sufficiently informed about radiation in general and MIR in particular. In a cross-sectional shopping mall survey ($N = 200$), participants reported that they associated the term, "*nuclear*," mostly with energy (65%) and war (25%), with only 7% associating the term with medicine. When asked what they thought caused most radiation-related cancers, most participants said sunrays (58.5%) or that they did not know (22.5%), with only 6% reporting that medical imaging was a cause of radiation cancers. Interestingly, however, nearly one-third of participants also believed that medical imaging is the leading source of exposure to radiation (Ludwig and Turner 2002). One potential outcome of poor knowledge is an inaccurate sense of risk. For instance, in the same study, less than a quarter of participants agreed with expert opinions that radiation exposure does not cause permanent biologic damage, and only 20% of the sample agreed with experts that living near a properly functioning nuclear reactor does not increase radiation exposure (Ludwig and Turner 2002). Alternatively, Busey and colleagues (2013) found that about one-third of randomly selected patients at a large academic medical center waiting to undergo non-urgent computed tomography (CT) and cardiac single photon emission computed tomography (SPECT), tests that deliver radiation, did not know they were being exposed to radiation at all, and of those who did, many vastly underestimated the associated health risks.

Similarly, Takakuwa et al. (2010) reported low levels of accurate risk knowledge in emergency room patients about to undergo medical imaging, with 41% incorrectly reporting that CT scans and chest radiographs gave the same amount of radiation and 25% reporting that CT scans used less radiation than radiographs. Further, a mere 7% of participants accurately described CT radiation doses as significantly higher than doses from chest x-rays. These authors also found that knowledge deficits and inaccuracies about MIR are higher in those with lower educational attainment, non-white race, and older age, which could be related to general knowledge base, access to care, and/or general attitudes towards the medical system or science. Further evidence reported by Repplinger and colleagues (2016) suggests that most emergency room patients' perceptions of medical imaging radiation were inaccurate. Among 500 such patients, 14% understood the relative difference between CT and chest x-ray, and 22% understood that MRI does not use ionizing radiation. However, 25% believed abdominal CTs increased cancer risk, while 55% believed that MRI increased cancer risk. While higher educational level and experience as a health-care professional were associated with more accurate responses, a significant majority of responses were incorrect for all questions.

Another study assessing knowledge in an emergency room setting aimed to identify demographic factors related to knowledge of radiation risk. McNierney-Moore and colleagues (2015) found evidence suggesting poor knowledge among participants ($N = 600$) regarding relative exposure from CT versus background radiation, lifetime risk of cancer following exposure, and relative exposure from CT versus chest x-ray. A multivariate logistic regression suggests no association between knowledge levels and education, gender, age, or ethnicity. However, compared to participants who earned less than $20,000 per year, participants with yearly income over $20,000 were more likely to demonstrate better knowledge. Knowledge of clinical guidelines about screening tests (e.g., at what age to begin or repeat mammography screening) is poor in general population samples (Kiviniemi and Hay 2012; Allen et al. 2013; Stoll et al. 2014) and could lead to over- or under-utilization of tests that deliver MIR.

Finally, Thornton and colleagues (2015) suggest high levels of uncertainty about which types of tests exposed the patient to radiation in a focus-group study in a cancer treatment setting. Concerns were raised about whether they had received the correct or best test and whether the use of multiple tests might possibly be redundant. This study indicated that insufficient knowledge contributed to patients feeling left out of the testing decision process and also contributed to distress about their medical care.

18.2.2 Attitudes and Beliefs

Patients' attitudes, such as evaluations and presumptions, towards communication about medical imaging are crucial for guiding efforts to improve the presentation of risk-benefit information. Research suggests that varied attitudes regarding MIR exist among patients (Ludwig and Turner 2002; Takakuwa et al. 2010; Hollada et al. 2014; Thornton et al. 2015). Hollada and colleagues (2014) reported that among 80 individuals undergoing research studies involving CT scans, those willing to participate in the study prioritized helping the medical community rather than MIR concerns, with those declining participation more concerned about MIR. Takakuwa et al. (2010) found that about three-quarters of patients waiting for CT scanning in an urban hospital setting undergoing CT scans either agreed (45%) or strongly agreed (28%) that a physician's ability to diagnose a condition using CT overshadows the importance of the risks radiation exposure from CT. Additionally, these patients were asked about their attitude towards shared decision-making for their CT as opposed to physician judgment. Interestingly, 62% of the sample either agreed or strongly agreed that it was more important that a physician discuss risks and benefits of a test than for the physician to simply recommend the best diagnostic test based on their judgment; black patients were more likely to agree than whites, and those with lower pain scores were significantly more likely to agree than those with higher pain scores.

Thornton and colleagues (2015) found that among cancer patients and parents of children with cancer there were expressions of indifference about testing risks, and they often justified this with the substantial distress they experienced in relation to the uncertainties of their illnesses. For these participants, the looming threat of chronic illness that could be alleviated with diagnostic information outweighed concerns about long-term risks of radiation exposure, and thus they deprioritized discussions of risks and benefits of imaging radiation with their physicians. Further, attitudes appeared to vary by severity of cancer treatment trajectory. For instance, some patients with late-stage cancer reported that shared decision-making for medical imaging tests was a low priority, considering that they viewed these tests as non-negotiable components of potentially life-saving experimental protocols. These patients cited their hope in the experimental treatment and their trust in the medical providers as factors that influence their attitudes. Yet some of these patients endorsed the availability of benefit-risk information for medical imaging, claiming it should be available to all patients as a part of routine care. In addition, some of those who had completed treatment and considered themselves cancer survivors reported growing interest and concern in MIR as they became aware of their need for lifelong surveillance.

18.2.3 Preferences

A growing literature has begun to identify a wide range of patient preferences regarding communication in the context of medical imaging tests (Picano 2004; Larson et al. 2007; Dauer et al. 2011; Jafari 2013; Thornton et al. 2015). Thornton and colleagues (2015) found that among the cohort of cancer patients they interviewed, many expressed a more nuanced interest in information about imaging and radiation well beyond the specific risks and benefits of the single test in question. Such information includes whether and how risk accumulates from multiple test studies over time, whether tests like an MRI could substitute for the radiation-exposing tests like CT, and whether longer waiting periods between scheduled scans would be an option to curb accumulation of possible radiation damage. They had preferences for information that would justify why, specifically, a CT scan was being ordered, and whether the image from positron emission tomography (PET) or magnetic resonance imaging (MRI) would be an appropriate alternative. Participants expressed confusion about different tests, and stated that this knowledge could be critical to their decision-making as to whether to undergo a test. Some participants asked for help keeping track of their radiation exposure with a schedule of planned tests or a personal history record of radiation exposure. Clearly, many cancer patients in this study understood that accumulation of radiation exposure potentially increases the associated health risks and that the decision to get a test may involve a complex array of variables. Patient preferences for resources to help them make the best decisions for their own health care were well articulated in this study of cancer patients. This study indicates that patients may well have specific preferences for information that goes beyond explication of the risks and benefits of any specific test, and may want more personalized information tailored to their own medical context and imaging history to help them make decisions about imaging tests over time.

In addition, Takakuwa et al. (2010) explored preferences towards two trade-offs for a hypothetical medical imaging study—test accuracy on the one hand and radiation exposure on the other—among emergency room patients waiting to undergo CT. Participants were asked whether they would prefer a test that exposes the individual to higher doses of radiation and risk but provides clearer, more definite diagnostic information. Participants narrowly supported the option for a better test with more radiation exposure, with no differences in terms of age, gender, or education. These findings indicate that participants place a slightly greater value on physicians' abilities to diagnose an illness than on MIR risks.

18.3 Addressing Patient Needs

Given that knowledge, attitudes, beliefs, and preferences towards communication vary by individual, some researchers are testing interventions to enhance and personalize risk communication practices. In an intervention study to improve knowledge, Larson and colleagues (2007) provided an informational leaflet to 100 parents of children who were undergoing non-emergent CT studies in a children's hospital. This brief double-sided leaflet provided information on one side in the form of concise text and simple images concerning what CT is, what x-rays are, and how an x-ray film differs from CT in terms of purpose and radiation dose. The other side of the handout included a small table that lists radiation doses of CT as compared to a standard daily background radiation estimate. While the handout clearly expresses risks of CT and radiation, it also provides information about benefits of CT and how the hospital attempts to minimize radiation doses. There were significant improvements in knowledge after exposure to the pamphlet. For example, knowledge that CT uses radiation significantly increased from 66% at pretest to 99% at posttest. Similarly, after participant exposure to the handout, knowledge that CT may increase lifetime risk of cancer increased from 13% at pretest to 86% at posttest. Most participants (59%) reported that the information in the pamphlet was either new to them or helped to clarify their understanding. Of note, most participants report the same level of concern for MIR as they did before exposure to the information, with only 14% reporting increased concern and 5% decreased concern. No parents opted out of the CT scan after exposure to the pamphlet. This informational pamphlet represents the potential that health communicators can present MIR risk information in a way that the individual will find comprehensible without raising unsubstantiated concerns or reducing test consent. This study provides evidence that individuals with minimal or no knowledge about medical imaging radiation or risks can benefit from a simple informational pamphlet that takes about five minutes to read and that concern about MIR does not become a major issue. This new information will allow these individuals to make informed decisions about medical imaging tests in the future.

Other efforts to address patient needs involve finding ways to actually reduce radiation dose. Such efforts involve shorter exposure times, lower radiation doses made possible with technological advances, and increased specificity in clinical decision guidelines regarding imaging tests, as well as recommended education programs that could help patients undergoing testing to remain directly involved in the decision-making process to self-regulate their imaging choices (Hricak et al. 2010). More appropriate use of imaging could involve replacing CT with other imaging modalities, such as for patients experiencing abdominal pain, chest trauma, and minor head injuries. For example, accurate determination of local tumor extent for patients with primary bone disease is not significantly different for CT and MRI (Panicek et al. 1997). The goal is to determine which diagnostic procedure is best for a given patient in a given situation, and in this context, if the use of ionizing radiation is warranted (Hricak et al. 2010). Certainly there are barriers to these efforts that will continue to present challenges, including test cost differential, longer wait time for MRI, and differential third-party payment (Kaste 2009). How well and for whom these communication practices work in terms of education and reduction of risks remain critical questions.

An emerging debate about the goals of doctor-patient dialogues regarding imaging risks questions the emphasis of the ALARA principle and the Image Gently campaign (Cohen 2016; Frush 2016). Cohen (2016) posits that radiation risks may be magnified by the nature of the dialogue, stating that the Image Gently campaign is the "cause of the CT cancer problem, not the solution." Cohen (2016) views efforts such as Image Gently as reactions to news media reports that sensationalize or misrepresent radiation risks that only serve to further increase concerns rather than relieve them. According to Cohen, Image Gently has been misguided in prioritizing its goal to inform and empower patients as opposed to the goal of quelling anxiety about imaging. Cohen claims that attempts to minimize radiation are wise but raises concerns about the effects of these particular approaches. Cohen notes misleading media headlines, diagnostic errors, concerns about nondiagnostic studies, and financial concerns as costs of taking an unjustified "better safe than sorry" approach. The investigation into the causes and scope of these concerns ultimately remains an empirical question, but Cohen (2016) asserts that the best route forward is to discontinue Image Gently and reduce focus on the ALARA principle in favor of an effort to dismiss anxieties about medical imaging radiation as unfounded.

The counterpoint to these views suggests that the Image Gently campaign has been largely beneficial to the improvement of understanding, using, and performing medical imaging tests. Frush (2016) states that Image Gently has taken an approach dedicated to "advocacy through positive and productive perspectives rather than assuming a 'harm and alarm' point of view." Frush goes on to claim that anecdotal evidence suggests that patients seem satisfied with balanced presentations of information and are willing to cope effectively with any raised concerns. Frush highlights the role of Image Gently in the development of guidelines for pediatric medical imaging in North America. Frush asserts that Cohen (2016) and other critics of Image Gently or the communication of the ALARA principle are crucial for the evolution of efforts to improve the use and quality of imaging services. In particular, Frush states that Image Gently and related campaigns should adapt by adjusting the focus of patient communication to address patient concerns without dismissing the importance of balanced information and understanding in patient-centered care. To achieve these goals, Frush (2016) suggests targeting specific needs. Frush and Lungren (2017) present the Think A-Head as an improvement to Image Gently targeted specifically to help families make informed decisions when children acquire head injuries. This campaign more directly addresses the concerns that families could experience while becoming more informed. The branding of the campaign itself illustrates these priorities with the tagline, "Keep calm and image gently." This debate highlights the need for further discussion of what constitutes patient-centered decision-making regarding imaging radiation.

18.4 Conclusions and Recommendations for Future Research

The growing importance of patient-centered medicine requires providers to be aware of patients' knowledge and preferences regarding medical recommendations and alternatives (Dauer et al. 2011). In the context of medical imaging, patient perspectives regarding enhanced communication of the risks and benefits of medical imaging will ensure that patients are well-informed about the rationale, risks, benefits, and alternatives to imaging tests, and that they feel engaged and empowered in the process by which a decision is made to recommend a specific imaging test to them. While current dialogues regarding communication of risks and benefits of medical imaging have highlighted patient safety and optimization of medical outcomes, and have centered around deliberation regarding medical-legal issues and medical provider responsibility (Picano 2004; Nievelstein and Frush 2012), there has been less direct consideration of the patient perspective in this dialogue. Much research in this area remains to be done to help patients acquire the knowledge and support they need to understand whether risks exist, and if so, how to manage them. With an ongoing debate about whether non-zero risks exist, medical professionals are faced with the challenge of better understanding patients' knowledge, attitudes, and preferences to create a patient-centered education.

While we have reviewed a series of useful descriptive studies outlining patient perspectives regarding MIR, there are few intervention studies identified to address patient needs to improve knowledge or satisfaction with communication regarding the risks and benefits of recommended medical imaging tests. The field also lacks studies examining system-level interactions around medical imaging that could clarify feasible yet distinct roles in aspects of patient engagement and communication for referring physicians, radiologists, and other radiology staff members. In particular, the field lacks descriptive studies clarifying the efficacy of existing efforts to educate patients regarding the risks and benefits of medical imaging. Questions including comprehension of comparison referents—such as background radiation and radiation received in an airplane flight—are commonly used to provide context about medical imaging exposures, but the field lacks studies examining how people interpret this information, and whether patients find that it adequately answers their questions and concerns about MIR. In addition, knowledge about medical imaging may be even more problematic in underserved and diverse populations (Viswanath et al. 2006). Opportunities to assess level of education and other demographic factors, as well as the role of health literacy and health numeracy, in determining distinct patient perspectives about medical imaging is a high priority. The development of interventions to address knowledge deficits in underserved populations also presents profound opportunities to "level the playing field" in patient-centered care regarding medical imaging. More research in this area could increase the availability of efficacious tools that could be disseminated into different treatment settings to improve assessment of patient perspectives and the use of such perspectives in doctor-patient communication strategies.

Finally, in addition to intervention research to introduce patients to basic risks and benefits of medical imaging at the time a test is recommended, patient preferences to have a method to keep track of radiation exposure over time (Thornton et al. 2015) present a unique opportunity to empower patients in their own clinical decision-making. While referring physicians have more of a "snapshot" context driven by the need for specific clinical diagnosis with each patient at a specific time, patients themselves may have a justifiably distinct, "longitudinal" context for risks and benefits of repeated medical imaging over their lives. The development of a feasible, comprehensible, acceptable strategy for patients to record cumulative radiation exposure information, utilizing physician input and facilitated by electronic medical records, would be a useful research direction to address this patient preference. Further research is also required to understand how medical professionals who argue that dose tracking undermines patient care may best address this patient preference. While many professionals argue that dose tracking is not appropriate or necessary, these professionals are likely to encounter patients who hold this preference. These considerations could guide patients towards knowledge of and satisfaction with the goals and principles guiding decisions.

There are many useful research agendas that could identify and enhance the patient perspective regarding communication about medical imaging. While there is general agreement that medical imaging radiation doses should be as low as possible and that patient communication is a good thing, there is much work to be done to identify the best and most feasible models for that communication within a patient-centered perspective. Further, patient attitudes, beliefs, and preferences for information may go well beyond the specific risks of one specific test—and may include rationale for one test over another—and test contribution to cumulative radiation exposure, as well as other patient preferences. With providers who are aware of patient needs and preferences and with patients who are well-informed and engaged, the use of medical imaging will continue a trajectory towards increasing positive impact on modern medicine while minimizing concerns of potential harms. With the growing inclusion of the patient voice in medical imaging decision-making, the dialogue will be further enhanced for the benefit of patients as well as their health-care providers.

References

Allen J, Bluethmann S, Sheets M et al. Women's responses to changes in U.S. preventive task force's mammography screening guidelines: Results of focus groups with ethnically diverse women. *BMC Public Health* 2013;13(1):1169. doi:10.1186/1471-2458-13-1169.

Baerlocher M, Detsky A. Discussing radiation risks associated with CT scans with patients. *JAMA* 2010;304:2170–2171.

Busey J, Soine L, Yager J et al. Patient knowledge and understanding of radiation from diagnostic imaging. *JAMA Intern Med* 2013;173(3):239–241. doi:10.1001/2013.jamainternmed.1013.

Cohen M. Point: Should the ALARA concept and image gently campaign be terminated? *JACR* 2016;13(10):1195–1198.

Dainiak, N. Radiation dose and stochastic risk from exposure to medical imaging. *Chest* 2013;144(5):1431–1433. doi:10.1378/chest.13-1064.

Dauer L, Thornton R, Hay J et al. Fears, feelings, and facts: Interactively communicating benefits and risks of medical radiation with patients. *AJR* 2011;196:756–761.

Frush D, Lungren M. The image gently think a-head campaign: Keep calm and image gently. *JACR* 2017;14(2):301–302.

Frush D. Counterpoint: Image gently: Should it end or endure? *JACR* 2016;13(10):1199–1202.

Goske M, Bulas D. Improving health literacy: Informed decision-making rather than informed consent for CT scans in children. *Pediatr Radiol* 2009;39:901–903.

Hollada J, Marfori W, Tognolini A et al. Successful patient recruitment in CT imaging clinical trials: What factors influence patient participation? *Acad Radiol* 2014;21(1):52–57. doi:10.1016/j.acra.2013.09.016.

Hricak H, Brenner D, Adelstein S et al. Managing radiation use in medical imaging: A multifaceted challenge. *Radiology* 2010;258(3):889–905.

International Commission on Radiation Protection. SPAR56, 23360–23474:1991. Standards for protection against radiation. Nuclear Regulatory Commission: Fed Regist 1991.

Jafari M, Daus A. Applying image gently SM and image wisely SM in nuclear medicine. *Health Phys* 2013;104(2 suppl 1):S31–S36. doi:10.1097/HP.0b013e3182764cd8.

Karsli T, Kalra M, Self J et al. What physicians think about the need for informed consent for communicating the risk of cancer from low-dose radiation. *Pediatr Radiol* 2009;39:917–925.

Kaste S. Imaging challenges: A US perspective on controlling exposure to ionizing radiation in children with cancer. *Pediatr Radiol* 2009;39(Suppl 1):S74–S79.

Khaliq W, Visvanathan K, Landis R, Wright SM. Breast cancer screening preferences among hospitalized women. *J Womens Health (Larchmt)* 2013;22(7):637–642. doi:10.1089/jwh.2012.4083.

Kiviniemi M, Hay J. Awareness of the 2009 US Preventive Services Task Force recommended changes in mammography screening guidelines, accuracy of awareness, sources of knowledge about recommendations, and attitudes about updated screening guidelines in women ages 40–49 and 50+. *BMC Public Health* 2012;12:899.

Larson D, Rader S, Forman H et al. Informing parents about CT radiation exposure: It's OK to tell them. *AJR* 2007;189:271–275.

Ludwig R, Turner L. Effective patient education in medical imaging: Public perceptions of radiation exposure risk. *J Allied Health* 2002;31(3):159–164.

Maloney E, Bylund C, Banerjee S et al. Newspaper coverage of the risks and benefits of medical radiation imaging. In: *Proceedings of the Annual Meeting of the International Communication Association*, Hilton Metropole Hotel, London, UK, 2013. http://citation.allacademic.com/meta/p640629_index.html (Accessed January 7, 2014).

McNierney-Moore A, Smith C, Guardiola J et al. Patient understanding of radiation risk from medical computed tomography—A comparison of Hispanic vs. non-Hispanic emergency department populations. *Peer J* 2015;3:e937. doi:10.7717/peerj.937.

Mettler F, Bhargavan M, Faulkner K et al. Radiologic and nuclear medicine studies in the United States and worldwide: Frequency, radiation dose, and comparison with other radiation sources—1950–2007. *Radiology* 2009;253(2):520–531.

National Council on Radiation Protection and Measurements. Ionizing radiation exposure of the population of the United States. National Council on Radiation Protection and Measurements, Bethesda, MD. NCRP Report number: 160, 2009.

Nievelstein R, Frush D. Should we obtain informed consent for examinations that expose patients to radiation? *Am J Roentgenol* 2012;199:664–669. doi:10.2214/AJR.11.8319.

Panicek D, Gatsonis C, Rosenthal D et al. CT and MR imaging in the local staging of primary malignant musculoskeletal neoplasms: Report of the radiology diagnostic oncology group. *Radiology* 1997;202(1):237–246.

Peters E, Burraston B, Mertz C. An emotion-based model of risk perception and stigma susceptibility: Cognitive appraisals of emotion, affective reactivity, worldviews, and risk perceptions in the generation of technological stigma. *Risk Anal* 2004;24(5):1349–1367.

Picano E. Informed consent and communication of risk from radiological and nuclear medicine examinations: How to escape from a communication inferno. *BMJ* 2004;329:849–851.

Repplinger M, Li A, Svenson J et al. Emergency department patients' perceptions of radiation from medical imaging. *WMJ* 2016;115:1.

Schonberg M, Hamel M, Davis R et al. Development and evaluation of a decision aid on mammography screening for women 75 years and older. *JAMA Intern Med* 2014;174(3):417–424. doi:10.1001/jamainternmed.2013.13639.

Slovic P, Peters E. Risk perception and affect. *Curr Dir Psychol Sci* 2006;15:322–325. doi:10.1111/j.1467-8721.2006.00461.x.

Slovic P. Perception of risk. *Science* 1987;236:280–285.

Stoll C, Roberts S, Cheng M et al. Barriers to mammography among inadequately screened women. *Health Educ Behav* 2014;42(1):8–15. doi:10.1177/1090198114529589.

Takakuwa K, Estepa A, Shofer F. Knowledge and attitudes of emergency department patients regarding radiation risk of CT: Effects of age, sex, race, education, insurance, body mass index, pain, and seriousness of illness. *Am J Roentgenol* 2010;195(5):1151–1158.

Thornton R, Dauer L, Shuk E et al. Patient perspectives and preferences for communication of medical imaging risks in a cancer care setting. *Radiology* 2015;275(2):545–552.

Viswanath K, Breen N, Meissner H et al. Cancer knowledge and disparities in the information age. *J Health Commun* 2006;11(1):1–17.

Additional Resources

Cohen M. ALARA, image gently and CT-induced cancer. *Pediatr Radiol* 2015;45:465–470.

Goske M, Applegate K, Bulas D et al. Image gently 5 years later: What goals remain to be accomplished in radiation protection for children? *AJR Am J Roentgenol* 2012;199(3):477–479.

Sidhu M, Goske M, Connolly B et al. Image gently, step lightly: Promoting radiation safety in pediatric interventional radiology. *AJR Am J Roentgenol* 2010;195(4):W299–W301. doi:10.2214/AJR.09.3938.

VI

Conclusion

19

Essential Messages: Dose, Benefit, and Risk in Medical Imaging

Lawrence T. Dauer,
Bae P. Chu, and
Pat B. Zanzonico

19.1 Introduction

The utilization of radiation in medicine for imaging (especially computed tomography [CT] and molecular imaging), as well as therapy, increased significantly over the past quarter century on a national and a global scale. The issues of perceived, potential, theoretical, and known risks associated with ionizing radiation exposure in medicine have come to the forefront of public and professional awareness, raising concern and controversy. Radiation-based imaging and therapy have, of course, revolutionized the way medicine is practiced. Imaging procedures using ionizing radiation carry relatively small individual risks that are usually justified by the medical need of patients, especially when diagnostic information is maximized and radiation dose is minimized. Therapies are increasingly being designed and administered to increase dose to the targeted treatment volume while minor dose to healthy tissues.

The appropriate use of radiation in medicine has been an important clinical objective since the advent of the X-ray. The use of justified and optimized radiation in medicine should certainly be viewed as a positive development. The responses to the increased use of ionizing radiation have included: practitioner-initiated approaches to improving quality, education, and utilization, data registries, legislative requirements and rulemaking guidance, and social media campaigns as grass-roots social marketing missions.

In 2012, the International Atomic Energy Agency and the World Health Organization, convened an international conference on radiation protection in medicine. The conference resulted in the "Bonn Call for Action," which suggested ten measures to improve radiation protection in medicine in the next decade (WHO 2014). Collectively, these proposed measures help to clarify the current challenges confronting enhancement of radiation responsibility in medicine:

1. Enhance the implementation of the principle of justification.
2. Enhance the implementation of the principle of optimization of protection and safety.
3. Strengthen manufacturers' role in contributing to the overall safety regime.
4. Strengthen radiation protection education and training of health professionals.
5. Shape and promote a strategic research agenda for radiation protection in medicine.
6. Increase availability of improved global information on medical exposures and occupational exposures in medicine.
7. Improve prevention of medical radiation incidents and accidents.

8. Strengthen radiation safety culture in health care.
9. Foster an improved radiation benefit-risk dialogue.
10. Strengthen the implementation of safety requirements globally.

This volume has provided a historical and current perspective as well as a comprehensive review and analysis of the doses, benefits (typically high when justified), and risks (generally very low when imaging is optimized) of essentially all current radiation-based diagnostic imaging procedures. As such, a number of the "Bonn Call for Action" activities have been addressed, including the enhancement of the principles of justification and optimization, the importance of the manufacturer's role, providing information and education on doses and risks for health professionals, identifying additional research needs, and fostering an improved radiation benefit-risk dialogue. As the previous chapters of this volume have indicated, significant advancement and improvement in these areas has already occurred over the last five years, and systems and processes are now in place for ongoing implementation of the foregoing principles in clinical practice.

The overall topic of this volume addresses several relevant and essential inquiries associated with dose, benefit, and risk in medical imaging (Chapter 1). Radiation dose index monitoring systems have improved documentation, collection, and reporting of imaging radiation dose and are used to collect modality-specific radiation dose information as well as other important data relevant to medical imaging procedures (Chapter 2). Such information, and the associated development of sophisticated information systems to acquire, collate, and analyze it could be characterized as "big data" applications. Facilities, institutions, and professional societies are beginning to appreciate the benefits of such data for quality initiatives, optimization, general radiation protection, protocol management, and beyond. These data also assist in survey studies, a means of tracking dose from imaging examinations and procedures by periodically soliciting the national (Chapter 3) and international (Chapters 3 and 16) medical community for data (e.g., procedure, modality, volume, and dose indicator information, among other patient-specific information), which can permit statistical analyses of current practice that can capture a broad range of data for future epidemiological studies. Periodic surveys also provide input for the development and periodic revision of recommendations for quality indicators, such as diagnostic reference levels and achievable doses (Chapter 4), for a given imaging modality and patient population. It is important to assess image quality and clinical usefulness systematically as well as radiation dose commensurate with clinical need.

Specific considerations for dose, benefit, and risk need to be made for each imaging modality. This is especially true for radiology imaging modalities, including dental (Chapter 5), mammography (Chapter 6), fluoroscopically guided interventions (Chapter 7), and CT applications (Chapter 8) that represent highly variable doses, benefits, and risks depending on parameter settings, organs within (and outside) of the field of view, and the imaging instrument, as well as the intended purpose of the imaging study. Although there has been recent improvement of both image quality and dose factors in each of these modalities, it is important to continue an ongoing dialogue between the medical community as users and vendors as developers to ensure overall optimization. This is especially important in the area of CT imaging because it represents the largest proportion of overall population exposure from medical imaging. Ongoing and future improvements are likely to result in even lower patient doses with improved image quality.

Present-day nuclear medicine imaging (Chapter 9) and the ongoing development of new radionuclides and molecularly targeted radiopharmaceuticals along with increasing utilization of hybrid imaging platforms (Chapter 10), yielding registered functional and anatomic images, present special challenges in balancing benefit and risk because of the attendant multi-modality radiation exposures. Such optimization is supported by special dosimetric information, radiopharmacokinetic data, and the use of reference and patient-specific calculational phantoms.

Considerations of dose, benefit, and risks are particularly acute in imaging of pediatric (Chapter 11) and of pregnant patients (Chapter 12). New and improving image reconstruction algorithms and approaches can result in significant reductions in administered activities and therefore dose in single photon emission computed tomography and positron emission tomography. For pregnant patients, the radiation doses to the conceptus following medical imaging of the mother—in nuclear medicine, CT, and in interventional fluoroscopy—are typically well below dose thresholds for tissue reactions (as they are for imaging non-pregnant patients as well), therefore emphasizing the importance of evaluating the residual risk of cancer and optimization when such imaging is required. The dose-response curve for cancer induction following in utero fetal exposure remains uncertain and is a focus of ongoing study in radiation epidemiological studies of cohorts exposed in utero.

Although the risks of radiation exposure have been extensively studied, there persists a significant uncertainty (perhaps as large as a factor of 2 or 3) in the low-dose region, that is, doses typically associated with medical imaging. Fundamental radiobiological (Chapter 13) and radiation epidemiological studies are essential for the development, application, and improvement of dose (i.e., low-dose)-effect models (Chapter 14) required for rationally assessing both justification and optimization in medical imaging procedures. Elucidation of the biological basis (Chapter 13) of radiation effects at low doses remains a significant challenge. Likewise, the increasing societal concern with medical radiation exposure demands not only an assessment of potential risks (albeit difficult in the low dose region) but also a clearer evaluation of the potential benefits. As such, quantitative risk-benefit analyses in medical imaging are essential (Chapter 15).

As public concern with medical radiation exposure increases, the prevalence of worry about the health harms of such exposure has become more evident. There has been heightened awareness among patients and physicians about the importance of holistic benefit-and-risk discussions in shared medical decision making (Chapter 17). Often, the mere mention of the word, "radiation," evokes fear in patients, families, and even health care professionals. Communicating benefits and risks in a comprehensible manner (Chapter 18) while presenting and discussing complex technical material with associated uncertainties is a challenge that, if not performed appropriately, could result in patients avoiding appropriate and medically necessary imaging because of misunderstood or unfounded fears. It is therefore important to recognize the psychological aspects of radiation risk communication, including: affect and reason, anxiety and decision-making, dread from unknown hazards pl outrage, anticipated regret and side-effect aversion, information source perceptions, and competence and care issues. Patient perspectives on medical imaging radiation is understudied but could guide point-of-care discussions. Clearly, patients believe that such information should be routinely available and that conversations with their personal physician, along with endorsed, readily available reference materials, could be ideal methods for such information exchange.

Reference

WHO (2014). Bonn call for action. 10 actions to improve radiation protection in medicine in the next decade. http://www.who.int/ionizing_radiation/medical_exposure/bonncallforaction2014.pdf.

Index

Note: Page numbers in italic and bold refer to figures and tables respectively.

Milton Keynes UK
Ingram Content Group UK Ltd.
UKHW051948071024
449327UK00026B/2222